T0331278

LIPOPROTEINS AS CARRIERS OF PHARMACOLOGICAL AGENTS

Targeted Diagnosis and Therapy

Editor

John D. Rodwell
Vice President, Research and Development
CYTOGEN Corporation
Princeton, New Jersey

Other Volumes in Preparation

LIPOPROTEINS AS CARRIERS OF PHARMACOLOGICAL AGENTS

EDITED BY
J. MICHAEL SHAW

Sterling Drug Inc.
Malvern, Pennsylvania

Marcel Dekker, Inc.   New York • Basel • Hong Kong

Library of Congress Cataloging-in-Publication Data

Lipoproteins as carriers of pharmacological agents / edited by J.
    Michael Shaw
            p.   cm. -- (Targeted diagnosis and therapy; 5)
        Includes bibliographical references and index.
        ISBN 0-8247-8505-3 (alk. paper)
        1. Lipoprotein drug carriers. I. Shaw, J. Michael
    II. Series.
        [DNLM: 1. Drug Carriers. 2. Lipoproteins--administration &
    dosage.   W1 TA579 v. 5 / QV 785 L764]
    RS201.L53L56   1991
    615'.19--dc20
    DNLM/DLC
    for Library of Congress   91-16479
                                                                      CIP

This book is printed on acid-free paper.

MARCEL DEKKER, INC.
270 Madison Avenue, New York, New York 10016

Current printing (last digit):
10 9 8 7 6 5 4 3 2 1

PRINTED IN THE UNITED STATES OF AMERICA

Introduction to the Series

Targeted Diagnosis and Therapy is a series intended to collect new knowledge generated in the research and development of self-directed diagnostic and therapeutic agents. The powerful tools of recombinant DNA and monoclonal antibody technologies have contributed immensely to our understanding of the concept of molecular recognition as well as protein structure-function relationships. This has yielded a view of the future that includes the use of a variety of new pharmaceutical products. These products will have the property of localizing to a predetermined site, with a consequent diagnostic or therapeutic effect. The clinical use of these products will include the treatment in vivo of malignant organs or tissues as well as elimination of specific cell types in ex vivo bone marrow-purging procedures. Each volume will focus on one product or strategy, and contain the relevant preclinical and/or clinical experience. The list of near-term subjects includes a variety of antibody conjugates, the interferons, the interleukins, tissue plasminogen activator, and gene therapy. Other volumes will deal with the next generation of agents, such as genetically engineered toxins and fusion proteins. It is expected that the series will be useful for basic researchers and clinicians alike.

New frontiers lie ahead. The opportunities for research and development of important new pharmaceutical products are considerable. It is hoped that Targeted Diagnosis and Therapy will assist in the efforts to achieve this goal.

John D. Rodwell
Editor

Preface

This volume brings together a wide range of talented scientists from seven countries with the objective of providing a more complete familiarity with and perspective on endogenous lipoprotein particles and their interaction with pharmacological agents. Lipoproteins are described in terms of their microemulsion characteristics and complex behavior in vivo as well as their transport across endothelial barriers and by receptor- and nonreceptor-mediated endocytosis at cell sites in various organs and peripheral tissues.

Each chapter either focuses on essential background information about lipoproteins or considers the interactions of pharmacological and diagnostic agents with lipoproteins. Relevant topics are reviewed and a wealth of previously published and original experimental studies is described. Readers from a variety of disciplines should find the text of interest since numerous topics in the biological, physical, medical, and pharmacological sciences are described. Chapter topics include (1) synthesis and interconversion of lipoproteins in biological compartments, (2) endocytosis and transcytosis of lipoproteins at the vascular endothelium, (3) complexing of pharmacological agents to lipoproteins, apoprotein-lipid microemulsions, and microemulsions, (4) diverse therapeutic and chemical agents that interact with lipoproteins, (5) lipoproteins and cancer, (6) modified lipoproteins as carriers of biological response modifiers, (7) receptor-dependent targeting to three cell types of the liver, (8) lipoproteins as carriers of radiopharmaceutical imaging agents, (9) pharmacological evaluation of sterylglucoside-lipoprotein complexes, and lastly (10) key issues for lipoprotein delivery and design of synthetic peptide-lipid carriers systems.

A remarkable literature and data base is available on lipoproteins, and several journals are devoted almost exclusively to topics concerned with their structure and function in normal and diseased states. The various classes of emulsionlike lipoproteins range in size from ~7 nm in diameter for high-density lipoproteins to over 1000 nm for chylomicrons and represent the highest percentage of naturally occurring particulate materials found in the extracellular biological fluids of man. References dating back over 20 years cite the interactions and transport of hydrophobic compounds with lipoproteins. The lipophilic core of lipoproteins and synthetic lipoprotein-like emulsions or micelles can potentially serve as a suitable domain and reservoir for any of a number of lipophilic or amphipathic compounds. This is especially significant when considering the vast number of drug candidates prepared by the pharmaceutical industry that are lipophilic or amphipathic in nature. Overall, the information derived from considering lipoproteins as potential carriers of drugs and diagnostic agents may be useful and applicable to the improved design of other particulate and novel carriers and the complex interactions and transport pathways they must undergo when administered parenterally.

J. Michael Shaw

Contents

Contributors

Martin K. Bijsterbosch, Ph.D. Division of Biopharmaceutics, Center for Bio-Pharmaceutical Sciences, Sylvius Laboratories, Leiden University, Leiden, The Netherlands

Raymond E. Counsell, Ph.D. Interim Chairman and Professor of Pharmacology and Medicinal Chemistry, Department of Pharmacology, The University of Michigan Medical School, Ann Arbor, Michigan

Laura E. DeForge, Ph.D. Postdoctoral Fellow, Department of Pathology, The University of Michigan Medical School, Ann Arbor, Michigan

Mark R. DeGalan, Ph.D., M.D.* Department of Pharmacology, The University of Michigan Medical School, Ann Arbor, Michigan

P. Chris De Smidt, Ph.D. Division of Biopharmaceutics, Center for Bio-Pharmaceutical Sciences, Sylvius Laboratories, Leiden University, Leiden, The Netherlands

---

Present Affiliation:
* Department of Radiology, Indiana University Medical School, Indianapolis, Indiana

Alexander T. Florence, Ph.D., D.Sc., F.R.S.E.   Professor of Pharmacy and Dean, School of Pharmacy, University of London, London, England

W. Stewart Futch, Jr., M.S., M.D.[*] Department of Animal Sciences, University of Illinois, Urbana, Illinois

Gavin W. Halbert, B.Sc., Ph.D., C.Chem. M.R.S.C., M.R.Pharm.S. Lecturer in Pharmaceutics, Department of Pharmacy, University of Strathclyde, Glasgow, Scotland

J. Kar Kruijt   Division of Biopharmaceutics, Center for Bio-Pharmaceutical Sciences, Sylvius Laboratories, Leiden University, Leiden, The Netherlands

Bo Lundberg, Ph.D.   Assistant Professor of Pharmaceutical Chemistry, Department of Biochemistry and Pharmacy, Åbo Akademi University, Åbo, Finland

Michèle Masquelier, Ph.D.   Chemist, Department of Clinical Pharmacology, Karolinska Hospital, Stockholm, Sweden

Roger S. Newton, Ph.D.   Director, Atherosclerosis Pharmacology, Pharmaceutical Research Division, Warner-Lambert Company, Ann Arbor, Michigan

Curt Peterson, M.D., Ph.D.   Associate Professor, Department of Clinical Pharmacology, Karolinska Hospital, Stockholm, Sweden

Mats Rudling, M.D.[†]   Department of Clinical Pharmacology, Karolinska Hospital, Stockholm, Sweden

Mark S. Rutherford, M.S., Ph.D.[‡] Department of Animal Sciences, University of Illinois, Urbana, Illinois

Lawrence B. Schook, M.S., Ph.D. Professor of Molecular Immunology, Department of Animal Sciences, University of Illinois, Urbana, Illinois

Susan W. Schwendner, M.S. Graduate Student Research Assistant, Department of Pharmacology, The University of Michigan Medical School, Ann Arbor, Michigan

---

Present Affiliations:

[*] Resident, Department of Internal Medicine, Eastern Virginia Graduate School of Medicine, Norfolk, Virginia

[†] Metabolism Unit, Department of Internal Medicine, Huddinge University Hospital, Huddinge, Sweden

[‡] Research Associate, Department of Biochemistry, St. Jude Children's Research Hospital, Memphis, Tennessee

Junzo Seki   Senior Research Associate of Pharmacy and Chemistry Department, Research Laboratories, Nippon Shinyaku Co., Ltd., Kyoto, Japan

J. Michael Shaw, Ph.D.[*]   Manager of Drug Delivery, Alcon Laboratories, Fort Worth, Texas

Kala V. Shaw, M.S.   Virginia-Maryland Regional College of Veterinary Medicine, Blacksburg, Virginia

Maya Simionescu, Ph.D.   Professor and Head of Section, Institute of Cellular Biology and Pathology, Bucharest, Romania

Nicolae Simionescu, M.D.   Professor and Director, Institute of Cellular Biology and Pathology, Bucharest, Romania

Kristina Söderberg   Department of Clinical Pharmacology, Karolinska Hospital, Stockholm, Sweden

David K. Spady, M.D.   Associate Professor of Medicine, Department of Internal Medicine, University of Texas, Southwestern Medical Center, Dallas, Texas

Makoto Sugiyama, Ph.D.   Manager of Pharmacy and Chemistry Department, Research Laboratories, Nippon Shinyaku Co., Ltd., Kyoto, Japan

Theo J. C. Van Berkel, Prof. Dr.   Head of the Division of Biopharmaceutics, Center for Bio-Pharmaceutical Sciences, Sylvius Laboratories, Leiden University, Leiden, The Netherlands

Sigurd Vitols, M.D., Ph.D.   Department of Clinical Pharmacology, Karolinska Hospital, Stockholm, Sweden

---

Present Affiliation:
[*] Section Head, Parenteral Drug Delivery, Sterling Drug Inc., Malvern, Pennsylvania

LIPOPROTEINS AS CARRIERS OF PHARMACOLOGICAL AGENTS

Lipoproteins in Biological Fluids and Compartments: Synthesis, Interconversions, and Catabolism

David K. Spady
University of Texas, Southwestern Medical Center, Dallas, Texas

I. INTRODUCTION

The plasma lipoproteins are emulsion or microemulsion particles that transport lipids through the vascular and extracellular fluid compartments of the body. The plasma lipoproteins have a common structural organization with a hydrophobic core of triglyceride and esterified cholesterol solubilized by a monomolecular film of amphipathic phospholipid, free cholesterol, and apoproteins whose hydrophilic domains project into the aqueous phase [1]. The major lipids transported through the plasma via lipoproteins are those present in the core, i.e., triglyceride and cholesterol [2,3]. For example, up to 200–300 g of triglyceride are transported out of the small intestine and liver each day. Depending on the metabolic state of the body, most of this triglyceride is delivered either to cardiac and skeletal muscle to be used as energy or to adipose tissue to be stored as fat. A much smaller mass of cholesterol is transported through the plasma via lipoproteins and, in contrast to triglyceride, the movement of cholesterol is primarily from peripheral tissues back to the liver and small intestine, the two organs through which cholesterol can be eliminated from the body.

This chapter presents an overview of the synthesis, intravascular processing, and catabolism of the major classes of lipoproteins. Special emphasis is given to recent information dealing with the quantitative importance of various organs for the uptake and degradation of lipoproteins and to the regulation of lipoprotein transport in these organs in vivo. In addition, the sites of catabolism of several modified and artificial lipoproteins are reviewed and the potential for selective delivery of lipoproteins to specific tissues is discussed.

II. CLASSIFICATION OF PLASMA LIPOPROTEINS

The plasma lipoproteins can be separated into various classes according to their density (ultracentrifugation), size (gel filtration), net surface charge (electrophoresis), and other surface properties (precipitation techniques, affinity columns). The most common classification of plasma lipoproteins is based on their density in salt solutions [2,4,5]. According to this classification, the major density fractions include the very low density lipoproteins (VLDL, $d < 1.006$ g/ml), intermediate density lipoproteins (IDL, $d = 1.006–1.019$ g/ml), low density lipoproteins (LDL, $d = 1.019–1.063$ g/ml), and high density lipoproteins (HDL, $d = 1.063–1.21$ g/ml). HDL particles are often further separated into HDL_2 ($d = 1.063–1.125$ g/ml) and HDL_3 ($d = 1.125–1.21$ g/ml). Chylomicrons, which appear in the plasma after a fatty meal, also have a density less than 1.006 g/ml. The relationship between these ultracentrifugally defined lipoprotein classes and their size and electrophoretic mobility is seen in Table 1 [2,4,6].

III. COMPOSITION OF PLASMA LIPOPROTEINS

Typical values for the composition of the major lipoprotein classes are summarized in Table 2 [2,5,6]. Chylomicrons are triglyceride-rich emulsion particles synthesized by the mucosal cells of the small intestine. Chylomicrons range in size from 800 Å to 10,000 Å in diameter with the largest particles being formed under conditions of high dietary triglyceride intake. Chylomicrons have the highest lipid/protein ratio and as a consequence have the lowest density ($d < 0.95$ g/ml) and will float to the surface of plasma stored overnight

Table 1 Physical Properties of Plasma Lipoproteins

Class	Density (g/ml)	Electrophoretic mobility	Diameter (Å)
Chylomicrons	0.95	–	800–10,000
VLDL	0.95–1.006	Pre–β	300–800
IDL	1.006–1.020	Slow pre–β	250–350
LDL	1.020–1.063	β	180–220
HDL_2	1.063–1.125	α	95–120
HDL_3	1.125–1.210	α	55–95

Source: Ref. 2, 4, and 6.

at 4°. Nascent chylomicron particles contain apoproteins A-I and B-48; however, once these particles enter the lymph and plasma, they quickly acquire apoproteins E and C [7–11]. Very low density lipoproteins are triglyceride-rich emulsion particles synthesized by the liver. They are somewhat smaller than chylomicrons and contain relatively less triglyceride and more cholesterol and protein. The major apoproteins of VLDL are apo B-100 and the C apoproteins. Low density lipoproteins are formed during the intravascular metabolism of VLDL. More than 95% of the protein content of LDL is apo B-100 and the core lipid is predominantly cholesteryl ester.

High density lipoproteins are the smallest of the human lipoproteins and have the highest protein/lipid ratio [13]. The main core lipid is cholesteryl ester. The major apoproteins of HDL are apo A-I and apo A-II with apo E, apo D, and the C apoproteins being present in smaller amounts.

IV. PLASMA APOPROTEINS

The apoproteins are important structural components of lipoproteins and are critically involved in the assembly, secretion, intravascular processing, and catabolism of lipoproteins. At least ten apoproteins have been identified in normal human serum including apo A-I, apo A-II, apo A-IV, apo B-100, apo B-48, apo C-I, apo C-II, apo C-III, apo D, and Apo E [14–16]. In addition, apo [a] is a component of lipoprotein [a], which has been demonstrated in more than 90% of the human population [16–18]. With the exception of apo B-48 and apo B-100, apoproteins are soluble in the aqueous environment of the plasma and readily transfer among lipoproteins of the same, and in some cases different classes. A common feature of these soluble apoproteins is a high degree of α-helical secondary structure in which the amino acid residues are distributed to form polar and nonpolar surfaces [4,19]. These amphipathic

Table 2 Composition of Plasma Lipoproteins (percent dry weight)

Class	Triglycerides	Cholesteryl esters	Cholesterol	Phospholipid	Protein
Chylomicrons	85	4	2	7	2
VLDL	53	13	7	19	8
IDL	23	30	9	19	19
LDL	8	40	8	22	22
HDL$_2$	8	20	5	30	37
HDL$_3$	4	14	3	25	54

Source: Ref. 2, 5, and 6.

Table 3 Plasma Apoproteins

Apoprotein	Plasma Concentration (mg/dl)	Molecular Weight	Amino Acids	Metabolic function
A-I	100–150	28,000	243	Activates LCAT
A-II	30–50	17,000	154	
A-IV	13–37	46,000	376	
B-100	70–100	549,000	4,536	Synthesis and secretion of VLDL; receptor-mediated catabolism of LDL
B-48	—	264,000	2,152	Synthesis and secretion of chylomicrons
C-I	4–6	6,500	57	Activates lipoprotein lipase
C-II	3–5	9,000	79	Inhibits hepatic uptake of triglyceride-rich lipoproteins; may inhibit C-II–activated lipoprotein lipase activity
C-III	12–14	8,000	79	
D	10	22,000		
E	3–5	34,000	299	Receptor-mediated catabolism of E-containing lipoproteins

Source: Ref. 2, 4, 12, 14, 15, and 16.

α-helical regions are thought to represent lipid-binding domains with the polar surface interacting with the aqueous environment of the plasma and the non-polar surface interacting with the hydrophobic core of the lipoprotein particle. In contrast, the B apoproteins are large hydrophobic proteins that contain major portions of ß structure and relatively little α-helical structure. As a consequence, the B apoproteins do not dissociate from lipoprotein particles and when delipidated are insoluble in aqueous buffers. The molecular weight, approximate plasma concentration, and function of the major plasma apoproteins are summarized in Table 3 [2,4,12,14–16]. Apo A-I and apo A-II are the major apoproteins of HDL. Synthesis of these proteins takes place predominantly in the liver and small intestine [7,9,20,21]. Apo A-I is a cofactor for lecithin-cholesterol acyltransferase (LCAT), a plasma enzyme that catalyzes the conversion of cholesterol and phosphatidylcholine to esterified cholesterol and lysophosphatidylcholine, respectively [22]. In addition, mixtures of apo A-I and phospholipid are capable of removing cholesterol from cell membranes [23]. In this manner HDL may facilitate the transfer of cholesterol from peripheral tissues back to the liver, a process referred to as reverse cholesterol transport [24]. The function of apo A-II has not been determined.

Apo B is present in plasma in two major forms. Normal VLDL, IDL, and LDL contain a 549,000 dalton species designated apo B-100 in a centile system of nomenclature, whereas the major species in chylomicrons has an apparent molecular weight of 264,000 daltons and is designated apo B-48 in the same nomenclature [25]. Patients with abetalipoproteinemia, a disorder in which apo B synthesis is genetically absent, have undetectable levels of apo B–containing lipoproteins (chylomicrons, VLDL, IDL, and LDL) in plasma [2]. Thus apoprotein B is absolutely required for the secretion of these particles. In humans apo B-100 is synthesized in the liver, whereas apo B-48 is synthesized in the small intestine [2,26]. Hepatic apo B-100 mRNA is about 14 kb in length and encodes a mature protein of 4536 amino acids [27,28]. Apo B-48 is collinear with the amino terminal 2152 amino acids of apo B-100 [29,30]. Furthermore, recent studies have shown that the gene that encodes apo B-48 in the small intestine is identical to the gene that encodes apo B-100 in the liver. In the intestine, however, apo B mRNA contains a single base change resulting in a translational stop at codon 2153 [30]. Apo B-100 is the determinant that is recognized by the LDL receptor [31,32]. The receptor–binding domain of apo B-100 is located in the carboxy terminal third of the protein; consequently, apo B-48 does not contain the receptor–binding domain and does not interact with the LDL receptor [30,33].

The C apoproteins are major components of triglyceride-rich lipoproteins and are also present on HDL. Synthesis is thought to take place primarily in the liver and small intestine. Apo C-II plays an important role in lipoprotein metabolism as an activator of lipoprotein lipase [34–36]. Lipoprotein lipase, an enzyme located in the capillary beds of several extrahepatic tissues, is responsible for hydrolyzing the triglycerides present in the core of chylomicrons and VLDL [37]. Apo C-II deficiency results in severe hypertriglyceridemia (similar to that seen with lipoprotein lipase deficiency), which can be corrected transiently by infusions of plasma or purified apo C-II. Apo C-III appears to inhibit apo C-II–activated lipoprotein lipase activity [37] and also inhibits the uptake of triglyceride-rich lipoproteins by the liver [38–40]. The physiological role of apo C-I is not known.

Apo E is a component of several classes of lipoproteins including chylomicrons, VLDL, IDL, and HDL. Apo E plays an important role in the catabolism of these particles through its interactions with the LDL and chylomicron receptors [41,42]. Genetic polymorphism of apo E has been demonstrated in humans, and variants of apo E that bind poorly to the LDL receptor are found in patients with familial type 3 hyperlipoproteinemia [2,14,43]. This disorder is characterized by accumulations of remnant particles, xanthomatosis, and premature atherosclerosis. The liver is the major site of apo E synthesis; however, 20–40% of total body apo E mRNA is extrahepatic, suggesting that a significant portion of circulating apo E maybe synthesized in peripheral tissues [44–47].

Apo[a] along with apo B are the protein components of lipoprotein[a]. Apo[a] is crosslinked to apo B in the same particle by disulfide bonds. The apo[a] component is a large protein of approximate M_r 650,000 that is distinctly different from apo B in composition and structure. The apo B component appears to be identical to that present in LDL. The function of apo[a] is unknown; presumably it is synthesized in the liver[16–18].

Over the last few years, a great deal of information has become available regarding the genetics and molecular biology of the plasma apoproteins. Complementary DNA and genomic clones for most of the apoproteins have been obtained and characterized and the apoprotein genes have been mapped within the human genome [14,27,27,48–60]. With this information in hand, the factors that regulate normal apoprotein gene expression can be examined in detail. Furthermore, it should now be possible to identify polymorphisms and mutations of the apoprotein genes that alter gene expression.

In addition to the native apoproteins described previously, a number of synthetic apoproteins have been designed. Synthetic peptides that contain amphipathic α-helical regions associate with lipid surfaces and displace the native exchangeable apoproteins of VLDL and HDL [61,62]. The factors that regulate the affinity of these peptides for lipid particles are currently being delineated. Other lipid-associating groups such as saturated fatty acyl chains have also been used to anchor synthetic peptides to lipid particles [62]. Thus by covalently linking a high affinity receptor–binding peptide to a lipid-associating group, it may be possible to target lipoprotein particles to specific tissues or cells.

V. DISTRIBUTION OF LIPOPROTEINS IN EXTRAVASCULAR FLUIDS

In order to interact with the parenchymal cells of intact organs, plasma lipoproteins must first be transported across the capillary wall into the interstitial fluid that is in contact with cells. The principal barrier to the passage of lipoproteins from plasma to the interstitial fluid is the capillary endothelium. The permeability of capillary endothelium to lipoproteins varies widely among different tissues and may be a major determinant of the rate of lipoprotein uptake by certain organs. The most permeable capillaries are the open sinusoids of the liver and spleen. In the liver, the endothelial cells lining the sinusoids contain large fenestrations of about 1000 Å in diameter [63,64]. These fenestrations readily allow the passage of all lipoproteins except large chylomicrons. Thus the concentration of a particular lipoprotein at the sinusoidal surface of hepatocytes is similar to that in plasma. A number of other tissues, including the small intestine, kidney, and endocrine organs, also contain fenestrated capillaries [63]. In these organs, however, the fenestrations involve only 2–30% of the endothelial surface, are much smaller in diameter (about 500 Å), and may be closed by a thin diaphragm. Capillary permeability to lipoproteins in these tissues is probably somewhat lower than that of the liver, especially for the larger lipoproteins such as VLDLs and chylomicrons.

Most of the large tissue compartments of the body, such as skeletal muscle, skin, and adipose tissue, contain nonfenestrated capillaries, which appear to offer significant resistance to the passage of all lipoproteins [63,65–67]. Thus the concentration of lipoproteins in peripheral lymph (which is thought to reflect the composition of interstitial fluid) relative to plasma equals about 0.03, 0.1, and 0.2 for VLDL, LDL, and HDL, respectively. Transport of plasma lipoproteins across nonfenestrated capillaries is mediated largely by vesicular transport, which, at least in the case of LDL, appears to be a nonsaturable, receptor–independent process [65,67].

The least permeable capillaries are those of the central nervous system, which are lined by a continuous tight–junction endothelium [63]. Indeed, no lipoproteins in the VLDL or LDL density range and no apo B can be detected in cerebrospinal fluid (CSF) [68]. Lipoproteins in the HDL density range are present in CSF and these lipoproteins contain apoproteins A-I and E [68,69]. The apo A-I and apo E are present on separate populations of lipoprotein particles. Several lines of evidence suggest that apo A-I in CSF is derived from plasma. First, the CSF/plasma concentration ratio of apo A-I (about 0.6%) is similar to that of albumin, a protein in CSF known to be derived from plasma. Furthermore, apo A-I mRNA cannot be detected in brain and apo A-I has not been seen in the secretory apparatus of brain cells. It is not known, however, if apo A-I or apo A-I–containing lipoproteins are actually transported from plasma into the interstitial fluid of the brain. Cerebrospinal fluid is derived both from the interstitial fluid that seeps off the surface of the brain and from the choroid plexus. The CSF secreted into the ventricles by the choroid plexus is derived largely from plasma. Thus choroid plexus, rather than brain interstitial fluid, may be the source of apo A-I in the CSF. The CSF/plasma concentration ratio of apo E (about 5%) is nearly 10 times that of apo A-I or albumin. The finding of high concentrations of apo E mRNA in brain and the presence of apo E in the Golgi apparatus of astrocytes suggest that apo E in CSF arises largely from synthesis within the brain [47,68,70].

VI. LIPOPROTEIN METABOLISM

At any point in time, the concentration of a particular lipoprotein in the plasma is determined by the rate at which it is introduced into the circulation relative to the rate at which it is removed from the plasma by the various tissues of the body. Lipoprotein uptake by the various organs of the body may, in turn, take place via receptor–dependent or receptor–independent pathways. In addition, lipoproteins may undergo a number of metabolic conversions before cellular uptake and degradation. Thus under steady–state conditions, the concentration of a lipoprotein in plasma represents the summation of several individual rate constants.

A. Metabolism of Triglyceride-Rich Lipoproteins

The primary function of the triglyceride-rich lipoproteins is the transport of triglyceride from the small intestine and liver to peripheral sites of utilization and storage. Chylomicrons are synthesized by the small intestine in response to fat absorption, whereas VLDLs are synthesized by the liver to transport endogenously synthesized triglyceride.

Synthesis of Triglyceride-Rich Lipoproteins

Following a fat-rich meal, lipids are hydrolyzed in the intestinal lumen and absorbed by enterocytes covering the tips of the villi. Within the enterocyte, absorbed fatty acids and monoglycerides are reesterified to triglyceride by enzymes located on the cytoplasmic surface of the smooth endoplasmic reticulum [71]. The newly formed triglyceride is sequestered within the lumen of the smooth endoplasmic reticulum in triglyceride-rich particles [72]. Absorbed cholesterol is also incorporated into the newly formed particles but generally is a minor component since cholesterol absorption amounts to less than 500 mg/day, whereas triglyceride absorption typically amounts to more than 100 g/day. Apoprotein synthesis is thought to take place on polyribosomes attached to the rough endoplasmic reticulum [73]. The newly synthesized apoproteins are transported to the smooth endoplasmic reticulum where they associate with the lipid particles. After passage through the Golgi apparatus, where further modifications such as glycosylation of the apoproteins take place, the nascent chylomicrons are concentrated in secretory vesicles which fuse with the basolateral membrane of the enterocyte and deliver the chylomicrons into the pericellular space to be picked up by the lymphatics of the villus core.

Apo B-48 is absolutely required for the secretion of chylomicrons. Thus, under conditions where protein synthesis has been inhibited or is genetically absent (abetalipoproteinemia), triglyceride droplets accumulate in the absorptive cells and chylomicron secretion ceases [2,74]. Ordinarily, the rate of lipid absorption (and thus the rate of chylomicron lipid secretion) varies widely throughout the day and from one day to the next. The small intestine responds to changes in lipid absorption primarily by altering the size and composition of the chylomicron particles rather than by altering the number of particles secreted. Thus during periods of rapid triglyceride absorption, chylomicron size, as well as the triglyceride/protein ratio, increases. In contrast, synthesis of apo B-48 is little affected by short or long term changes in triglyceride flux across the intestine [7,75].

Just as chylomicrons are synthesized by the small intestine to transport exogenous fat, so VLDLs are synthesized by the liver to transport endogenous lipid. In the hepatocyte, as in the enterocyte, the apoproteins are synthesized in the rough endoplasmic reticulum and translocated to the smooth endoplasmic reticulum where the enzymes involved in lipid synthesis are located. The nascent VLDLs are then transported to the Golgi apparatus and sequestered into secretory vesicles which fuse with the sinusoidal membrane of the hepatocyte and exocytose the nascent lipoproteins into the plasma [71,76].

The secretion of VLDL triglyceride is regulated by the availability of triglyceride within the hepatocyte [2,77–79]. Hepatic triglyceride may be derived from several sources depending on the metabolic state of the organism. After a fatty meal a portion of the absorbed triglyceride is delivered to the liver via chylomicron remnants. On the other hand, after a high carbohydrate meal, carbohydrate delivered to the liver in excess of that which can be stored as glycogen is oxidized to acetyl CoA and used for fatty acid synthesis. Even during the postabsorptive state, albumin–bound free fatty acids derived from lipolysis in the adipose tissue provide the liver with a constant source of free fatty acids that may be used for triglyceride synthesis. As is the case with chylomicrons, changes in the availability of triglyceride or cholesterol tend to produce changes in the size and composition of VLDL rather than changes in the number of particles secreted [79,80]. With prolonged fasting, however, the number of VLDL particles secreted as well as the triglyceride content of VLDL are reduced. Individuals lacking the ability to synthesize apo B-100 are completely unable to secrete VLDL, indicating that apo B-100 is absolutely required for the secretion of VLDL by the liver [2,74].

Catabolism of Triglyceride-Rich Lipoproteins

The catabolism of chylomicrons and VLDL begins with the hydrolysis of core triglyceride by lipoprotein lipase, an enzyme found in the vascular beds of several extrahepatic tissues including heart, skeletal muscle, and adipose tissue [37,81,82]. The enzyme is synthesized in the parenchymal cells of these tissues and is then transported to its functional site on the luminal surface of capillary endothelial cells. Under a variety of nutritional states, the activity of lipoprotein lipase in adipose tissue is inversely correlated with the activity in heart and skeletal muscle. Such coordinate regulation allows chylomicron and VLDL triglyceride to be delivered preferentially to heart and skeletal muscle for oxidation during the fasted state but to the adipose tissue for storage when animals are in the fed state. Thus lipoprotein lipase plays an important role in regulating the rates at which lipoprotein triglycerides are provided to different tissues of the body for use in metabolic pathways. Regulation of lipoprotein lipase appears to be articulated by multiple hormonal actions that affect both the synthesis and degradation of the enzyme. For example, insulin enhances lipoprotein lipase activity in adipose tissue both in vitro and in vivo. On the other hand, epinephrine and glucagon inhibit lipoprotein lipase activity in the adipose tissue but stimulate it in cardiac and skeletal muscle. The mechanism by which a particular hormone enhances lipase activity in one tissue while reducing it in another is not known. One possibility is the existence of tissue–specific forms of the enzyme. However, at least in the rat and guinea pig, the enzyme appears to be similar, if not identical, in all tissues examined in terms of molecular weight and kinetic properties [82,83].

The C apoproteins serve two important functions during the hydrolysis of chylomicrons and VLDL [37–40]. First, large amounts of apo C-III relative to apo E inhibit the uptake of triglyceride–rich particles by the liver, thus allowing time for interaction of these particles with lipoprotein lipase. Second, apo C-II activates lipoprotein lipase. Indeed, deficiency of apo C-II leads to severe hypertriglyceridemia similar to that seen with lipoprotein lipase deficiency [35,36].

Interaction of chylomicrons and VLDL with lipoprotein lipase results in the liberation of large amounts of free fatty acids that are either taken up in adjacent tissues or recirculated via albumin. As triglyceride in the core of the chylomicron and VLDL is removed, the particle shrinks in size and some of its excess surface material (including phospholipid and the A and C apoproteins) buds off as discs or vesicles and is transferred to other lipoproteins, particularly HDLs [6,13].

It is at this point that the catabolism of chylomicrons and VLDL diverges. Following hydrolysis of 80–90% of the core triglyceride and transfer of the excess surface material to HDL, the chylomicron particle (which is now called a chylomicron remnant) becomes recognized by a saturable high velocity transport system on hepatocytes and is rapidly cleared by the liver [84,85]. Apo E is the determinant responsible for interaction of the chylomicron remnant with these high affinity receptors on hepatocytes [86]. Presumably, the decrease in the ratio of apo C-III to apo E is the event that leads to remnant recognition by the liver [38].

Due to the presence of apo E, chylomicron remnants bind to the LDL receptor with high affinity [87]. Nevertheless, several lines of evidence suggest that the uptake of chylomicron remnants by the liver is mediated by a receptor pathway distinct from the LDL receptor pathway. First, rabbits and patients who genetically lack functional LDL receptors appear to have normal hepatic clearance of chylomicron remnants [88]. Second, a number of pharmacological and dietary manipulations have been described that alter hepatic LDL receptor activity but have little effect on the handling of chylomicron remnants by the liver [89,90]. Third, binding of chylomicron remnants to liver membranes is only partially and variably competed by LDL or by antibodies directed against the binding site of the LDL receptor [91,92]. Finally, a 500,000 dalton liver membrane protein has recently been described as being distinct from the LDL receptor, but having close structural and biochemical similarities to this receptor. Examination of the amino acid sequence derived from a cDNA clone indicates that this protein contains an extracellular domain resembling four copies of the LDL receptor, a transmembrane segment, and a cytoplasmic tail containing two copies of a possible signal for clustering into coated pits [193]. Although little is known about the physiology of this protein, its structure suggests that it could bind and internalize lipoproteins containing apo E or apo B with high efficiency. In the mouse, this molecule is most highly expressed in the liver; however, significant amounts of mRNA for this molecule are also found in lung, brain, intestine, and muscle. Why extrahepatic tissues would have a receptor that presumably binds chylomicron remnants is not known. However, chylomicron remnants also bind to LDL receptors on a variety of cell types in vitro, whereas in vivo these particles are cleared quantitatively by the liver. Presumably the large size of chylomicron remnants prevents their passage through the capillary endothelium of extrahepatic tissues.

Little is known about the regulation of chylomicron remnant transport. In the rat, chylomicron uptake may be affected by the type of triglyceride present in the core of the particle [93] but appears not to be regulated by cholesterol feeding or ethinyl estradiol, two manipulations known to alter hepatic LDL receptor activity [89,90]. Following uptake by the liver, chylomicron remnants are delivered to the lysosomal compartment, where they are completely degraded.

Hydrolysis of VLDL triglyceride also results in the formation of a remnant particle. This remnant, like that formed by the action of lipoprotein lipase on the chylomicron, is rapidly and quantitatively taken up by the liver [94,95]. However, an alternative pathway exists for VLDL in that this particle may be metabolized through an intermediate density lipoprotein fraction to low density lipoproteins [95,96]. During the conversion of VLDL to LDL, nearly all of the triglyceride and all of the apoproteins except apo B-100 are removed. This transformation apparently takes place in the vascular space and may simply represent continued activity of hepatic and lipoprotein lipase [95]. In some species, such as the rat and rabbit, VLDLs are metabolized predominantly via the remnant pathway. In these species the rate of LDL production as well as circulating LDL levels are low [94]. In humans, however, most VLDL is metabolized to LDL [96]. What regulates the proportion of VLDL that is metabolized via these two routes is not known, although the presence or absence of apo E on newly synthesized VLDL particles may play a role [97].

Like chylomicron remnants, VLDL remnants contain apo E and thus bind with high affinity to the LDL receptor. In the case of VLDL remnants, however, it is clear that uptake by the liver is mediated by the LDL receptor pathway. Thus the hepatic uptake of VLDL remnants (but not chylomicron remnants) is markedly impaired in rabbits that genetically lack functional LDL receptors [98]. Although VLDL remnants bind to LDL receptors on a variety of cell types in vitro, VLDL remnants are cleared almost quantitatively by the liver in vivo [94], probably because most LDL receptors are located in the liver and because the size of VLDL remnants restricts their passage through the capillary endothelium of other tissues.

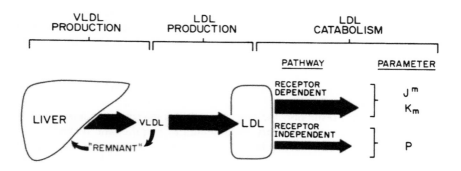

Figure 1 General scheme for the metabolism of VLDL and LDL. As described in the text, the liver synthesizes and secretes VLDL, which may then undergo one of two fates. A fraction of VLDL is metabolized to remnants that are rapidly cleared by the liver via the LDL receptor pathway while the remainder of VLDL is converted into LDL. Thus the rate of LDL production is determined both by the rate of VLDL production and by the rate of conversion of VLDL to LDL. Low density lipoprotein is removed from plasma by the various organs of the body by receptor-dependent and receptor-independent pathways. The rate of LDL catabolism by either of these pathways can be described by classical kinetic parameters. For the receptor-dependent pathway, these parameters include the maximal transport rate (J^m) and the concentration of LDL necessary to achieve half-maximal transport rates (K_m). The receptor-independent pathway can be described in terms of a proportionality constant (P).

B. Metabolism of LDL

Synthesis of LDL

As discussed previously, LDLs are formed during the metabolism of VLDLs. Since VLDL may be metabolized to remnants, which are rapidly cleared by the liver, or converted to LDL, the rate of LDL production depends both on the rate of VLDL formation by the liver and on the proportion of VLDL that is converted to LDL (Fig. 1). It appears that the liver secretes VLDL particles at a relatively constant rate, although the composition of the particle may be altered dramatically depending on the availability of triglyceride and cholesterol in the liver. Thus changes in rates of LDL production appear to be due primarily to changes in the rate of conversion of VLDL to LDL. If, as discussed earlier, VLDL remnants are cleared from plasma primarily by hepatic LDL receptors, then a decrease in hepatic LDL receptor activity would lead to an accumulation of VLDL, which, in turn, could increase the rate of conversion of these particles into LDL. Indeed, there exists an inverse relationship between LDL receptor activity and LDL production rates in rabbits and humans who genetically lack functional LDL receptors, and the changes in rates of LDL production are the result of changes in the proportion of VLDL converted to LDL [98]. The rate of LDL production may also depend on the apoprotein composition of VLDL. For example, in the rabbit, about 10% of VLDL secreted by the liver contains no apo E and these apo E-free VLDLs are approximately three times as likely to be converted to LDL than are apo E-containing VLDLs [97].

Catabolism of LDL

The turnover of LDL is considerably slower than that of VLDL and chylomicrons. In normal humans the plasma half-life of LDL is 3–4 days. With the development of radiolabeled markers for LDLs that are retained by tissues after uptake, it has become possible to study in quantitative terms the uptake of LDL from plasma by the various tissues of the body. Among the most useful markers are [^{14}C]sucrose and radioiodinated tyramine cellobiose [99,100]. These radiolabeled molecules can be conjugated to LDL without significantly altering its biological behavior [101,102]. During catabolism of LDL these molecules become trapped in the lysosomal compartment because of the absence in mammalian lysosomes of the enzymes necessary to degrade them. Using a primed-continuous infusion of LDL labeled with such radiolabeled molecules, it has been possible to quantitate rates of LDL transport in all of the major organs of experimental animals under conditions where the rate of LDL uptake is a linear function of the time of infusion, where leakage of the radiolabel from tissues is small relative to rates of uptake, and where circulating LDL levels can be experimentally adjusted to any desired value

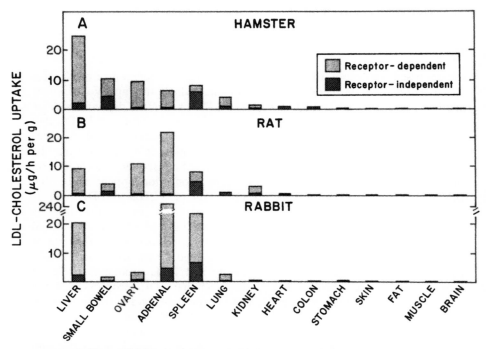

Figure 2 Rates of LDL cholesterol uptake per gram of tissue in the hamster, rat, and rabbit. In each organ LDL uptake is expressed as micrograms of LDL cholesterol removed from the plasma per hour per gram of tissue and is divided into receptor-dependent and receptor-independent components.

[101–105]. These methods have now been applied to the study of several animal species including the rat, hamster, rabbit, mouse, and dog. Representative studies in the hamster, rat, and rabbit are illustrated in Fig.2 [101,103,-104]. In all three species, rates of LDL uptake, when expressed per gram of tissue, are highest in the endocrine organs, liver, small intestine, and spleen. Somewhat lower rates of transport are observed in the kidney, lung, colon, heart, and stomach. Importantly, extremely low rates of LDL transport are found in the major tissue compartments of the body such as skeletal muscle, adipose tissue, skin, and brain. As illustrated in Fig. 3, when organ weight is taken into account, the liver becomes the overwhelmingly important site of LDL catabolism in all species and accounts for 60–80% of total LDL turnover. The small intestine is the only other organ that contributes significantly to LDL turnover. Information concerning rates of LDL uptake in the various organs of humans is not available. However, circulating LDL levels fell 50% in a

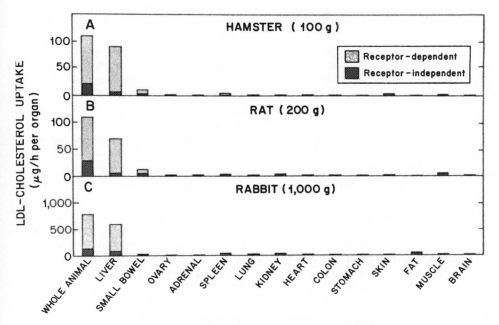

Figure 3 Rates of LDL cholesterol uptake per whole organ in the hamster, rat, and rabbit. The values represent the same data seen in Fig. 2 but are expressed as micrograms of LDL cholesterol removed from the plasma per hour per whole organ. In addition, the rate of LDL cholesterol uptake by the whole animal is shown. Uptake rates of LDL are again divided into receptor-dependent and receptor-independent components. (From Ref. 3.)

patient with a genetic defect in the LDL receptor pathway who received a normal liver transplant, suggesting that the liver also accounts for the majority of LDL turnover in humans [106].

Mechanism of LDL Transport. Low density lipoprotein is cleared from plasma by two basic pathways, as illustrated in Fig. 1. Uptake of LDL by tissues may be mediated by LDL receptors (receptor–dependent transport) or by a less well–understood process (or process) that is independent of the LDL receptor pathway (receptor–independent transport). Specific receptors that bind LDL were first described on the surface of cultured human fibroblasts and have subsequently been demonstrated on a variety of cell types [107–110]. These receptors mediate the binding and internalization of lipoproteins containing apo B-100 or apo E [41]. Following binding of LDL to its receptor, the LDL-receptor complex is incorporated into coated pits and internalized into endosomes [108]. Apparently due to a drop in the pH of the endosome, the LDL dissociates from its receptor. The LDL is then delivered to the lysosomal compartment where it is completely degraded while the LDL receptor is returned to the cell surface. The entire cycle takes 10–20 min and continues whether or not ligand is present in the media.

The LDL receptor is a glycoprotein of apparent M_r 160,000 as determined by SDS gel electrophoresis. Complementary DNA and genomic clones for the LDL receptor have been isolated and analyzed and several natural occurring mutations of the LDL receptor gene have been identified [110,111]. These mutations disrupt receptor function and give rise to familial hypercholesterolemia (FH). Cultured fibroblasts from individuals who inherit one abnormal gene (heterozygous FH) express approximately half the normal number of LDL receptors, whereas fibroblasts from subjects with two abnormal genes (homozygous FH) express no functional LDL receptors [112]. As a consequence, plasma LDL levels are approximately twice normal in heterozygous FH and six to ten times normal in homozygous FH.

By comparing LDL turnover rates in normal subjects with LDL turnover rates in patients who lack functional LDL receptors, it is possible to determine the contribution of receptor–dependent and receptor–independent mechanisms to LDL turnover in humans. These studies indicate that about two-thirds of total LDL turnover is mediated by LDL receptors and about one-third by receptor–independent processes [113]. The contribution of receptor–dependent and receptor-independent processes to total LDL turnover can also be determined in normal individuals by comparing the turnover rates of native LDL and LDL modified to prevent its recognition by the LDL receptor [114,115]. This approach is based on the observation that the arginine and lysine residues of apo B are essential for the binding of LDL to its receptor [116]. Techniques used for modifying LDL include glucosylation, reductive methylation, and treatment with cyclohexanedione; however, glucosylation and methylation most effectively eliminate receptor binding of LDL and give the lowest turnover rates in vivo [101,117]. Studies in a variety of animal species including humans indicate that these modified LDL preparations are cleared from plasma approximately 20–35% as rapidly as unmodified LDL, indicating that 65–80% of total LDL turnover is mediated by LDL receptors [101,104, 113,115,117–119].

Through the use of a primed-continuous infusion of methylated LDL, receptor–independent LDL transport has been measured in all of the major tissues of several animal species as illustrated in Fig. 2 [101,103,104]. Receptor–independent transport can be demonstrated in all tissues, although, with the exception of the spleen, rates are generally quite low. By subtracting receptor–independent uptake from total LDL uptake in a particular tissue, the receptor–dependent component of total LDL uptake is obtained. As is apparent, the high rates of LDL transport seen in the liver and endocrine organs are mediated largely by LDL receptors (>90%). In contrast, the extremely low rates of uptake in the large tissue compartments such as skin, heart, skeletal muscle, and fat can be accounted for entirely by receptorindependent pathways. Thus in all species examined it can be calculated that

80-90% of whole body LDL receptor activity is located in the liver (Fig. 3). On the other hand, receptor-independent transport is widely distributed among all tissues of the body with no single organ accounting for more than 20% of total receptor-independent LDL turnover. Data regarding the distribution of receptor-dependent and receptor-independent transport among the various human tissues are not available, although indirect evidence suggests that LDL receptor activity is located largely in the liver, as is the case in the other species that have been examined [106].

Kinetics of LDL Transport. The distribution of receptor-dependent and receptor-independent transport among the various organs of the body as described previously applies only to animals with normal plasma LDL concentrations. If the concentration of LDL in plasma is raised or lowered, the relative importance of receptor-dependent and receptor-independent LDL uptake in each organ will vary depending on the kinetic characteristics of these two transport processes in each organ. By adding mass amounts of unlabeled LDL to the infusions of radiolabeled LDL, it has been possible to measure rates of LDL transport under circumstances where plasma LDL levels have been abruptly raised and maintained throughout the experimental period (4-6 h). Thus the rates of receptor-dependent and receptor-independent LDL transport in the various organs of normal animals can be determined as a function of circulating LDL concentrations [105]. Figure 4 shows an example of such a study in the hamster. Rates of total and receptor-independent LDL uptake were measured in the liver and small intestine in normal hamsters under conditions where circulating LDL levels were acutely varied from normal (25 mg/dl) to 500 mg/dl. In both organs, total LDL uptake increases as a curvilinear function of the plasma LDL concentration and demonstrates saturation kinetics. In contrast, receptor-independent LDL uptake increases as a linear function of the plasma LDL concentration and shows no evidence for saturation even at LDL levels 20 times normal. At any particular LDL level, the rate of receptor-dependent LDL uptake equals the difference between total and receptor-independent LDL transport. Figure 5 shows the kinetic curves for total (A) and receptor-independent (B) LDL uptake for several other tissues of the normal hamster.

The relationship between total LDL uptake and plasma LDL concentrations can be analyzed using true nonlinear regression analysis and the appropriate transport equations to obtain the various transport parameters for both receptor-dependent and receptor-independent LDL uptake. As illustrated in Fig. 1, receptor-dependent transport is generally described in terms of a maximal transport rate (J^m) and the concentration of LDL necessary to achieve half-maximal transport rates (K_m) while receptor-independent transport is described in terms of a proportionality constant (P). The results of this type of analysis in the normal hamster are summarized in Fig. 6. Saturable receptor-dependent LDL uptake can be identified in five organs: the liver, adrenal gland, small intestine, spleen, and kidney. The plasma LDL cholesterol

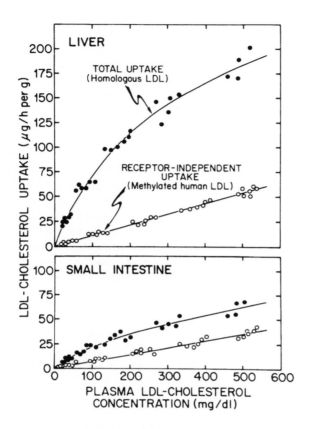

Figure 4 Kinetics of LDL uptake by the liver and small intestine of the hamster. Uptake rates for both homologous LDL (receptor-dependent and receptor-independent uptake) and methylated human LDL (receptor-independent uptake) are shown as a function of the plasma concentration of LDL cholesterol in the same animal. Rates of LDL uptake were measured in control animals whose plasma LDL cholesterol concentrations during the 4-hour experimental period were abruptly raised and maintained at values ranging from normal (25 mg/dl) to about 500 mg/dl. (Data derived from experiments detailed in Ref. 105.)

concentration necessary to achieve half-maximal uptake rates (K_m) is approximately 100 mg/dl in all five tissues. In contrast, the maximal transport rate (J^m) for the receptor-dependent pathway varies widely from about 140 μg/h per gram in the liver to < 10 μg/h per gram in the kidney. The proportionality constant (P) for receptor–independent uptake equals the micrograms of LDL cholesterol taken up per hour per gram of tissue per milligram per deciliter of LDL cholesterol in plasma and ranges from 0.032 in spleen to 0.003 in skeletal muscle. The proportionality constant for receptor–independent transport can

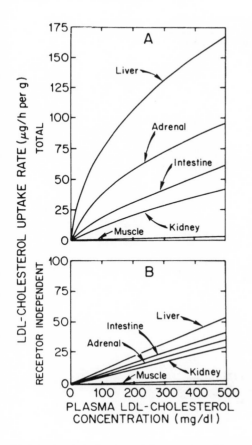

Figure 5 Kinetic curves for LDL uptake in various tissues of the hamster. (A) Relationship between total LDL cholesterol uptake by various tissues of the body and the concentration of LDL cholesterol in plasma. (B) The same relationship for the receptor-independent component of total LDL uptake. The lines represent the best-fit curves for individual data points obtained as described in Fig. 4. In these examples, both total and receptor-independent LDL uptake are expressed as micrograms of LDL cholesterol removed from plasma per hour per gram of tissue. (From Ref. 120.)

also be obtained from the slope of the relationship between methylated LDL uptake and plasma LDL concentrations (Fig. 5B) and these values are essentially identical to those obtained from an analysis of the kinetic curves for total (homologous) LDL uptake.

These types of studies have now been performed in three species and the following generalizations can be made. First, saturable receptor–dependent LDL transport can be demonstrated in the endocrine organs, liver, small in-

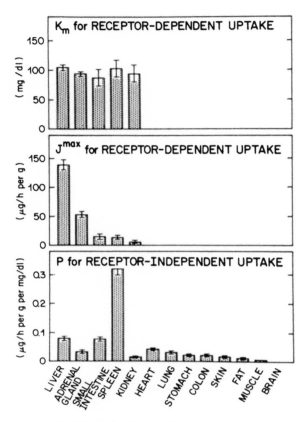

Figure 6 Kinetic constants for receptor-dependent and receptor-independent LDL transport in the hamster. (Data derived from experiments detailed in Ref. 105.)

testine, spleen, and kidneys of all three species. Maximal transport rates for the receptor–dependent pathway vary widely but are highest in the endocrine organs and liver when expressed per gram of tissue. Despite these wide variations in the maximal transport rate, the K_m for the receptor–dependent transport process is approximately 100 mg/dl in all tissues. Receptor–independent transport is not saturable even at high LDL concentrations.

The LDL uptake rates seen in Fig. 5 can be multiplied by the whole organ weight to give the kinetic curves for LDL uptake by the whole tissue. In addition, the kinetic curves for LDL uptake in all of the individual tissues of the body can be summed to yield the kinetic curves for total body LDL transport. Such curves for total body LDL transport in the hamster are seen in Fig. 7. In panel A the solid curve represents the total rate of LDL cholesterol

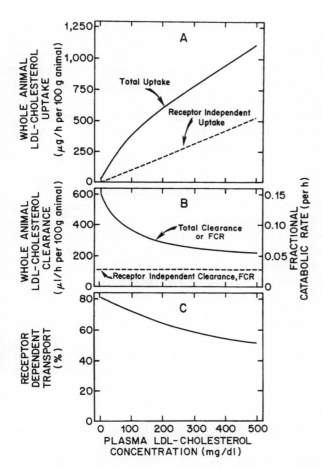

Figure 7 Kinetic curves for LDL transport in the whole hamster weighing 100 g. (A) Rate of LDL cholesterol removal from the plasma by all tissues of the body expressed as micrograms of LDL cholesterol taken up per hour per 100 g animal. This uptake process is also shown for the receptor-independent component of total LDL uptake. (B) The same data presented as whole-animal clearance rates or fractional catabolic rates. Again, both the total and receptor-independent components are shown. (C) Percentage of total LDL uptake that is receptor dependent at various concentrations of plasma LDL cholesterol. These curves are constructed under conditions in which J^m and K_m for the receptor-dependent process and P for the receptor-independent process are kept constant and only the production rate is changed. (From Ref. 105.)

removal from plasma, which, in a steady state, must equal the rate of LDL production. The dashed line shows the proportion of total LDL transport that is receptor independent. Panel B illustrates these same data expressed either as whole animal clearance rates or as fractional catabolic rates (FCRs). Panel C shows the percentage of total LDL removal from plasma that is receptor dependent at any plasma LDL cholesterol concentration. The receptor-independent component of LDL transport is a linear function of the plasma LDL cholesterol concentration (panel A) and therefore is a constant when expressed either as a clearance rate or as an FCR (panel B).

It should be noted that the plasma LDL clearance rate, FCR, and percentage of LDL removed from plasma by the receptor-dependent process decrease markedly as the plasma LDL cholesterol concentration is raised. These changes take place even though there has been no change in LDL receptor activity in the body. Thus changes in LDL clearance or FCR cannot be interpreted as a change in LDL receptor activity since anything that changes the plasma concentration of LDL (such as a change in the rate of LDL production) will also alter these two parameters [105,120]. Data on LDL clearance or FCR can be interpreted only if superimposed upon the kinetic curves for normal LDL transport in that species. For these reasons, interpretation of human LDL turnover data can be extremely difficult. Low density lipoprotein turnover studies have been performed in normal as well as hyperlipidemic patients subjected to a wide variety of dietary and drug treatments. However, since kinetic curves for normal LDL transport in humans are not available, it is generally impossible to determine whether a decrease in the FCR for LDL represents downregulation of receptor-dependent LDL transport or overproduction of LDL with saturation of receptor-dependent LDL transport.

Regulation of Receptor-Dependent LDL Transport In Vivo. The availability of standard kinetic curves for normal LDL transport in several species has made it possible to study in detail the regulation of receptor-dependent LDL transport in these species under in vivo conditions. Since the vast majority of receptor-dependent LDL transport takes place in the liver, most studies dealing with the regulation of LDL receptor activity in vivo have focused on this organ. In addition to genetic defects in the LDL receptor pathway, a large number of environmental factors have been shown to regulate receptor-dependent LDL transport in the liver. It is now apparent that the composition of the diet is one of the most important determinants of hepatic LDL receptor activity. Studies dealing with the effects of dietary cholesterol and triglycerides on hepatic LDL receptor activity in the hamster are illustrated in Fig. 8. The normal kinetic curves for total and receptor-independent transport in the liver of control hamsters are shown in the shaded areas. The area between the two curves represents the amount of receptor-dependent LDL transport at each plasma LDL cholesterol concentration. Superimposed on these kinetic curves for normal LDL transport are data obtained in hamsters fed 0.12% cholesterol in the presence or absence of 20% safflower oil or 20% hydrogenated coconut oil for 1 month [102,121]. When the diet is supplemented with cholesterol alone, plasma LDL cholesterol levels increase about twofold and the rate of hepatic LDL cholesterol uptake is significantly less than would be expected in a control animal at the same LDL cholesterol level. Indeed, receptor–dependent LDL uptake has been reduced by nearly 50%. When the diet is further supplemented with safflower oil, receptor–dependent LDL transport increases to near normal values and plasma LDL levels fall slightly. In contrast, when hydrogenated coconut oil rather than safflower oil is added to the diet, receptor–dependent LDL uptake is markedly reduced and plasma LDL cholesterol levels dramatically rise. The effects of these dietary manipulations are summarized in Fig. 9, where whole liver LDL receptor activity is shown in panel A as a percentage of control, while plasma LDL cholesterol concentrations are shown in panel B. As is apparent, dietary cholesterol produces a dose-dependent suppression of hepatic LDL receptor activity with reciprocal elevations of plasma LDL cholesterol concentrations. At each level of dietary cholesterol, polyunsaturated triglyceride somewhat reduces, while the saturated triglyceride markedly augments the deleterious effect of dietary cholesterol on

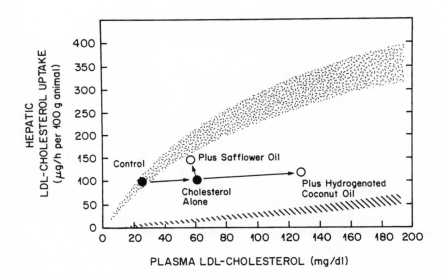

Figure 8 The effect of dietary cholesterol and triglycerides on hepatic LDL cholesterol uptake. The shaded areas represent the kinetic curves for total (stippled) and receptor-independent (hatched) LDL uptake determined in control hamsters. The individual points superimposed on these normal kinetic curves show the mean values for animals maintained on the indicated diets for one month. (Data derived from experiments detailed in Ref. 102.)

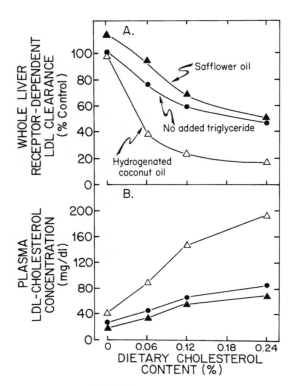

Figure 9 Whole liver LDL receptor activity and plasma LDL cholesterol concentration in animals fed diets enriched with 0.06, 0.12, or 0.24% cholesterol with and without 20% safflower oil or hydrogenated coconut oil. Whole liver LDL receptor activity is expressed as a percentage of control and represents the rate of receptor-dependent LDL uptake in the whole liver of experimental animals relative to the rate of receptor-dependent uptake that would be seen in the whole liver of control animals at the same LDL cholesterol concentration. (From Ref. 102.)

hepatic LDL receptor activity and plasma LDL cholesterol concentrations. In contrast to receptor–dependent LDL transport, receptor–independent LDL transport remains largely unaffected by these dietary fats.

Thus the quantity and type of dietary fat appear to have a profound effect on hepatic LDL receptor activity. Indeed, suppression of hepatic LDL receptor activity by a diet rich in cholesterol and saturated fat may be responsible, at least in part, for the progressive rise in plasma LDL levels that occurs with aging in Western societies. Recent studies in this laboratory have shown that plasma LDL levels and hepatic LDL receptor activity remain constant throughout the entire life span of male hamsters when maintained on a low cholesterol, low triglyceride diet.   In contrast, when animals are maintained on a diet enriched in saturated triglyceride and a small amount of cholesterol, plasma LDL cholesterol levels gradually rise into the range commonly seen in Western man and this rise is due, at least in part, to downregulation of receptor-dependent LDL transport in the liver.

Other experimental manipulations have been shown to regulate hepatic LDL receptor activity in vivo.   For example, expansion of the enterohepatic pool of bile acids by feeding cholic acid, chenodeoxycholic acid, or deoxycholic acid apparently increases cholesterol availability within the liver either by enhancing cholesterol absorption or by suppressing bile acid synthesis, a major pathway for cholesterol elimination from the body.   The result is suppression of de novo cholesterol synthesis in the liver and downregulation of hepatic LDL receptor activity [122].   Conversely, interfering with bile salt absorption by the administration of bile salt sequestrants such as cholestyramine enhances the rate of conversion of cholesterol to bile salts in the liver [103].   The result is a compensatory increase in de novo cholesterol synthesis in the liver and upregulation of hepatic LDL receptors.   It should be noted that the primary response to a change in cholesterol availability within the hepatocyte is an appropriate adjustment in the rate of hepatic cholesterol synthesis.   Only if this compensatory mechanism is inadequate is the LDL receptor pathway regulated [123].   Thus species that have a great capacity to stimulate and suppress hepatic cholesterol synthesis, such as the rat, are less susceptible to changes in cholesterol flux across the liver.

Because of the relationship between plasma LDL levels and atherosclerosis, a major emphasis has been placed on the development of drugs to lower LDL levels.   At the present time, mevinolin appears to be a very promising LDL-lowering agent both in terms of efficacy and lack of side effects [124,125]. Mevinolin is a competitive inhibitor of hydroxymethylglutaryl CoA (HMG CoA) reductase, the rate-limiting enzyme of cholesterol biosynthesis [124,125]. Suppression of de novo cholesterol synthesis leads to a reduced content of cholesterol within the hepatocyte, which in turn leads to upregulation of LDL receptor activity.   Mevinolin does not lower plasma cholesterol levels in most small experimental animals since these animals can circumvent the competitive blockade of HMG CoA reductase by synthesizing large quantities of this enzyme.

Little is known regarding the regulation of LDL receptor activity in extrahepatic tissues under in vivo conditions.   It appears that LDL receptor activity can be regulated in the endocrine organs by hormonal manipulations that alter steroid hormone synthesis.   In addition, extrahepatic LDL receptors appear to be downregulated by prolonged hypercholesterolemia.

There is also little information regarding the regulation of receptor-independent LDL transport.   Receptor-independent LDL transport is thought to represent bulk fluid phase endocytosis and it is certainly possible that this process could be altered, especially by dietary manipulations in which cell membranes become enriched in particular fatty acids. However, in recent studies in our laboratory we have been unable to demonstrate regulation of receptor–independent LDL transport.

Finally, receptor–dependent LDL transport may be affected by polymorphisms and mutations of apoprotein B, the constituent of LDL recognized by the LDL receptor. Several individuals have been described in whom the FCR of autologous LDL was significantly lower than the FCR of homologous LDL, suggesting that the LDL of these individuals has a diminished affinity for LDL receptors [126]. In addition, apo B gene variants appear to be associated with the determination of plasma cholesterol and LDL levels in normal and hyperlipidemic patients [127]. The degree to which polymorphisms of apo B contribute to the variability in plasma LDL levels observed in the general population is not known.

Regulation of LDL Receptor Activity in Malignantly Transformed Tissue. All cells of the body require at least some new cholesterol to support membrane synthesis and turnover. After malignant transformation, requirements for cholesterol may increase, especially during periods of rapid proliferation and growth of the tumor. In vivo, tumors derive cholesterol via two tightly regulated pathways, de novo synthesis and receptor–dependent LDL uptake. Many cell lines derived from malignant tissue express LDL receptor activity in vitro [128–131]. Thus the possibility exists that chemotherapeutic agents could be delivered to tumors via LDL if LDL receptor activity in the tumor was high relative to other tissues of the body. However, there is little evidence that malignantly transformed tissue expresses high levels of LDL receptor activity in vivo. For example, using the Lewis rat renal carcinoma model, we found that after malignant transformation, LDL receptor activity disappeared entirely and the cancer acquired the cholesterol needed for growth by a fivefold increase in the rate of de novo cholesterol synthesis [132,133]. This loss of receptor–dependent LDL transport was probably not due to a decrease in blood flow or capillary permeability since receptor–independent LDL transport was essentially identical in normal and malignant renal tissue. If similar results are seen with other malignant tumors, selective delivery of drugs to malignantly transformed tissue via the LDL receptor pathway may prove to be difficult.

Modified LDL

As described previously, in normal animals and humans about three-quarters of LDL turnover is mediated by LDL receptors, 80–90% of which are located in the liver. Recognition of LDL by the LDL receptor may be blocked by methylation or glucosylation of LDL. In this case, LDL is taken up only through receptor–independent pathways which are distributed throughout the body rather than predominantly in the liver [101,104]. A number of other modifications of LDL have been reported whereby LDL becomes recognized by receptors other than the LDL receptor. The most relevant modifications include those that allow LDL to be recognized and catabolized by the galactose-specific receptor pathway present in hepatocytes or by the scavenger receptor pathway present in macrophages and endothelial cells.

Uptake of LDL by the Galactose-Specific Receptor. Low density lipoprotein derivatized with 200–250 lactose residues per LDL particle becomes a ligand for the galactose-specific receptor [134]. This receptor pathway is a high capacity transport system located almost exclusively on the parenchymal cells of the liver which mediates the binding, internalization, and lysosomal catabolism of asialofetuin. Lactosylated LDL is rapidly and quantitatively cleared by the liver at rates up to 20-fold greater than the rate of LDL transport. Furthermore, this pathway is not regulated by alterations in cholesterol balance across the liver. Thus LDL whose apo B has been extensively lactosylated could serve as a vehicle for rapidly and quantitatively delivering drugs to the lysosomal compartment of hepatocytes.

Lactosaminated Fab fragments of IgG are also rapidly cleared by the galactose-specific receptor in the liver [135]. Lactosaminated Fab fragments directed at a particular plasma protein will therefore mediate the rapid hepatic clearance of that protein. Administration of Fab fragments directed against apo B has been shown to cause rapid clearance of LDL by the galactose-specific receptor in the liver. This is another approach by which lipoproteins may be targeted specifically to the hepatocyte in vivo.

Uptake of LDL by the Scavenger Receptor. Under in vitro conditions peripheral blood monocytes and macrophages derived from a variety of tissues express high affinity binding sites that recognize a diverse group of polyanionic macromolecules including modified LDL, maleylated albumin, and certain polysaccharides (dextran sulfate and fucoidin) and polynucleotides (polyinosinic acid and polyguanylic acid) [136–138]. Charge modification of critical lysine residues of the LDL protein by acetylation, maleylation, succinylation, or treatment with malondialdehyde results in recognition of the particle by the scavenger receptor and concomitant loss of recognition by the LDL receptor [137,139]. In addition, incubation of LDL with endothelial cells or with copper also produces LDL that is recognized by the scavenger pathway [140]. In contrast to the LDL receptor pathway, the scavenger receptor pathway is not regulated by the cholesterol content of the cell. Thus macrophages incubated with modified LDL may accumulate large amounts of cholesteryl esters and eventually assume the appearance of foam cells. In addition to peripheral blood monocytes and tissue macrophages, endothelial cells also express scavenger receptors. In fact, when chemically modified LDL is administered to animals in vivo, the vast majority is rapidly cleared by endothelial cells in the liver [141,142]. It appears that the modified LDL is then translocated from the sinusoidal endothelial cells to the parenchymal cells of the liver [141]. Thus it may be possible to selectively deliver drugs to the reticuloendothelial system via LDL that has been chemically modified so as to be recognized by the scavenger receptor pathway (see Chapter 6). In vivo, however, a large portion of such LDL may ultimately be catabolized by parenchymal cells of the liver. It should be emphasized that modified LDLs used to determine receptor–independent LDL transport (methylated LDL and glucosylated LDL) are not recognized by the scavenger receptor pathway.

C. Metabolism of HDL

Cholesterol is an essential structural component of all cells. The major source of cholesterol for most peripheral tissues is de novo synthesis. Indeed, in most species, including humans, most of the cholesterol synthesized in the body is synthesized in peripheral tissues [3]. Several peripheral tissues also acquire significant amounts of cholesterol from LDL uptake. Regardless of the source, cholesterol acquired by nonsteroidogenic peripheral tissues must ultimately return to the liver to be eliminated from the body. High density lipoprotein is thought to play a major role in this movement of cholesterol from peripheral tissues back to the liver.

Synthesis and Intravascular Metabolism of HDL

Information regarding the origin of plasma HDL is derived largely from organ perfusion and cell culture studies. The isolated perfused rat liver secretes disc-shaped lipid bilayer particles that are approximately 190 Å in diameter and 46 Å thick [143]. Lipid analysis shows these nascent HDL particles to be enriched in phospholipid and, to a lesser extent, free cholesterol and to be essentially devoid of esterified cholesterol and triglyceride. Apo E is the major apoprotein present on nascent HDL of hepatic origin [143–145]. This is in contrast to plasma HDL where apo A-I and A-II are the predominant apoproteins. The isolated perfused rat intestine also secretes discoidal HDL particles, although apo A-I rather than apo E appears to be the major apoprotein [146]. In addition to secretion by the liver and small intestine, it has been suggested that discoidal HDL particles may arise from excess surface material generated during the lipolysis of triglyceride-rich lipoproteins [147–149].

Regardless of the source, nascent HDLs are thought to play an important role in the initiation of reverse cholesterol transport by serving as acceptors for free cholesterol from cell membranes [150,151]. Transfer of free cholesterol between cell membranes and HDL is bidirectional and appears to occur by diffusion through the aqueous medium [150–153]. Net movement between cell membranes and HDL occurs when the cholesterol/phospholipid ratio in plasma membranes is greater than that on the surface of HDL. Several cell types have been shown to possess specific high affinity binding sites that recognize HDL particles containing apo A-I and apo A-II but not particles containing apo B or apo E [154–159]. It is unclear whether the transfer of cholesterol between cell membranes and HDL is facilitated by the binding of HDL to these cell surface receptors [160–164].

The net movement of free cholesterol from cell membranes to plasma is maintained by the enzyme LCAT, the principal cholesterol-esterifying enzyme of plasma. This enzyme transfers the sn-2 fatty acid of phosphatidylcholine to the 3-ß-hydroxy group of cholesterol [22,24]. By virtue of its hydrophobicity, the newly formed cholesteryl ester moves into the interior of the particle converting the bilayer disc into spherical HDL_3 and creating a concentration gradient for the further transfer of membrane cholesterol into plasma [152]. When LCAT is genetically absent cholesterol is no longer transported into plasma. As a consequence, cholesterol accumulates in tissues and discoidal HDLs accumulate in the plasma [165,166].

Lecithin-cholesterol acyltransferase is a glycoprotein of apparent M' 63,000 that is synthesized and secreted into plasma by the liver [167]. Its activity is greatly enhanced by apo A-I [168–171]. In the presence of optimal amounts of apo A-I, the further addition of other apoproteins is inhibitory. However, when suboptimal amounts of apo A-I are present, apo A-II, apo C-II, and apo C-III enhance activity though not to the same extent as optimal amounts of apo A-I alone [172].

High density lipoprotein formed through the action of LCAT may undergo a number of interconversions in the plasma. Under in vitro conditions, lipolysis of triglyceride-rich lipoproteins leads to the transfer of surface lipids and proteins as well as core triglyceride to HDL, resulting in the conversion of HDL_3 to HDL_2 [173]. The conversion of HDL_3 to HDL_2 by this mechanism is probably also dependent on the presence of LCAT and lipid transfer protein activities [174]. Thus there appears to be an important metabolic relationship between triglyceride-rich lipoproteins and HDL_2. Indeed, HDL_2 levels increase after a triglyceride–rich meal and the rise in HDL_2 levels is directly related to the rate of chylomicron metabolism [175,176]. HDL_2 may also be converted back to HDL_3. In this process cholesteryl ester in HDL_2 is exchanged for triglyceride in LDL via lipid transfer proteins resulting in the formation of triglyceride-rich HDL_2 particles that are largely depleted of cholesteryl ester. Lipolysis of HDL_2 triglyceride by hepatic or lipoprotein lipase then results in the regeneration of HDL_3 [177,178].

Catabolism of HDL

Studies dealing with the catabolism of HDL are complicated by a number of factors, including (a) the heterogeneity of HDL with respect to size, density, and composition (b) the interconversion of HDL subpopulations, and (c) the fact that individual components of HDL turn over at different rates. As discussed previously, HDL is thought to function primarily in the transport of cholesterol from peripheral tissues back to the liver. Several mechanisms have been described by which HDL cholesterol may be delivered to the liver. In some species, such as rabbits and human beings, the plasma contains high levels of lipid transfer activity, which mediates the transfer of HDL cholesterol to VLDL in exchange for triglyceride [179–192]. Very low density lipoprotein, along with the newly acquired cholesteryl ester, is then converted to remnant particles or to LDL, both of which are largely cleared by the liver. In species that lack lipid transfer proteins, such as the rat, HDL_2 appears to be converted into HDL_1 [183–185], which is a large cholesteryl ester–rich lipoprotein found in the density range of 1.030–1.095 g/ml that contains predominantly apoprotein E. As a consequence, these particles are recognized by the LDL receptor and again cleared largely by the liver.

High density lipoprotein may also transfer free cholesterol to the liver through the action of hepatic lipase. Hepatic lipase and lipoprotein lipase are responsible for the lipase activity in postheparin plasma. Hepatic lipase is located primarily along liver sinusoids and can be distinguished from lipoprotein lipase on the basis of its differing cofactor and salt requirements. In addition, hepatic lipase is particularly active in hydrolyzing HDL phospholipid. The degradation of HDL phospholipid by hepatic lipase increases the concentration gradient of free cholesterol between the surface of HDL and the hepatocyte plasma membranes. This may lead to a net transfer of HDL free cholesterol to the liver or at least prevent the movement of hepatic cholesterol to the HDL particle [186,187].

Finally, selective uptake of HDL cholesteryl esters has been demonstrated in several tissues of the rat [188]. In this species, HDL provides cholesterol to the endocrine organs for steroid hormone synthesis and is the main source of cholesterol for the adrenal gland [189]. The endocrine organs contain high affinity receptors that bind HDL. Core cholesteryl esters are then removed from the HDL particles and selectively taken up by parenchymal cells at rates fourfold to sevenfold higher than the rate of apo A-I uptake [188,190]. Selective uptake of HDL cholesteryl esters may also occur in the liver [188,191]. The quantitative importance of these various pathways of HDL cholesterol catabolism in vivo is not known and probably varies from species to species.

VII. CONCLUDING REMARKS

Water-insoluble lipids are transported through the fluid compartments of the body as emulsion or microemulsion particles. In some cases these particles are targeted to certain tissues of the body by apoproteins present on their surface that interact with high affinity cell surface receptors. The best studied lipoprotein receptor pathway is the LDL receptor pathway, which mediates the uptake of lipoproteins containing apo B or apo E. When expressed per gram of tissue, rates of receptor–dependent LDL transport are highest in the liver and endocrine organs; lower rates are observed in the small intestine, spleen, kidney, and lung. However, when expressed per whole organ, the liver accounts for more than 80% of total body LDL receptor activity in all species studied to date. Receptor–dependent transport is saturable and can be regulated by a number of dietary and pharmacological manipulations. Tissues also take up LDL via receptor–independent transport, which, in contrast to receptor–dependent transport, is not saturable and probably not regulable. The kinetic curves for receptor–dependent and receptor–independent LDL uptake in the various organs of the body have been experimentally determined in several species [105]. This information makes it possible to predict how changes in tissue receptor activity or circulating LDL concentrations will affect the rates of LDL uptake in the various organs of the body. For example, when hamsters are fed a low cholesterol, low triglyceride diet, plasma LDL cholesterol equals 25 mg/dl and LDL cholesterol uptake per gram of tissue is 3 to 7 times higher in the liver than in the lung, small intestine, spleen, and endocrine organs and more than 20 times higher than in the remaining tissues of the body. However, when hamsters are fed a diet enriched with saturated triglyceride and small amounts of cholesterol, hepatic LDL receptor activity is suppressed and plasma LDL levels increase to 75–150 mg/dl. Under these conditions, LDL uptake in the liver is reduced relative to that in the extrahepatic tissues; indeed, several extrahepatic organs may take up LDL at higher rates than does the liver (when expressed per gram of tissue). On the other hand, LDL uptake in the liver, relative to the extrahepatic organs, is increased by dietary or pharmacological manipulations that selectively increase hepatic LDL receptor activity. Although not as well studied as the LDL receptor pathway, the chylomicron remnant receptor pathway exhibits even greater hepatic selectivity. This pathway mediates the rapid and quantitative removal of chylomicron remnants by the liver through a high velocity receptor pathway that recognizes apo E. This pathway appears not to be regulated by experimental manipulations known to alter the LDL receptor pathway. Finally, through mechanisms that are not completely understood, cholesteryl esters present in the core of HDL also ultimately return largely to the liver, although in some species the endocrine organs also take up HDL cholesterol at high rates.

From this overview of lipoprotein transport several strategies by which lipoproteins might be used to target hydrophobic molecules to specific organs can be considered. Clearly, native lipoproteins could be used to deliver hydrophobic drugs to the liver. The native lipoprotein most rapidly and selectively taken up by the liver is the chylomicron remnant. In addition, native LDL could be modified so as to be taken up rapidly and quantitatively by the galactose-specific receptor on parenchymal liver cells. More sophisticated strategies may be required to selectively target lipid particles to individual extrahepatic tissues. In this regard, it should be possible to develop synthetic apoproteins designed to interact with cell surface receptors located preferentially in the organ of interest. In addition, by linking monoclonal antibodies to lipid vesicles or microemulsion particles, either directly or via lipid-associating molecules, it may be possible to target these particles to any antigen or receptor in the body [192]. Such strategies may be particularly important in targeting lipid particles to malignant tumors since there is little evidence for selective uptake of native lipoproteins by malignant tissue in vivo.

ACKNOWLEDGMENT

I wish to thank Imogene Robison for her assistance in preparing this chapter. The author, as well as some of the original research reported in this chapter, is supported by NIH grants HL 38049 and AM 01221 and by a grant-in-aid from the American Heart Association.

REFERENCES

1.   Edelstein C, Kezdy FJ, Scanu AM, Shen BW, Apolipoproteins and the structural organization of plasma lipoproteins: human plasma high density lipoproteins-3. J Lipid Res 1979; 20:143–153.
2.   Havel RJ, Goldstein JL, Brown MS, Lipoproteins and lipid transport. In: Bondy PK, Rosenberg LE, Eds. Metabolic control and disease. Philadelphia: W. B. Saunders, 1980:393–494
3.   Turley SD, Dietschy JM, The metabolism and excretion of cholesterol by the liver. In: Shafritz DA, Eds. The liver: biology and pathobiology, 2nd ed. Arias IM, Jakoby WB, Popper H, Schachter D, New York:Raven Press, 1988:617-641.
4.   Gotto AM Jr, Pownall HJ, Havel RJ. Introduction to the plasma lipoproteins. In: Segrest JP, Albers JJ, eds. Methods in enzymology, vol. 128. Orlando, FL:Academic Press, 1986:3–41.
5.   Bragdon JH, Havel RJ, Boyle E, Human serum lipoproteins. I. Chemical composition of four fractions. J Lab Clin Med 1956; 48:36–42.
6.   Patsch JR, Gotto AM Jr. Metabolism of high density lipoproteins. In: Gotto AM Jr., ed. Plasma lipoproteins. Amsterdam:Elsevier, 1987:299–333.
7.   Davidson NO, Magun AM, Brasitus TA, Glickman RM. Intestinal apolipoprotein A-I and B-48 metabolism: effects of sustained alterations in dietary triglyceride and mucosal cholesterol flux. J Lipid Res 1987; 28:388–402.
8.   Kostner G, Holasek A. Characterization and quantitation of the apolipoproteins from human chyle chylomicrons. Biochemistry 1972; 11:12171223.
9.   Imaizumi K, Havel RJ, Fainaru M, Vigne J-I. Origin and transport of the A-I and arginine-rich apolipoproteins in mesenteric lymph of rats. J Lipid Res 1978; 19:1038–1046.
10.   Imaizumi K, Fainaru M, Havel RJ. Composition of proteins of mesenteric lymph chylomicrons in the rat and alterations produced upon exposure of chylomicrons to blood serum and serum proteins.   J Lipid Res 1978; 19:712–722.

11. Havel RJ. Lipoprotein biosynthesis and metabolism. Ann NY Acad Sci 1980; 348:16–29.
12. Kane JP, Sata T, Hamilton RL, Havel RJ. Apoprotein composition of very low density lipoproteins of human serum. J Clin Invest 1975; 56:1622–1634.
13. Tall AR, Small DM. Plasma high-density lipoproteins. N Engl J Med 1978; 299:1232–1236.
14. Breslow JL. Lipoprotein genetics and molecular biology. In: Gotto AM Jr, ed. Plasma lipoporteins. Elsevier, 1987:Amsterdam: 359–397.
15. McConathy WJ, Alaupovic P. Studies on the isolation and partial characterization of apolipoprotein D and lipoprotein D of human plasma. Biochemistry 1976; 15:515–520.
16. McConathy WJ, Alaupovic P. Isolation and characterization of other apolipoproteins. In:Segrest JP, Alkers JJ, eds. Methods of enzymology, vol. 128. Orlando, FL:Academic Press, 1986:297–310.
17. Gaubatz JW, Chari MV, Nava ML, Guyton JR, Morrisett JD. Isolation and characterization of the two major apoproteins in human lipoproteins[a]. J Lipid Res, 1987; 28:69–79.
18. Morrisett JD, Guyton JR, Gaubatz JW, Gotto AM Jr. Lipoprotein[a]: structure, metabolism and epidemiology. Gotto AM Jr, ed. In:Plasma lipoproteins. Amsterdam:Elsevier, 1987:129–152.
19. Segrest JP, Jackson RL, Morrisett JD, Gotto AM Jr. A molecular theory of lipid–protein interactions in the plasma lipoproteins. FEBS Lett 1974; 38:247–253.
20. Wu A-L, Windmueller HG. Relative contributions by liver and intestine to individual plasma apolipoproteins in the rat. J Biol Chem 1979; 254:7316–7322.
21. Green PHR, Glickman RM, Sandek CD, Blum CB, Tall AR. Human intestinal lipoproteins. Studies in chyluric subjects. J Clin Invest 1979; 64:233–242.
22. Jonas A. Lecithin cholesterol acyltransferase. In:Gotto AM Jr, ed. Plasma lipoproteins. Amsterdam:Elsevier, 1987:299–333.
23. Jackson RL, Gotto AM, Stein O, Stein Y. A comparative study on the removal of cellular lipids from Landsüchtz ascites cells by human plasma apolipoproteins. J Biol Chem 1975; 250:7204–7209.
24. Glomset JA. The plasma lecithin:cholesterol acyltransferase reaction. J Lipid Res 1968; 9:155–167.
25. Kane JP, Hardman DA, Paulus HE. Heterogeneity of apolipoprotein B:isolation of a new species from human chylomicrons. Proc Natl Acad Sci USA 1980; 77:2465–2469.
26. Malloy MJ, Kane JP, Hardman DA, Hamilton RL, Dalal KB. Normotriglyceridemic abetalipoproteinemia. J Clin Invest 1981; 67:1441–1450.
27. Law SW, Grant SM, Higuchi K, Hospattankar A, Lackner K, Lee N, Brewer HB Jr. Human liver apolipoprotein B-100 cDNA:Complete nucleic acid and derived amino acid sequence. Proc Natl Acad Sci USA 1986; 83:8142–8146.
28. Blackhart BD, Ludwig EM, Pierotti VR, Caiati L, Onasch MA, Wallis SC, Powell L, Pease R, Knott TJ, Chu M-L, Mahley RW, Scott J, McCarthy BJ, Levy-Wilson B. Structure of the human apolipoprotein B gene. J Biol Chem 1986; 261:15364–15367.

29.   Hardman DA, Kane JP. Isolation and characterization of apolipoprotein B-48.
       In: Segrest JP, Albers JJ, eds. Methods in enzymology, vol. 128. Orlando,
       FL:Academic Press, 1986:262–272.
30.   Powell LM, Wallis SC, Pease RJ, Edwards YH, Knott TJ, Scott J. A novel
       form of tissue-specific RNA processing produces apolipoprotein-B48 in
       intestine. Cell 1987; 50:831–840.
31.   Brown MS, Goldstin JL. A receptor-mediated pathway for cholesterol
       homeostasis. Science 1986; 232:34–47.
32.   Marcel YL, Hogue M, Theolis R Jr, Milne RW. Mapping of antigenic
       determinants of human apolipoprotein B using monoclonal antibodies against
       low density lipoproteins. J Biol Chem 1982, 257:13165–13168.
33.   Knott TJ, Pease RJ, Powell LM, Wallis SC, Rall SC Jr, Innerarity TL,
       Blackhart B, Taylor WH, Marcel Y, Milne R, Johnson D, Fuller M, Lusis AJ,
       McCarthy BJ, Mahley RW, Levy-Wilson B, Scott J. Complete protein sequence
       and identification of structural domains of human apolipoprotein B. Nature
       1986; 323:734–738.
34.   LaRosa JC, Levy RI, Herbert P, Lux SE, Fredrickson DS. A specific
       apoprotein activator for lipoprotein lipase. Biochem Biophys Res Commun
       1970; 41:57–62.
35.   Breckenridge WC, Little JA, Steiner G, Chow A, Poapst M. Hypertriglyce-
       ridemia associated with deficiency of apolipoprotein C-II. N Engl J Med 1978;
       298:1265–1273.
36.   Miller NE, Rao SN, Alaupovic P, Noble N, Slack J, Brunzell JD, Lewis B.
       Familial apolipoprotein CII deficiency: plasma lipoproteins and apolipoproteins
       in heterozygous and homozygous subjects and the effects of plasma infusion.
       Eur J Clin Invest 1981; 11:69–76.
37.   Garfinkel AS, Schotz MC. Lipoprotein lipase. In:Gotto AM Jr, ed. Plasma
       lipoproteins. Amsterdam:Elsevier, 1987:95–128.
38.   Windler E, Chao Y-S, Havel RJ. Regulation of the hepatic uptake of
       triglyceride-rich lipoproteins in the rat. Opposing effects of homologous
       apolipoprotein E and individual C apoproteins. J Biol Chem 1980; 255:8303-
       8307.
39.   Quarfordt SH, Michalopoulos G, Schirmer B. The effect of human C
       apolipoproteins on the in vitro hepatic metabolism of triglyceride emulsions in
       the rat. J Biol Chem 1982; 257:14642–14647.
40.   Windler E, Havel RJ. Inhibitory effects of C apolipoproteins from rats and
       humans on the uptake of triglyceride-rich lipoproteins and their remnants by
       the perfused rat liver. J Lipid Res 1985; 26:556–565.
41.   Innerarity TL, Mahley RW. Enhanced binding by cultured human fibroblasts
       of apo-E-containing lipoproteins as compared with low density lipoproteins.
       Biochemistry 1978; 17:1440–1447.
42.   Pitas RE, Innerarity JL, Mahley RW. Cell surface receptor binding of
       phospholipid protein complexes containing different ratios of receptor-active
       and -inactive E apoprotein. J Biol Chem 1980; 255:5454–5460.
43.   Zannis VI, Breslow JL. Human very low density lipoprotein apolipoprotein E
       isoprotein polymorphism is explained by genetic variation and posttranslational
       modification. Biochemistry 1981; 20:1033–1041.

44. Newman TC, Dawson PA, Rudel LL, Williams DL. Quantitation of apolipoprotein E mRNA in the liver and peripheral tissues of nonhuman primates. J Biol Chem 1985; 260:2452–2457.

45. Williams DL, Dawson PA, Newman TC, Rudel LL. Apolipoprotein E synthesis in peripheral tissues of nonhuman primates. J Biol Chem 1985; 260:2444–2451.

46. Blue M-L, Williams DL, Zucker S, Khan SA, Blum CB. Apolipoprotein E synthesis in human kidney, adrenal gland, and liver. Proc Natl Acad Sci USA 1983; 80:283–287.

47. Elshourbagy NA, Liao WS, Mahley RW, Taylor JM. Apolipoprotein E mRNA is abundant in the brain and adrenals, as well as in the liver, and is present in other peripheral tissues of rats and marmosets. Proc Natl Acad Sci USA 1985; 82:203–207.

48. Law SW, Lackner KJ, Hospattankar AV, Anchors JM, Sakaguchi AY, Naylor SL, Brewer HB Jr. Human apolipoprotein B-100: cloning, analysis of liver mRNA, and assignment of the gene to chromosome 2. Proc Natl Acad Sci USA 1985; 82:8340–8344.

49. Shoulders CC, Kornblihtt AR, Munro BS, Baralle FE. Gene structure of human apolipoprotein A1. Nucl Acids Res 1983; 11:2827–2837.

50. Karathanasis SK, Zannis VI, Breslow JL. Isolation and characterization of the human apolipoprotein A-I gene. Proc Natl Acad Sci USA 1983; 80:6147–6151.

51. Cheung P, Kao F-T, Law ML, Jones C, Puck TT, Chan L. Localization of the structural gene for human apolipoprotein A-I on the long arm of human chromosome 11. Proc Natl Acad Sci USA 1984; 81:508–511.

52. Moore MN, Kao F-T, Tsao Y-K, Chan L. Human apolipoprotein A-II: nucleotide sequence of a cloned cDNA, and localization of its structural gene on human chromosome 1. Biochem Biophys Res Commun 1984; 123:1–7.

53. Tsao Y-K, Wei C-F, Robberson DL, Gotto AM Jr, Chan L. Isolation and characterization of the human apolipoprotein A-II gene. Electron microscopic analysis of RNA:DNA hybrids, nucleotide sequence, identification of a polymorphic MspI site, and general structural organization of apolipoprotein genes. J Biol Chem 1985; 260:15222–15231.

54. Elshourbagy NA, Walker DW, Boguski MS, Gordon JI, Taylor JM. The nucleotide and derived amino acid sequence of human apolipoprotein A-IV mRNA and the close linkage of its gene to the genes of apolipoproteins A-I and C-III. J Biol Chem 1986; 261:1998–2002.

55. Tata F, Henry I, Markham AF, Wallis SC, Weil D, Grzeschik KH, Junien C, Williamson R, Humphries SE. Isolation and characterisation of a cDNA clone for human apolipoprotein CI and assignment of the gene to the chromosome 19. Hum Genet 1985; 69:345–349.

56. Knott TJ, Robertson ME, Priestley LM, Urdea M, Wallis S, Scott J. Characterisation of mRNAs encoding the precursor for human apolipoprotein CI. Nucl Acids Res 1984; 12:3909–3915.

57. Jackson CL, Bruns GAP, Breslow JL. Isolation and sequence of a human apolipoprotein CII cDNA clone and its use to isolate and map to human chromosome 19 the gene for apolipoprotein CII. Proc Natl Acad Sci USA 1984; 81:2945–2949.

58. Wei C-F, Tsao Y-K, Robberson DL, Gotto AM Jr, Brown K, Chan L. The structure of the human apolipoprotein C-II gene. Electron microscopic analysis of RNA:DNA hybrids, complete nucleotide sequence, and identification of 5' homologous sequences among apolipoprotein genes. J Biol Chem 1985; 260:15211–15221.

59. Protter AA, Levy-Wilson B, Miller J, Bencen G, White T, Seilhamer JJ. Isolation and sequence analysis of the human apolipoprotein CIII gene and the intergenic region between the apo AI and apo CIII genes. DNA 1984; 3:449–456.

60. Das HK, McPherson J, Bruns GAP, Karathanasis SK, Breslow JL. Isolation, characterization, and mapping to chromosome 19 of the human apolipoprotein E gene. J Biol Chem 1985; 260:6240–6247.

61. Segrest JP, Chung BH, Brouillette CG, Kanellis P, McGahan R. Studies of synthetic peptide analogs of the amphipathic helix. J Biol Chem 1983; 258:2290–2295.

62. Pownall HJ, Massey JB, Sparrow JT, Gotto AM Jr. Lipid–protein interactions and lipoprotein reassembly. In: Lipoproteins, Gotto AM Jr, ed. Plasma lipoproteins. Amsterdam:1987:95–128.

63. Renkin EM, Relation of capillary morphology to transport of fluid and large molecules: a review. Acta Physiol Scand 1979; 463:81–91.

64. DeZanger R, Wisse E, The filtration effect of rat liver fenestrated sinusoidal endothelium on the passage of chylomicrons to the space of Disse. In: Knook DL, Wisse E, eds. Sinusoidal liver cells. New York:Elsevier Biomedical Press, 1982:69–72.

65. Sloop CH, Dory L, Roheim PS. Interstitial fluid lipoproteins. J Lipid Res 1987; 28:225–237.

66. Reichl D, Myant NB, Pflug JJ. Concentration of lipoproteins containing apolipoprotein B in human peripheral lymph. Biochim Biophys Acta 1977; 489:98–105.

67. Vasile E, Simionescu N. Transcytosis of low density lipoprotein through vascular endothelium. In:Glomerular dysfunction and biopathology of vascular wall. Tokyo:Academic Press, 1985:87–101.

68. Pitas RE, Boyles JK, Lee SH, Hui D, Weisgraber KH. Lipoproteins and their receptors in the central nervous system. Characterization of the lipoproteins in cerebrospinal fluid and identification of apolipoprotein B, E(LDL) receptors in the brain. J Biol Chem 1987; 262:14352–14360.

69. Roheim PS, Carey M Forte T, Vega GL. Apolipoproteins in human cerebrospinal fluid. Proc Natl Acad Sci USA 1979; 76:4646–4649.

70. Pitas RE, Boyles JK, Lee SH, Foss D, Mahley RW. Astrocytes synthesize apolipoprotein E and metabolize apolipoprotein E-containing lipoproteins. Biochim Biophys Acta 1987; 917:148–161.

71. Bell RM, Ballas LM, Coleman RA. Lipid topogenesis. J Lipid Res 1981; 22:391–403.

72. Sabesin SM, Frase S. Electron microscopic studies of the assembly, intracellular transport, and secretion of chylomicrons by rat intestine. J Lipid Res 1977; 18:496–511.

73. Christensen MJ, Rubin CE, Cheung MC, Albers JJ. Ultrastructural immuno-localization of apolipoprotein B within human jejunal absorptive cells. J Lipid Res 1983; 24:1229–1242.
74. Mallay MJ, Kane JP. Hypolipidemia. Med Clin North Am 1982; 66:469–484.
75. Davidson NO, Kollmer ME, Glickman RM. Apolipoprotein B synthesis in rat small intestine: regulation by dietary triglyceride and biliary lipid. J Lipid Res 1986; 27:30–39.
76. Jones AL, Ruderman NB, Herrera MG. Electron microscopic and biochemical study of lipoprotein synthesis in the isolated perfused rat liver. J Lipid Res 1967; 8:429–446.
77. Ide T, Ontko JA. Increased secretion of very low density lipoprotein triglyceride following inhibition of long chain fatty acid oxidation in isolated rat liver. J Biol Chem 1981; 256:10247–10255.
78. Wilcox HG, Heimberg M. Secretion and uptake of nascent hepatic very low density lipoprotein by perfused livers from fed and fasted rats. J Lipid Res 1987; 28:351–360.
79. Patsch W, Tamai T, Schonfeld G. Effect of fatty acids on lipid and apoprotein secretion and association in hepatocyte cultrues. J Clin Invest 1983; 72:371–378.
80. Johnson FL, St Clair RW, Rudel LL. Studies on the production of low density lipoproteins by perfused livers from nonhuman primates J Clin Invest 1983; 72:221–236.
81. McLean LR, Demel RA, Socorro L, Shinomiya M, Jackson RL. Mechanism of action of lipoprotein lipase. In: Albers JJ, Segrest JP, eds. Methods in enzymology, vol. 129. Orlando, FL:Academic Press, 1986:738–763.
82. Speake BK, Parkin SM Robinson DS. Lipoprotein lipase in the physiological systems. Biochem Soc Trans 1985; 13:29–31.
83. Semb H, Olivecrona T. Lipoprotein lipase in guinea pig tissues: molecular size and rates of synthesis. Biochim Biophys Acta 1986; 878:330–337.
84. Bergman EN, Havel RJ, Wolfe BM, Bohmer T. Quantitative studies of the metabolism of chylomicron triglycerides and cholesterol by liver and extrahepatic tissues of sheep and dogs. J Clin Invest 1971; 50:1831–1839.
85. Sherrill BC, Dietschy JM. Characterization of the sinusoidal transport process responsible for uptake of chylomicrons by the liver. J Biol Chem 1978; 253:1859–1867.
86. Sherril BC. Rapid hepatic clearance of the canine lipoproteins containing only the E apoprotein by a high affinity receptor. Identity with the chylomicron remnant transport process. J Biol Chem 1980; 255:1804–1807.
87. Floren C-H, Albers JJ, Kudchodkar BJ, Bierman EL. Receptor–dependent uptake of human chylomicron remnants by cultured skin fibroblasts. J Biol Chem 1981; 256:425–433.
88. Kita T, Goldstein JL, Brown MS, Watanabe Y, Hornick CA, Havel RJ. Hepatic uptake of chylomicron remnants in WHHL rabbits: a mechanism genetically distinct from the low density lipoprotein receptor. Proc Natl Acad Sci USA 1982; 79:3623–3627.
89. Cooper AD, Yu PYS. Rates of removal and degradation of chylomicron remnants by isolated perfused rat liver. J Lipid Res 1978; 19:635–643.

90.   Arbeeny CM, Rifici VA. The uptake of chylomicron remnants and very low density lipoprotein remnants by the perfused rat liver. J Biol Chem 1984; 259:9662–9666.

91.   Carrella M, Cooper AD. High affinity binding of chylomicron remnants to rat liver plasma membranes. Proc Natl Acad Sci USA 1979; 76:338–342.

92.   Cooper AD, Nutik R, Chen J. Characterization of the estrogen-induced lipoprotein receptor of rat liver. J Lipid Res 1987; 28:59–68.

93.   Kortz WJ, Schirmer BD, Mansback CM II, Shelburne F, Toglia MR, Quarfordt SH. Hepatic uptake of chylomicrons and triglyceride emulsions in rats fed diets of differing fat content. J Lipid Res 1984; 25:799–804.

94.   Faergeman O, Sata T, Kane JP, Havel RJ. Metabolism of apoprotein B of plasma very low density lipoproteins in the rat. J Clin Invest 1975; 56:1396–1403.

95.   Havel RJ. The formation of LDL: mechanisms and regulation. J Lipid Res 1984; 25:1570–1576.

96.   Sigurdsson G, Nicoll A, Lewis B. Conversion of very low density lipoprotein to low density lipoprotein. J Clin Invest 1975; 56:1481–1490.

97.   Yamada N, Shames DM, Stoudemire JB, Havel RJ. Metabolism of lipoproteins containing apolipoprotein B-100 in blood plasma of rabbits: heterogeneity related to the presence of apolipoprotein E. Proc Natl Acad Sci USA 1986; 83:3479–3483.

98.   Kita T, Brown MS, Bilheimer DW, Goldstein JL. Delayed clearance of very low density and intermediate density lipoproteins with enhanced conversion of low density lipoprotein in WHHL rabbits. Proc Natl Acad Sci USA 1982; 79:5693–5697.

99.   Pittman RC, Attie D, Carew TE, Steinberg D. Tissue sites of degradation of low density lipoprotein: application of a method for determining the fate of plasma proteins. Proc Natl Acad Sci USA 1979; 76:5345–5349.

100.   Glass CK, Pittman RC, Keller GA, Steinberg D. Tissue sites of degradation of apoprotein A-I in the rat J Biol Chem 1983; 344:7161–7167.

101.   Spady DK, Turley SD, Dietschy JM. Receptor–independent low density lipoprotein transport in the rat in vivo. Quantitation, characterization, and metabolic consequences. J Clin Invest 1985; 76:1113–1122.

102.   Spady DK, Dietschy JM. Interaction of dietary cholesterol and triglycerides in the regulation of hepatic low density lipoprotein transport in the hamster. J Clin Invest 1988; 81:300–309.

103.   Spady DK, Bilheimer DW, Dietschy JM. Rates of receptor–dependent and –independent low density lipoprotein uptake in the hamster. Proc Natl Acad Sci USA 1983; 80:3499–3503.

104.   Spady DK, Huettinger M, Bilheimer DW, Dietschy JM. Role of receptor-independent low density lipoprotein transport in the maintenance of tissue cholesterol balance in the normal and WHHL rabbit. J Lipid Res 1987; 28:32–41.

105.   Spady DK, Meddings JB, Dietschy JM. Kinetic constants for receptor–dependent and receptor–independent low density lipoprotein transport in the tissues of the rat and hamster. J Clin Invest 1986; 77:1474–1481.

106.   Bilheimer DW, Goldstein JL, Grundy SM, Starzl TE, Brown MS. Liver

transplantation to provide low-density-lipoprotein receptors and lower plasma cholesterol in a child with homozygous familial hypercholesterolemia. N Engl J Med 1984; 311:1658–1664.

107. Brown MS, Goldstein JL. Familial hypercholesterolemia: defective binding of lipoproteins to cultured fibroblasts associated with imparied regulation of 3-hydroxy-3-methylglutaryl coenzyme A reductase activity. Proc Natl Acad Sci USA 1974; 71:788–792.

108. Goldstein JL, Brown MS. The low-density lipoprotein pathway and its relation to atherosclerosis. Annu Rev Biochem 1977; 46:897–930.

109. Brown MS, Kovanen T. Regulation of plasma cholesterol by lipoprotein receptors. Science 1981; 212:628–635.

110. Tolleshaug H, Hobgood KK, Brown MS, Goldstein JL. The LDL receptor locus in familial hypercholesterolemia: multiple mutations disrupt transport and processing of a membrane receptor. Cell 1983; 32:941–951.

111. Goldstein JL, Brown MS, Anderson RGW, Russell DW, Schneider WJ. Receptor-mediated endocytosis: concepts emerging from the LDL receptor system. Ann Rev Cell Biol 1985; 1:1–39.

112. Goldstein JL, Brown MS. Binding and degradation of low density lipoproteins by cultured human fibroblasts. Comparison of cells from a normal subject and from a patient with homozgous familial hypercholesterolemia. J Biol Chem 1974; 249:5153–5162.

113. Bilheimer DW, Stone NJ, Grundy SM. Metabolic studies in familial hyper-cholesterolemia. J Clin Invest 1979; 64:524–533.

114. Shepherd J, Bicker S, Lorimer AR, Packard CJ. Receptor-mediated low density lipoprotein catabolism in man J Lipid Res 1979; 20:999–1006.

115. Kesaniemi YA, Witztum JL, Steinbrecher UP. Receptor-mediated catabolism of low density lipoprotein in man. Quantitation using glycosylated low density lipoprotein. J Clin Invest 1983; 71:950–959.

116. Mahley RW, Weisgraber KH, Melchior GW, Innerarity TL, Holcombe KS. Inhibition of receptor-mediated clearance of lysine and arginine-modified lipoproteins from the plasma of rats and monkeys. Proc Natl Acad Sci USA 1980; 77:225–229.

117. Steinbrecher UP, Witztum JL, Kesaniemi YA, Elam RL. Comparison of glucosylated low density lipoprotein with methylated or cyclohexanedione-treated low density lipoprotein in the measurement of receptor-independent low density lipoprotein catabolism. J Clin Invest 1983; 71:960–964.

118. Dietschy JM, Spady DK. Regulation of low density lipoprotein uptake and degradation in different animal species. In:Parnham MJ, Brune K, eds. Agents and actions supplements, vol. 16, Cologne Atherosclerosis Conference No. 2: Lipids. Stuttgart:Birkhauser Verlag, 1984:177–190.

119. Bilheimer DW, Watanabe Y, Kita T. Impaired receptor-mediated catabolism of low density lipoprotein in the WHHL rabbit, an animal model of familial hypercholesterolemia. Proc Natl Acad Sci USA 1982; 79:3305–3309.

120. Meddings JB, Dietschy JM. Regulation of plasma levels of low-density lipoprotein cholesterol: interpretation of data on low-density lipoprotein turnover in man. Circulation 1986; 74:805–814.

121. Spady DK, Dietschy JM. Dietary saturated triacylglycerols suppress hepatic low

density lipoprotein receptor activity in the hamster. Proc Natl Acad Sci USA 1985; 82:4526–4530.

122. Spady DK, Stange EF, Bilhartz LE, Dietschy JM. Bile acids regulate hepatic low density lipoprotein receptor activity in the hamster by altering cholesterol flux across the liver. Proc Natl Acad Sci USA 1986; 83:1916–1920.

123. Spady DK, Turley SD, Dietschy JM. Rates of low density lipoprotein uptake and cholesterol synthesis are regulated independently in the liver. J Lipid Res 1985; 26:465–472.

124. Alberts AW, Chen J, Kuron G, Hunt V, Huff J, Hoffman C, Rothrock J, Lopez M, Joshua H, Harris E, Patchett A, Monaghan R, Currie S, Stapley E, Albers-Schonberg G, Hensens O, Hirshfield J, Hoogsteen K, Liesch J, Springer J. Mevinolin: a highly potent competitive inhibitor of hydrozymethylglutaryl-coenzyme A reductase and a cholesterol lowering agent. Proc Natl Acad Sci USA 1980; 77:3957–3961.

125. Bilheimer DW, Grundy SM, Brown MS, Goldstein JL. Mevinolin and colestipol stimulate receptor-mediated clearance of low density lipoprotein from plasma in familial hypercholesterolemia heterozygotes. Proc Natl Acad Sci USA 1983; 80:4124–4128.

126. Vega GL, Grundy SM. In vivo evidence for reduced binding of low density lipoproteins to receptors as a cause of primary moderate hypercholesterolemia. J Clin Invest 1986; 78:1410–1414.

127. Talmud PJ, Barni N, Kessling AM, Carlsson P, Darnfors C, Bjursell G, Galton D, Wynn V, Kirk H, Hayden MR, Humphries SE. Apolipoprotein B gene variants are involved in the determination of serum cholesterol levels: a study in normo- and hyperlipidaemic individuals. Atherosclerosis 1987; 67:81–89.

128. Faust J, Goldstein JL, Brown MS. Receptor-mediated uptake of low density lipoprotein and utilization of its cholesterol for steroid synthesis in cultured mouse adrenal cells. J Biol Chem 1977; 252:4861–4871.

129. Ho YK, Smith RG, Brown MS, Goldstein JL. Low-density lipoprotein (LDL) receptor activity in human acute myelogenous leukemia cells. Blood 1978; 52:1099–1114.

130. Barnard GF, Erickson SK, Cooper AD. Lipoprotein metabolism by rat hepatomas. Studies of the etiology of defective dietary feedback inhibition of cholesterol synthesis. J Clin Invest 1984; 74:173–184.

131. Vitols S, Gahrton G, Ost A, Peterson C. Elevated low density lipoprotein receptor activity in leukemic cells with monocytic differentiation. Blood 1984; 63:1186–1193.

132. Clayman RV, Bilhartz LE, Spady DK, Buja LM, Dietschy JM. Low density lipoprotein-receptor activity is lost in vivo in malignantly transformed renal tissue. FEBS Lett 1986; 196:87–90.

133. Clayman RV, Bilhartz LE, Buga LM, Spady DK, Dietschy JM. Renal cell carcinoma in the Wistar-Lewis rat: a model for studying the mechanisms of cholesterol acquisition by a tumor in vivo. Cancer Res 1986; 46:2958–2963.

134. Attie AD, Pittman RC, Steinberg D. Metabolism of native and of lactosylated human low density lipoprotein: evidence for two pathways for catabolism of exogenous proteins in rat hepatocytes. Proc Natl Acad Sci USA 1980; 77:5923–5927.

135. Bernini F, Tanenbaum SR, Sherrill BC, Gotto AM Jr, Smith LC. Enhanced catabolism of low density lipoproteins in rat by lactosaminated Fab fragment. J Biol Chem 1986; 261:9294–9299.

136. Goldstein JL, Ho YK, Basu SK, Brown MS. Binding site on macrophages that mediates uptake and degradation of acetylated low density lipoprotein, producing massive cholesterol deposition. Proc Natl Acad Sci USA 1979; 76:333–337.

137. Brown MS, Basu SK, Falck JR, Ho YK, Goldstein JL. The scavenger cell pathway for lipoprotein degradation: specificity of the binding site that mediates the uptake of negatively-charged LDL by macrophages. J Supramol Struct 1980; 13:67–81.

138. Haberland ME, Fogelman AM. Scavenger receptor-mediated recognition of maleyl bovine plasma albumin and the demaleylated protein in human monocyte macrophages. Proc Natl Acad Sci USA 1985; 82:2693–2697.

139. Haberland ME, Olch CL, Folgelman AM. Role of lysines in mediating interaction of modified low density lipoproteins with the scavenger receptor of human monocyte macrophages. J Biol Chem 1984; 259:11305–11311.

140. Steinbrecher UP, Parthasarathy S, Leake DS, Witztum JL, Steinberg D. Modification of low density lipoprotein by endothelial cells involves lipid peroxidation and degradation of low density lipoprotein phospholipids. Proc Natl Acad Sci USA 1984; 81:3883–3887.

141. Blomhoff R, Drevon CA, Eskild W, Helgerud P, Norum KP, Berg T. Clearance of acetyl low density lipoprotein by rat liver endothelial cells. J Biol Chem 1984; 259:8898–8903.

142. Nagelkerke JF, Barto KP, van Berkel TJC. In vivo and in vitro uptake and degradation of acetylated low density lipoprotein by rat liver endothelial, Kupffer, and parenchymal cells. J Biol Chem 1983; 258:12221–12227.

143. Hamilton RL, Williams MC, Fielding CJ, Havel RJ. Discoidal bilayer structure of nascent high density lipoproteins from perfused rat liver. J Clin Invest 1976; 58:667–680.

144. Felker TE, Fainaru M, Hamilton RL, Havel RJ. Secretion of the arginine-rich and A-I apolipoproteins by the isolated perfused rat liver. J Lipid Res 1977; 18:465–473.

145. Marsh JB. Apoproteins of the lipoproteins in a nonrecirculating perfusate of rat liver. J Lipid Res 1976; 17:85–90.

146. Green PHR, Tall AR, Glickman RM. Rat intestine secretes discoid high density lipoprotein. J Clin Invest 1978; 61:528–534.

147. Eisenberg S, Patsch JR, Sparrow JT, Gotto AM, Olivecrona T. Very low density lipoprotein. Removal of apolipoproteins C-II and C-III during lipolysis in vitro. J Biol Chem 1979; 254:12603–12608.

148. Deckelbaum RJ, Eisenberg S, Fainaru M, Barenholz Y, Olivecrona T. In vitro production of human plasma low density lipoprotein-like particles. A model for very low density lipoprotein catabolism. J Biol Chem 1979; 254:6079–6087.

149. Chajek T, Eisenberg S. Very low density lipoprotein Metabolism of phospholipids, cholesterol, and apolipoprotein C in the isolated perfused rat heart. J Clin Invest 1978; 61:1654–1665.

150. Johnson WJ, Bamberger MJ, Latta RA, Rapp PE, Phillips MC, Rothblat GH. The bidirectional flux of cholesterol between cells and lipoproteins. Effects of phospholipid depletion of high density lipoprotein. J Biol Chem 1986; 261:5766–5776.

151. Rothblat GH, Phillips MC. Mechanism of cholesterol efflux from cells. Effects of acceptor structure and concentration. J Biol Chem 1982; 257:4775–4782.

152. Fielding CJ. The origin and properties of free cholesterol potential gradients in plasma, and their relation to atherogenesis. J Lipid Res 1984; 25:1624–1628.

153. Bierman EL, Oram JF. The interaction of high-density lipoproteins with extrahepatic cells. Am Heart J 1987; 113:549–550.

154. Miller NE, Weinstein DB, Steinberg D. Binding, internalization, and degradation of high density lipoprotein by cultured normal human fibroblasts. J Lipid Res 1977; 18:438–450.

155. Biesbroeck R, Oram JF, Albers JJ, Bierman EL. Specific high-affinity binding of high density lipoproteins to cultured human skin fibroblasts and arterial smooth muscle cells. J Clin Invest 1983; 71:525–539.

156. Fong BS, Rodriques PO, Salter AM, Yip BP, Despres J-P, Angel A, Gregg RE. Characterization of high density lipoprotein binding to human adipocyte plasma membranes. J Clin Invest 1985; 75:1804–1812.

157. Suzuki N, Fidge N, Nestel P, Yin J. Interaction of serum lipoproteins with the intestine. Evidence for specific high density lipoprotein-binding sites on isolated rat intestinal mucosal cells. J Lipid Res 1983; 24:253–264.

158. Hoeg JM, Demosky SJ Jr, Edge SB, Gregg RE, Osborne JC Jr, Brewer HB Jr. Characterization of a human hepatic receptor for high density lipoproteins, Arteriosclerosis 1985; 5:228–237.

159. Rifici VA, Eder HA. A hepatocyte receptor for high-density lipoproteins specific for apolipoprotein A-I, J Biol Chem 1984; 259:13814–13818.

160. Karlin JB, Johnson WJ, Benedict CR, Chacko GK, Phillips MC, Rothblat GH. Cholesterol flux between cells and high density lipoproteins. J Biol Chem 1987; 262:12557–12564.

161. Tabas I, Tall AR. Mechanism of the association of HDL$_3$ with endothelial cells, smooth muscle cells, and fibroblasts. J Biol Chem 1984; 259:13897–13905.

162. Slotte JP, Oram JF, Bierman EL. Binding of high density lipoproteins to cell receptors promotes translocation of cholesterol from intracellular membranes to the cell surface. J Biol Chem 1987; 262:12904–12907.

163. Brinton EA, Oram JF, Chen C-H, Albers JJ, Bierman EL. Binding of high density lipoprotein to cultured fibroblasts after chemcial alteration of apoprotein amino acid residues. J Biol Chem 1986; 261:495–503.

164. Graham DL, Oram JF. Identification and characterization of a high density lipoprotein-binding protein in cell membranes by ligand blotting. J Biol Chem 1987; 262:7439–7442.

165. Glomset JA. The metabolic role of lecithin: cholesterol acyltransferase: perspectives from pathology. In:Paoletti R, Kritchevsky D, eds Advances in lipid research, vol 2. New York:Academic Press,1973:1–65.

166. Stokke KT, Bjerve KS, Blomhoff JP, Oystese B, Flatmark A, Norum KR, Gjone E. Familial lecithin: cholesterol acyltransferase deficiency studies on lipid composition and morphology of tissues. Scand J Clin Lab Invest 1974; 33:93–100.

167. Osuga T, Portman OW. Origin and disappearance of plasma lecithin:cholesterol acyltransferase. Am J Physiol 1971; 220:735–741.
168. Fielding LJ, Shore VG, Fielding PE. A protein cofactor of lecithin:cholesterol acyltransferase. Biochem Biophys Res Comm 1972; 46:1493–1498.
169. Soutar AK, Garner CW, Baker HN, Sparrow JT, Jackson RL, Gotto AM, Smith LC. Effect of the human plasma apolipoproteins and phosphatidycholine acyl donor on the activity of lecithin:cholesterol acyltransferase. Biochemistry 1975; 14:3057–3064.
170. Yokoyama S, Murase T, Akanuma Y. The interaction of apolipoproteins with lecithin:cholesterol acyltransferase. Biochim Biophys Acta 1978; 530:258–266.
171. Furukawa Y, Nishida T. Stability and properties of lecithin-cholesterol acyltransferase. J Biol Chem 1979; 254:7213–7219.
172. Nishida HI, Nakanishi T, Yen EA, Arai H, Yen FT, Nishida T. Nature of the enhancement of lecithin–cholesterol acyltransferase reaction by various apolipoproteins. J Biol Chem 1986; 261:12028–12035.
173. Patsch JR, Gotto Am Jr, Olivecrona T, Eisenberg S. Formation of high density lipoprotein$_2$-like particles during lipolysis of very low density lipoproteins in vitro. Proc Natl Acad Sci USA 1978; 75:4519–4523.
174. Diepliner H, Zechner R, Kostner GM. The in vitro formation of HDL$_2$ during the action of LCAT: the role of triglyceride-rich lipoproteins. J Lipid Res 1985; 26:273–282.
175. Tall AR, Blum CL, Forester GP, Nelson CA. Changes in the distribution and composition of plasma high density lipoproteins after ingestion of fat. J Biol Chem 1982; 257:198–207.
176. Nikkila EA, Taskinen M-R, Sane T. Plasma high-density lipoprotein concentration and subfraction distribution in relation to triglyceride metabolism. Am Heart J 1987; 113:543–548.
177. Deckelbaum RJ, Eisenberg S, Oschry Y, Gronot E, Sharon I, Bengtsson-Olivecrona G. Conversion of human plasma high density lipoprotein-2 to high density lipoprotein-3. J Biol Chem 1986; 261:5201–5208.
178. Shinomiya M, Sasaki N, Barnhart RL, Shirai K, Jackson RL. Effect of apolipoproteins on the hepatic lipase-catalyzed hydrolysis of human plasma high density lipoprotein$_2$-triacylglycerols. Biochim Biophys Acta 1982; 713:292–299.
179. Zilversmit DB, Lipid transfer proteins. J Lipid Res 1984; 25:1563–1569.
180. Barter PJ, Hopkins GJ, Ha YC. The role of lipid transfer proteins in plasma lipoprotein metabolism. Am Heart J 1987; 113:538–542.
181. Fielding CJ, Factors affecting the rate of catalyzed transfer of cholesteryl esters in plasma. Am Heart J 1987; 113:532–537.
182. Tall AR. Plasma lipid transfer proteins J Lipid Res 1986; 27:361–376.
183. Gavish D, Oschry Y, Eisenberg S. In vivo conversion of human HDL$_3$ to HDL$_2$ and apoE-rich HDL$_1$ in the rat: effects of lipid transfer protein. J Lipid Res 1987; 28:257–267.
184. Koo C, Innerarity TL, Mahley RW. Obligatory role of cholesterol and apolipoprotein E in the formation of large cholesterol-enriched and receptor-active high density lipoproteins. J Biol Chem 1985; 260:11934–11943.
185. Eisenberg S, Oschry Y, Zimmerman J. Intravascular metabolism of the cholesteryl ester moiety of rat plasma lipoproteins. J Lipid Res 1984; 25:121–128.

186.   Grosser J, Schrecker O, Greten H. Function of hepatic triglyceride lipase in lipoprotein metabolism. J Lipid Res 1981; 22:437–442.
187.   Bamberger M, Glick JM, Rothblat GH. Hepatic lipase stimulates the uptake of high density lipoprotein cholesterol by hepatoma cells. J Lipid Res 1983; 24:869–876.
188.   Glass C, Pittman RC, Civen M, Steinberg D. Uptake of high-density lipoprotein-associated apoprotein A-I and cholesterol esters by 16 tissues of the rat in vivo and by adrenal cells and hepatocytes in vitro. J Biol Chem 1985; 260:744–750.
189.   Andersen JM, Dietschy JM. Kinetic parameters of the lipoprotein transport systems in the adrenal gland of the rat determined in vivo. J Biol Chem 1981; 256:7362–7370.
190.   Reaven E, Chen Y-D, Spicher M, Azhar S. Morphological evidence that high density lipoproteins are not internalized by steroid-producing cells during in situ organ perfusion. J Clin Invest 1984; 74:1384–1397.
191.   Stein Y, Dabach Y, Hollander G, Halperin G, Stein O. Metabolism of HDL-cholesteryl ester in the rat, studied with a nonhydrolyzable analog, cholesteryl linoleyl ether. Biochim Biophys Acta 1983; 752:98–105.
192.   Smirnov VN, Domogatsky SP, Dolgov VV, Hvatov VB, Klibanov AL, Koteliansky VE, Muzykantov VR, Repin VS, Samokhin GP, Shekhonin BV, Smirnov MD, Sviridov DD, Torchilin VP, Chazov EI. Carrier-directed targeting of liposomes and erythrocytes to denuded areas of vessel wall. Proc Natl Acad Sci USA 1986; 83:6603–6607.
193.   Herz J, Hamann U, Rogne S, Myklebost O, Gausepohl H, Stanley KK. Surface location and high affinity for calcium of a 500-kd liver membrane protein closely related to the LDL-receptor suggest a physiological role as lipoprotein receptor. EMBO J 1988; 7:4119–4127.

2

Cellular Interactions of Lipoproteins with the Vascular Endothelium: Endocytosis and Transcytosis

Nicolae Simionescu and Maya Simionescu
Institute of Cellular Biology and Pathology, Bucharest, Romania

I. INTRODUCTION

The successful delivery of drugs to desired sites of action implies complex interactions and passage through cell barriers among which the vascular endothelium is of paramount importance. This ubiquitous cell monolayer lining up all blood vessels is not only highly differentiated for rapid, extensive, and selective exchange of molecules between plasma and interstitial fluid, but it is also endowed with a broad and sophisticated spectrum of metabolic activities.

At the interface between blood and endothelial cell surface, a plasma molecule can undergo two types of interaction:

1. Surface interaction, which may be reciprocal, leading to metabolic effects.
2. Uptake into the endothelial cell to be used for its own metabolic needs (endocytosis) or transport across the endothelial cell to be delivered to the surrounding tissues (transcytosis).

Site-specific drug delivery aimed at the restricted distribution of a drug–carrier complex to the capillary bed of a designated target tissue (or organ) is considered a first-order targeting [38,101,116,157].

The use of lipoproteins as natural carriers for delivery of lipophilic drugs raises several issues regarding the endothelial transport of lipoproteins in different tissues and in various physiopathological conditions. Knowledge in this field is very scarce, and in this review we consider some insights from the cell biology standpoint, based in part on the work carried out in our laboratory.

Whenever relevant evidence is available, we refer to the following points:

Metabolic interactions between a given class of lipoproteins and components of endothelial cell surface;
Transcytosis of lipoprotein across different types of endothelia;
The fate of lipoproteins in the extracellular matrix;
The drainage of lipoproteins into the interstitial fluid and lymph.

Because of the potential use of lipoprotein–drug complexes in treating atherosclerosis, a special emphasis is places on lipoprotein transport and fate in the hyperlipoproteinemic atherogenesis.

II. GENERAL ROLE OF ENDOTHELIAL CELLS IN LIPOPROTEIN METABOLISM

A. Metabolic Interactions

Once secreted by the small intestine and the liver, lipoproteins (LPs) undergo rapid transformations in plasma either by physical transfer of lipid and apoprotein components or by enzymatic catalysis by lecithin-cholesterol acyltransferase (LCAT) and lipoprotein lipase (LPL). The latter is associated with the surface of endothelial cells, especially in the capillaries of adipose tissue, muscle, lung, and liver. This enzymatic hydrolysis involves the triglyceride-rich chylomicrons and very low density lipoproteins (VLDLs) and results in the formation of cholesterol esters from cholesterol and phosphatidylcholine (the effect of LCAT) and the removal of triglycerides further converted into fatty acids (FA) and glycerol (the effect of LPL). Although the mechanisms underlying the FA transfer across the capillary endothelium remain hypothetical, it can be assumed that, as in other cells [138], FAs cross the plasmalemma and the cytosol by free diffusion. Fatty acids are stored in adipocytes as triglycerides and are used by the myocytes for energy-generating oxidative phosphorylation. The glycerol moiety is not taken up by endothelium [35].

Upon this first step of enzymatic degradation, chylomicrons and VLDL become triglyceride-poor remnants. The chylomicron remnant is taken up by a specific receptor found exclusively in hepatocytes, whereas the VLDL remnant, also referred to as IDL (intermediate density lipoprotein), is subsequently converted into LDL (low density lipoprotein), the main cholesterol-carrying lipoprotein in humans and several animal species. The LDL is continuously removed from the circulation via a receptor-mediated mechanism taking place mainly (~75%) in the liver and by a combined receptor-mediated and receptor-independent mechanism occurring in many extra-hepatic tissues, including vascular endothelium.

In the liver, the cholesterol resulting from the degradation of chylomicron remnants is either secreted into the intestine as bile acids or is utilized by packing with triglyceride in VLDL. These are secreted into the sinusoid capillaries from where they reach the blood of the inferior vena cava. Since all LPs drained by lymph are discharged into the blood of the superior vena cava system, it means that the whole amount of LP of both the endogenous pathway (secreted by the liver) and the exogenous pathway (secreted by the intestine) is passed first through the pulmonary capillary network. The extent and nature of the LP modifications at this level are poorly understood. However, this special distribution may be fruitfully exploited when LP–drug complexes are targeted to the lung.

The cholesterol from extrahepatic cells is removed by high density lipoprotein (HDL), which, through the action of LCAT and a cholesterol-transport protein, is delivered to IDL [14,45,51]. In some species, such as the rat, in which HDL is the major plasma lipoprotein, HDL may play a role analogous to that of LDL in humans.

A more detailed account of LP state and metabolic conversions in different fluids and compartments of the body is given in Chapter 1.

The LP–endothelial interactions in the special case of the modified LDL are discussed in Section V.

B. Transport Processes

Whether administered orally or parenterally, once inside the vasculature, LPs (and their associated drugs) are transported into the extravascular compartments under a strict structural and physiological control residing at the endothelial cell (EC) level.

In the plasma and when transported, the complex, dynamic, and unstable LPs generally behave as hydrophilic macromolecules. They can be taken up by endothelial cells and conveyed to the lysosomal compartment for hydrolytic catabolism (endocytosis), or they can be translocated across the cell, bypassing the lysosomes, and discharged into the interstitial fluid (transcytosis) [120,121] (Fig. 1). From the latter, LP can be either drained into the initial lymphatic capillaries or taken up by other cells. The cellular and molecular mechanisms involved in this dual process have been only partially elucidated so far. Endothelial cells have been shown to contain cell surface receptors, which can be instrumental in the intracellular uptake and degradation of a certain amount of circulating LP. But unlike other cells, ECs appear to be provided with the machinery necessary to mediate and control the transcellular delivery of LP to the other cells. This transcytotic capacity renders ECs key entity for understanding and designing ways for general or site-specific drug delivery using LPs as natural vehicles.

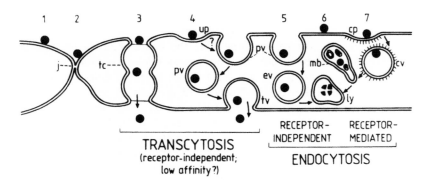

Figure 1 Diagrammatic representation (not at scale) of lipoprotein (LP) interactions with endothelium and the possible transport pathways and processes involved. (1) Upon LP direct contact or binding to the luminal plasma membrane, the two partners can affect each other's properties and interchange cholesterol. (2) LPs do not normally penetrate endothelial junctions. (3) Occasionally, particles of the size of LDL or VLDL can be seen within transendothelial channels or/and at their abluminal openings. (4) ß-LP may be attached to uncoated pits (up) or label plasmalemmal vesicles (pv) which in time course experiments can be detected associated with the luminal front, internalized or open to the abluminal surface, apparently discharging their content into the subendothelial space. The processes under (3) and (4) define transcytosis, which seems to be largely receptor-independent or possibly mediated by low affinity binding sites (so far poorly characterized) (tv, transcytotic vesicles). (5) It is assumed (but not yet proved) that some plasmalemmal vesicles transport LDL and VLDL particles by fluid phase endocytosis to lysosomes (ly). Such vesicles behave as endocytotic vesicles (ev), which perform a receptor-independent uptake, distinct from receptor-mediated endocytosis (6). In the latter LP particles randomly bound to plasma membrane receptors may cluster in coated pits (cp) (7) and be delivered via coated vesicles (cv) to multivesicular bodies (mb) and lysosomes (ly).

Endocytosis

In normal conditions the LP–EC surface interactions are largely reversible. Studies in situ and in vitro have shown that ECs express receptors for LDL, ß-VLDL and HDL which mediate a specific uptake of the cognate LP. However, endocytosis by ECs of various classes of LP in vivo has not yet been convincingly quantitated. In addition ECs are among the few cell types provided with acetyl-LDL (or scavenger) receptors [84,85,99,169], which are not subjected to downregulation [4,12–14]; this characteristic together with their restricted distribution may represent an advantage for their use in drug targeting. A fraction of LP endocytosis may occur through a receptor-independent pathway [134–136,151,162–165].

The endocytic activity of arterial endothelium appears to be markedly enhanced in experimental hyperlipoproteinemia. In advanced stages with elevated levels of plasma LDL in hamster [92] or ß-VLDL in rabbit [72,123,

124], endothelial cells of lesion-prone areas of the aortic arch [72,123] or cardiac valves [33] accumulate large amounts of cholesterol esters and turn into foam cells.

Transcytosis

As with any other hydrophilic macromolecule, LP transport from the plasma through vascular endothelium is governed by three groups of factors [120,121]:

1. Plasma driving forces (the balance between the hydrostatic pressure and osmotic pressure).
2. The physicochemical properties of the permeant LP molecule (size, shape, charge, and chemistry) and their concentration gradient in plasma vs. interstitial fluid (generating gradients for diffusion exchange).
3. The surface properties and specific activities of the endothelial cells relevant for interactions with LP molecules.

Lipoprotein permeant species can be taken up by the endothelial structures involved in transport either in fluid phase or by adsorption. Whereas the specific adsorption by a cognate receptor has been well documented, the contribution of a nonspecific (electrostatic) LP uptake by ECs remains to be clarified.

By their size, ranging between 10 nm and 30 nm for HDL, LDL, and some VLDL particles [45], such macromolecular complexes have access to the plasmalemmal vesicles of endothelial cells [119,121–124,162–165]. Occasionally chylomicrons or large VLDL particles can be seen engulfed by large vacuoles but the fate of these (intracellular or transcellular delivery) is uncertain. Due to their large dimensions, all LPs are normally excluded from the intercellular junctions of endothelium [162–165].

In advanced stages of atherosclerotic plaque formation, LP particles can be detected in the arterial intima as components of a bulk plasma extravasation [59,60,80,123,124].

Vascular endothelium exhibits locally differentiated structural and functional properties which represent various opportunities for LP transport. A first order of differentiation is represented by three major morphologic types of endothelia encountered: (a) continuous endothelia found primarily in the large vessels and capillaries of lung, heart, somatic tissues, and nervous system; (b) fenestrated endothelia occurring in most visceral capillaries; and (c) discontinuous or sinusoidal endothelia of the liver, spleen, and bone marrow capillaries (Fig. 2).

For transcytosis of LP, the continuous endothelium offers a very large population of plasmalemmal vesicles (rather uniform, of ~70 nm diameter), which can vary greatly in number from a few thousand per cell in myocardial capillaries to only a few dozen in the blood–brain barrier. Although a general

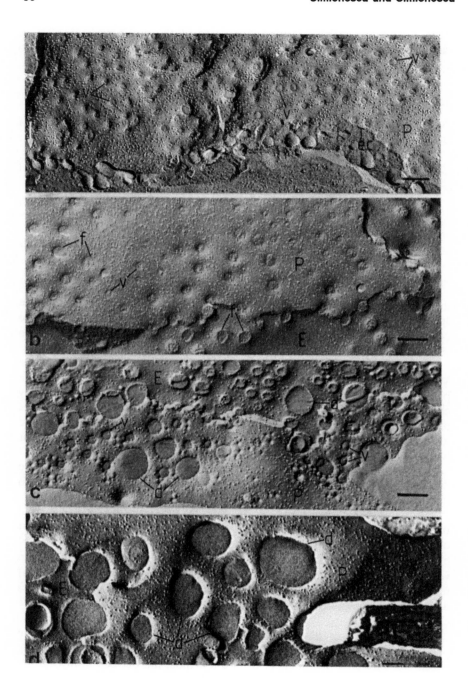

consensus on the role of these structures in macromolecular transport has not yet been reached [118,119,121], there is strong evidence that plasmalemmal vesicles either can function as isolated units shuttling from one cell surface to another or can form transiently patent transendothelial channels [42,118–120]. In continuous endothelia, the intercellular occluding junctions display a characteristic high tightness in brain capillaries.

Fenestrated endothelia, in addition to vesicles and channels, contain numerous 70 nm diaphragmed openings, but there is no good evidence that LPs transcytose through these fenestrae.

The discontinuous endothelium of liver sinusoid capillaries contains undiaphragmed openings (~100 nm in diameter), which have been considered to sort chylomicrons according to size [25,86]. The capillaries of adrenal cortex represent an intermediary type between fenestrated and discontinuous, containing both fenestrae and large discontinuities (Fig. 2). In the sinusoid capillaries, such as those of the liver, although blood cells do not reach the parenchymal cells, plasma does since it percolates freely through the space of Disse. Therefore, for macromolecules (including lipoproteins) plasma of the sinusoid capillaries bathes the main cell types of the liver: endothelial, Kupffer cells, and hepatocytes. In addition, these endothelial cells have the capacity to rapidly catabolize acetyle-LDL and to form fatty acids (from cholesteryl esters), which are further transferred to hepatocytes [84,85].

A second order of specialization is represented by the existence on the endothelial cell surface of biochemically differentiated microdomains generated by the preferential distribution of some glycoconjugates and charged groups. These microdomains of defined electric charge and chemistry are characteristically associated with the structures involved in transendothelial exchanges [121]. They may influence, especially by electrostatic interactions, the uptake of normal or modified LP.

Figure 2 Freeze-fracture replicas of capillary endothelia illustrating the main structures potentially involved in transendothelial passage of macromolecules. These structures appear as opening of plasmalemmal vesicles (v), fenestrae (f), or discontinuities (d). (A) Myocardial capillary (P face): numerous papillae (~25 nm diameter) (v) represent the fractured necks of plasmalemmal vesicles open on the cell surface. In the lower part of the field, a cross-fracture of endothelial cell (ec) shows continuity between vesicle membrane and its opening (*arrows*). (B) Pancreatic capillary; two kinds of openings can be seen representing either necks of plasmalemmal vesicles (v) or fenestrae (f) of a larger diameter than the former. Both features appear as papillae on the P face (P) and craters on the E face (E). (C) Adrenal cortex capillary displaying a large variety of openings, ranging from the size of plasmalemmal vesicles (v) to large discontinuities, ~100 nm to 150 nm in diameter (d); they appear as papillae on the P face (P) and craters on the E face (E). (D) Liver sinusoid capillary: area containing numerous oval discontinuities ~100 nm to 300 nm in diameter (d) which have the pattern of a papilla on the P face (P) and crater on E face (E). Bars = 100 nm.

A third order of local differentiations is the occurrence in certain endothelia of organ-specific antigens, binding sites, and receptors for certain plasma constituents, LP included. Such specific binding sites and receptors may operate in receptor-mediated transcytosis of some plasma components and the drugs they may carry.

In normal conditions, transcytosis of macromolecules including LP largely prevails over endocytosis of the same molecular species.

III. CHYLOMICRONS

A. Synthesis-Secretion Metabolic Pathways

Dietary fatty acids and cholesterol absorbed by the intestinal mucosal cells are used for the synthesis of triglycerides and phospholipids and of cholesteryl esters, respectively. Together with newly synthesized apolipoproteins A and B–48, they are assembled into large polymorphic particles, the nascent chylomicrons. Once secreted within the interstitial fluid, these particles acquire apoproteins C and E by transfer from HDL and in exchange with some apo A. Permeating the loose intercellular junctions of initial lymphatic capillaries, chylomicrons have ready access to the lacteals of the intestinal villi, and via the thoracic duct they are drained into the blood of the superior vena cava system. In the plasma, chylomicrons (CHs) continue to acquire apo E and C and cholesteryl ester from HDL (Fig. 3).

B. Interaction with Endothelium and Lipoprotein Lipase

Within circulation, the triglyceride-rich CHs rapidly interact with the lipoprotein lipase, which is firmly associated with the endothelial cell surface (ECS), especially in capillaries.

Lipoprotein lipase (LPL) is a widely distributed enzyme synthesized as a proenzyme by many parenchymal cells of extrahepatic tissues, but not by endothelial cells. By an unknown mechanism, the inactive form of LPL is translocated to the luminal surface of capillary endothelium, where it is bound to the heparan sulfate chains. Immunocytochemically, LPL was indirectly detected on the luminal front of capillary endothelium [96]. Lipoprotein lipase becomes functional upon activation by specific interaction with apo C-II contained by CH [40,91], VLDL, and HDL particles.

On contact with the immobilized CH, in a few minutes LPL catalyzes the hydrolysis of CH–triglycerides at a rate of ~300 g triglycerides per day (estimated in human) [45]. As hydrolysis proceeds, the released free fatty acids (FFAs) are taken up by the endothelial cells via a still unclear process postulated to be by transport across membrane [35] or by lateral diffusion in the plane of the membrane [114,115]. In monocytes, FFAs are used as the

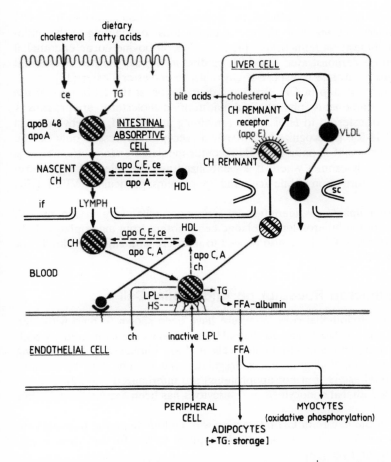

Figure 3 General metabolic pathways of chylomicrons (CH) and their interactions with endothelium. (The components and steps involved are discussed in the text.) The nascent chylomicron contains triglycerides, cholesteryl esters, phospholipids, and apolipoproteins B-48 and A. ce, cholesteryl ester; FFA, free (unesterified) fatty acid; HDL, high density lipoprotein; HS, heparan sulfate proteoglycan; if, interstitial fluid; LDL, lipoprotein lipase; ly, lysosome; PCET, parenchymal cells of extrahepatic tissues; TG, triglyceride; VLDL, very low density lipoprotein. Diagram not drawn to scale.

main source of metabolic energy, whereas in adipocytes they are reesterified and stored as triglycerides (Fig. 3).

It has been suggested that the CH and VLDL triglycerides are hydrolyzed during a series of attachments and detachments from LPL at a definite number of sites along the endothelial glycocalyx [91].

Lipoprotein lipase also facilitates the transfer of cholesteryl esters from CH and chylomicron remnants to the endothelium, thus contributing to the endothelial cell cholesterol homeostasis [40]. Studies on bovine aortic endothelial cells in culture demonstrated that LPL binding to the cell surface is essential for the transmembrane transport of chylomicron cholesteryl ester [19].

During the first step of CH enzymatic degradation at the ECS, part of the CH surface components (phospholipid, unesterified cholesterol, apoproteins A and C) are transferred to HDL. The remaining part of CH, the chylomicron remnant (CH-R), is recognized by a specific receptor expressed only on the sinusoidal surface of hepatocytes. By receptor-mediated endocytosis, CH-R is internalized to lysosomes where it is catabolized and the released cholesterol is used for the formation of bile acids and lipoproteins, particularly VLDL (Fig. 3).

Lipoprotein lipase was identified by immunocytochemistry in the smooth muscle cells of the atherosclerotic plaque being absent from macrophages [63].

High amounts of CH have also proved to interfere with capillary blood flow in diabetics [43].

C. Chylomicron Receptor on Endothelial Cell

Bovine aortic endothelial cells in culture have been shown to bind with high affinity CH. It has been claimed that CH binding sites are expressed on both actively growing and contact-inhibited cells and are unrelated to LPL binding sites. Chylomicron binding is not downregulated and upon binding triglycerides and cholesteryl esters (but not apoproteins) are internalized [29–32].

No real endothelial endocytosis or transcytosis has been ascribed to CH thus far.

Because of the nature of CH interaction with endothelium, the potential use of CHs as carriers for pharmaceuticals appears to be, for the time being, restricted to the liver [38].

IV. Very Low Density Lipoproteins

A. Synthesis-Secretion Metabolic Pathways

Nascent very low density lipoprotein (VLDL), the other triglyceride-rich lipoprotein, is synthesized by the liver. Most of its triglycerides (TGs) originate from the free fatty acids of the adipose tissue, which are packed in the core of the particle together with the cholesteryl ester resulting from the lysosomal hydrolysis of various LPs taken up by the hepatocyte. The particle contains unesterified cholesterol, phospholipids, apo B-100, apo E, and apo C.

The VLDL secreted in the plasma bathing the space of Disse enter through endothelial discontinuities into the sinusoid capillary. In the plasma there is

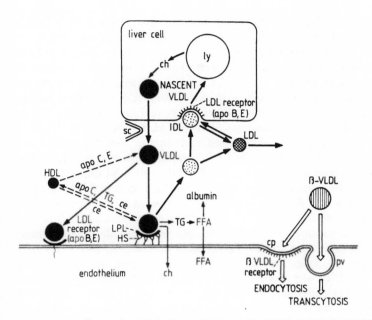

Figure 4 General metabolic pathways of VLDL and their interactions with endothelium. (The metabolic steps involved are discussed in the text.) ce, cholesteryl ester; cp, coated pit; FFA, free (unesterified) fatty acid; HDL, high density lipoprotein; HS, heparan sulfate proteoglycan; IDL, intermediate density lipoprotein; LDL, low density lipoprotein; LPL, lipoprotein lipase; ly, lysosome; pv, plasmalemmal vesicle; sc, sinusoid capillary; TG, triglyceride; VLDL, very low density lipoprotein. Diagram not drawn to scale.

an early transfer of cholesteryl ester, phospholipid, and apo C from VLDL to HDL in exchange for more apo C-I, II, and III and apo E (Fig. 4).

The liver secretes several distinct species of VLDL, the size of which is generally inversely related to the TG load of each VLDL [88]. In humans, plasma contains several VLDL subpopulations, ranging in diameter from 30 nm to 80 nm, which are involved in multiple compartmented but dynamic and interrelated pathways of metabolism. In general, large VLDLs are cleared from plasma like the CH, whereas small VLDLs are converted to IDL and LDL.

B. Interaction with Endothelium and Lipoprotein Lipase

The initial VLDL metabolic interaction with endothelial cell surface is basically the same as for CH: binding the LPL followed by TG hydrolysis (see Section III). As hydrolysis proceeds, apo C, cholesteryl esters, and phospholipids are transferred to HDL. The resulting VLDL remnant, also called IDL, is largely depleted of G, apo C, and phospholipids, but it retains apo B and apo E. In humans, about half of the IDL particles are taken up by the liver LDL receptors, which, via receptor-mediated endocytosis, deliver the particles to lysosomes for degradation (Fig. 4). The LDL receptor has a higher affinity for IDL than for LDL. The other half of the circulating IDL undergoes further loss of TGs, apo E, and phospholipids to become LDL, in which among apoproteins only apo B-100 is retained. Unlike CH, CH remnants, VLDL< and VLDL remnants, LDL is catabolized in both the liver and peripheral tissues (see Section V). The hepatic LDL receptor (the apo B/E receptor) recognizes several ligands: IDL, LDL, apo E-HDL, and ß-VLDL.

In humans, VLDL particles larger than 45 nm bind to liver LDL receptor only if they contain apo E in an accessible position, as is the case with the hypertriglyceridemic subjects [45].

Data on the interactions of normal VLDL with vascular endothelium, especially in vivo, are rather scarce. Perfusion of homologous rate [^{125}I]VLDL through isolated rat lungs, in the presence of heparin, resulted in the proteolysis of apo C-II, with low uptake of radioactivity by the lung parenchyma [137]. Since the pulmonary vascular bed is the first to contact the newly synthesized VLDL, the lung endothelium has the potential to modify the VLDL apoproteins prior to their peripheral catabolism.

Primary cultures of rat liver endothelial cells have been shown to take up and degrade human [^{125}I]VLDL in a saturable fashion. The uptake was inhibited by human VLDL, LDL, HDL, and CH, and by purified apo E and C-III:1. THe latter also bound to liver endothelial cells with a dissociation constant of 0.5 μM [46].

A receptor type of binding was also demonstrated for human VLDL on porcine aortic endothelium. The binding was Ca^{2+} time- and temperature-dependent, saturable and reversible, and was competed by VLDL, LDL, and HDL. THese observations suggested that aortic endothelial cells bind some VLDL with a higher affinity than they do LDL. The triglyceride-rich LP may contribute to the atherogenic process by local production of significant amounts of remnants on the ECS of arteries [24].

There is some evidence that normal VLDL or some metabolic products thereof are capable of penetrating arterial endothelium, though at a much lower rate than LDL and IDL [90]. After perfusion with radiolabeled lipoproteins, the binding and uptake of VLDL by the aorta was found to be at a lower rate than that of LDL and HDL [140].

Although lipoproteins of the size of VLDL have been visualized in various tissues, including capillary endothelium [122], the transendothelial transport of normal VLDL and IDL has not been examined in detail yet. Moreover, it has been claimed that in capillaries isolated from rat fat tissue fluorescent-labeled VLDLs are excluded from vesicular ingestion [108]. These findings remain to be checked at cellular level. Endothelial ingestion and transcytosis of LP particles of this size was, however, demonstrated for the abnormal VLDL, ß-VLDL both in normal and hypercholesterolemic animals, as well as by cultured endothelial cells [163–166].

C. Abnormal VLDL

Animals (dog, rabbit, pigeon) fed large amounts of cholesterol develop a special type of VLDL rich in cholesterol esters, beta migrating, and containing much apo E but little apo C. These ß-VLDL particles can also appear in humans with type III hyperlipoproteinemia. In hypertriglyceridemic subjects, another class of abnormal LP is detected, the hypertriglyceridic VLDL (HTG-VLDL).

Together with CHs, the ß-VLDL particles are recognized by a special, genetically distinct ß-VLDL receptor, which is expressed only by cells of the reticuloendothelial (mononuclear phagocytic) system such as the endothelial cells, macrophages, and foam cells of the xanthoma or atherosclerotic plaques. These receptors do not bind normal VLDL, LDL, or HDL. ß-VLDL receptors are also expressed by endothelial cells of the Watanabe heritable hyper-lipidemic (WHHL) rabbits [4]. Endothelial cells and monocyte-macrophages are thus far the only cells known to express distinct receptors for LDL, acetyl-LDL (scavenger), and ß-VLDL [4]. Upon receptor-mediated endocytosis, the ß-VLDL is hydrolyzed in lysosomes and the released cholesterol is reesterified to oleate, which is stored within the cytoplasm as lipid inclusions the accumulation of which is the morphologic hallmark of a foam cell. Normal VLDL does not induce formation of such lipid-laden cells.

Generally HTG-VLDLs follow the same intracellular pathway but in lysosome TGs are hydrolyzed and finally stored as intracytoplasmic inclusions, which, unlike those produced by cholesteryl esters, are not birefringent. Foam cells isolated from xanthomas of diabetics with HTG contain predominantly TG. ß-VLDL receptor is downregulated by cholesterol loading but not by TG loading. Cultured rabbit aortic endothelial cells preincubated with ß-VLDL showed an increased permeability to LDL and albumin without altering the transendothelial electrical resistance. Since the same results were obtained on endothelial cell monolayers from WHHL rabbits, it was concluded that LDL receptor was not involved in the modulation of this transport [87]. Endothelial cell exposure to ß-VLDL reportedly enhanced their production of a monocyte chemotactic factor [77] and altered monocyte adherence to endothelium [71].

ß-VLDL receptor appears as an expression of a defense mechanism found exclusively in cells of the mononuclear phagocytic system, capable of cleaning up arterial atherosclerotic plaques of these abnormal LPs, which are known to have deleterious effects on endothelial cells.

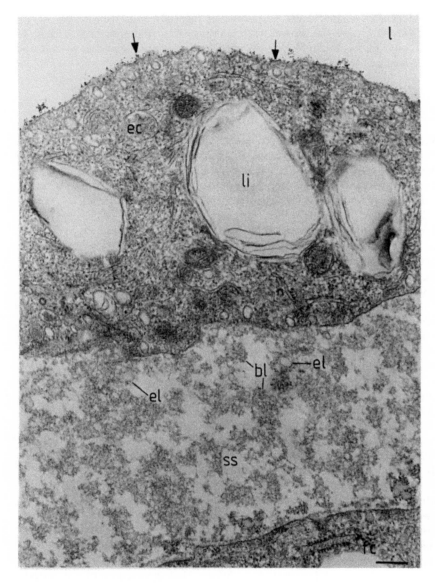

Figure 5 Segment of the aortic arch intima from a hypercholesterolemic rabbit at 8 weeks of diet. The specimen was reacted with *Ricinus communis* followed by lactosaminated bovine serum albumin–gold complex to detect the cell surface galactosyl residues. Although the endothelial cell (ec) is loaded with large lipid inclusions (li), the labeling density of the tracer used (*arrows*) is quasinormal. The subendothelial space (ss) is filled with a floccular material (probably plasma insudate) intermixed with basal laminalike material (bl) and extracellular liposomes (el).

D. ß-VLDL: Endothelial Interactions and Transport in Hypercholesterolemic Animals

It was demonstrated that diet-induced hypercholesterolemia in rabbit is due to overproduction of ß-VLDL accompanied by its high influx and accumulation in the aortic wall (more than 15-fold over control). The lipid deposition was not paralleled by an enhanced accumulation of HDL or albumin [22].

Using as an experimental model the hyperlipoproteinemic rabbit, we examined the ß-VLDL transport across endothelium and subsequent deposition in the aortic intima and the cardiac valves. A special emphasis was put on the early prelesional events detectable at the ultrastructural level at the onset of the atherogenetic process.

In the lesion-prone area of the aortic arch and coronary artery, changes in the distribution and density of ECS charged groups, sialoconjugates, and oligosaccharides during progressive stages of atherosclerotic plaque formation were investigated cytochemically. The findings revealed a remarkable resistance of the EC coat even to very high levels of serum cholesterol. In lesional areas including fatty streaks, cationic sites and galactyosyl- and N-acetylglucosaminyl moieties were not altered (Fig. 5), but mannosyl residues increased in density. Significant ECs alterations appeared only in advanced stages, particularly on ECs heavily loaded with lipid inclusions [41,70].

In the first two weeks of diet, endothelium was structurally intact and no platelets or monocytes were seen attached to ECS. While plasma ß-VLDL cholesterol was increased up to 15-fold compared with the normal animals, the arterial intima of hyperlipoproteinemic rabbits displayed progressively enhanced uptake and deposition of dietary [^{3}H]cholesterol, [^{125}I]ß-VLDL, and fluorescent conjugate ß-VLDL–1,1'-dioctadecyl-3,3,3',3'-tetramethylindocarbocyanine (DiI). The uptake of [^{125}I]ß-VLDL by the aortic wall of hyperlipoproteinemic animals was 57.2 ± 6 ng [^{125}I]ß-VLDL protein per milligram of wet tissue, about eightfold higher than the control rabbits (12.2 ± 1.6 ng protein/mg wet tissue). In the aortic arch the uptake was almost double that counted in the descending thoracic aorta. These data were generally confirmed by light and electron microscopic autoradiography (unpublished observations).

According to the stage of intimal lesions, ß-VLDL–gold particles were detected in relatively few coated pits and vesicles and were more numerous in plasmalemmal vesicles in different locations. Concomitantly, a progressive subendothelial accumulation of densely packed unilamellar or multilamellar vesicles, referred to as extracellular liposomes (ELs), was noticed (Fig. 6) [123,124]. Incubation with filipin revealed that ELs contain unesterified cholesterol (Figs. 7 and 8) [123]. By immunocytochemistry and incubation with tomatine it was also seen that ELs are frequently surrounded by apoprotein B, which colocalized with phospholipids and apo B (Fig. 9) [80]. The ß-VLDL–gold complex perfused in situ was transcytosed across endothelium by plasmalemmal vesicles and in time was found associated with extracellular liposomes. At more advanced stages, the lipid deposition appeared both extra-

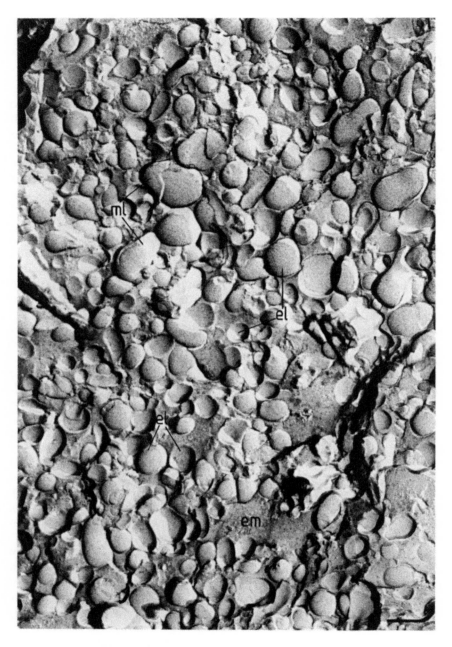

Figure 6 Hyperlipoproteinemic rabbit at 2 weeks of cholesterol-rich diet: the grossly normal aortic arch contains in the intima densely packs extracellular liposomelike vesicles (el), which in freeze fracture preparations commonly show a bi- or multilamellar structure (ml) with smooth surfaces devoid of particles. em, extracellular matrix. Bar = 200 nm.

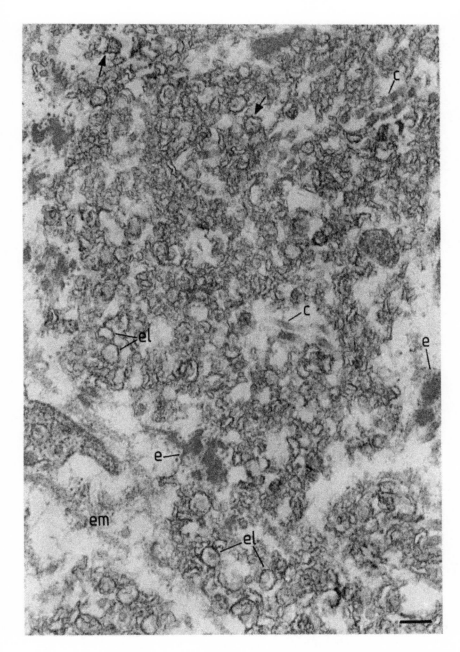

Figure 7 Specimen similar to that in Fig. 6, but incubated with filipin, which binds specifically to unesterified cholesterol. Such complexes appear as characteristic deformations in the lamellae of extracellular liposomes (el) (*arrows*). e elastin; c, collagen; em, extracellular matrix. Bar = 200 nm.

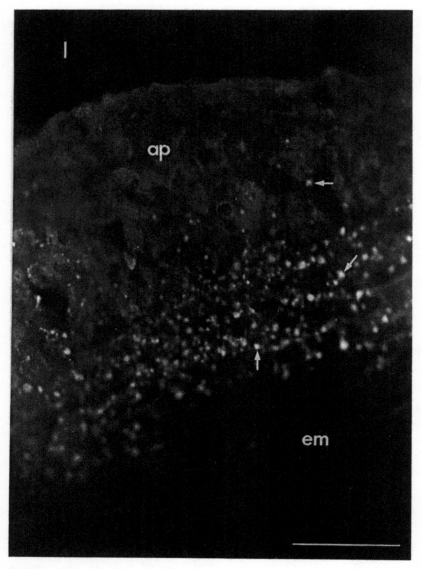

Figure 8 Atherosclerotic plaque (ap) of the aortic arch of a hyperlipidemic rabbit at 8 weeks of cholesterol-rich diet. The specimen was treated briefly (45 s) with propylene glycol, washed with water, incubated with 0.02% filipin in phosphate-buffered saline for 30 min, and mounted in glycerol. Filipin binds specifically to 3-β-hydroxysterols and has an intrinsic fluorescence. Cholesterol–filipin complexes appear as fluorescent spherical particles (*arrows*) located mainly extracellularly.   em, extracellular matrix; 1, vessel lumen. Bar = 100 μm. (Courtesy of F. Lupu.)

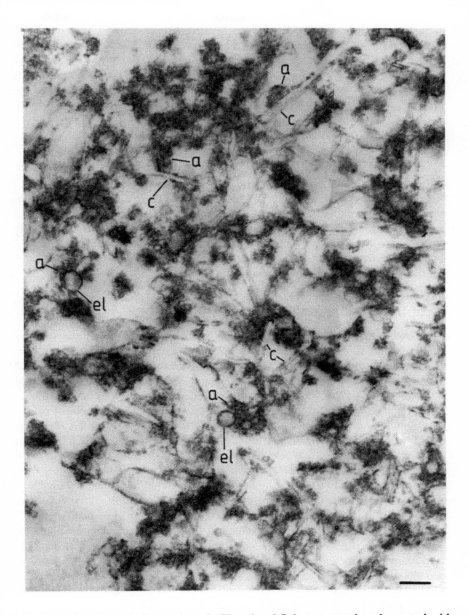

Figure 9 Specimen similar to those in Figs. 6 and 7, but cryosectioned, reacted with an antibody to apoprotein B conjugated to horseradish peroxidase, and incubated with tomatine (known to bind specifically to 3-β-hydroxysterols). Extracellular liposomes (el) are closely surrounded by apo B-HRP reaction product (a), which colocalizes with the spicules that reveal the unesterified cholesterol–tomatine complexes (c). Bar = 200 nm.

cellularly and intracellularly transforming in foam cells the monocyte-derived macrophages, smooth muscle, and endothelial cells [72]. Similar ultrastructural events and changes developed in the cardiac valves [33]. These observations indicate that at the inception of atherogenesis, ß-VLDL or/and its components are progressively accumulated extracellularly and later also intracellularly. The process seems to be mainly the result of enhanced transcytosis of these abnormal VLDLs across arterial and valvular endothelium.

V. LOW DENSITY LIPOPROTEINS

The use of drug–low density lipoprotein complexes has taken advantage of the receptor-mediated uptake of low density lipoprotein (LDL) not only by normal but also by many abnormal cells. However, the selectivity of such delivery depends largely on a better understanding of LDL interactions with endothelial cells of various vascular beds. Thus it may be possible to design ways of more specific targeting of drug–LDL complexes.

A. Formation Metabolic Pathways

Very low density lipoprotein remnant or intermediary density lipoprotein is largely removed from the plasma by the LDL receptor of the liver where part of it is converted directly to LDL by a still unknown mechanism. The crucial loss from VLDL is that of apo E, a process that has been ascribed to the action of hepatic LDL; thus the resulting LDL retains only the apo B-100.

The liver represents the major site of LDL removal from plasma by a predominantly receptor-mediated pathway [13,134]. The hepatic LDL receptor has a higher affinity for IDL than for LDL, which is due to high IDL content in apo E. Hepatocytes have additional recognition sites for LDL other than LDL receptor; they perform a less efficient endocytosis.

Low density lipoprotein is also taken up by extrahepatic tissues, as a function of the rate of transcapillary transport and the LDL receptor activity on the surface of those cells (Fig. 10). In addition, both in liver and extrahepatic tissues, receptor-independent endocytosis of LDL takes place at rates that vary with the organ [2,134–136]. The receptor-independent pathway is especially prominent in the receptor-negative FH homozygotes and the WHHR, in which virtually all plasma LDL is degraded in the liver and extrahepatic tissues by a receptor-independent process. However, in normal conditions, in most species, including humans, the receptor-mediated clearance of plasma LDL in vivo amounts to 70–80% of the whole uptake.

In humans, LDL particles are 18–25 nm in diameter and transport 60% of cholesterol, of which about three-fourths is esterified.

Chemically modified LDL (acetylated, acetoacetylated, or maleylated) at its arginine and lysine residues (which increases the negative charge of the pro-

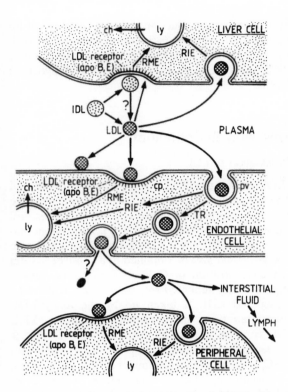

Figure 10 General metabolic pathways of LDL and their interactions with endothelium. Upon partial degradation of IDL within plasma or on hepatocytes, the latter can take up the resulting LDL particles by the LDL receptors and deliver them to lysosomes (ly) by receptor-mediated endocytosis (RME); some particles can be internalized in fluid phase by receptor-independent endocytosis (RIE) also reaching the lysosomal compartment in which, upon hydrolysis, the unesterified cholesterol (ch) is released into the cytoplasm. At the level of endothelial cells, LDL particles can be nonspecifically attached to the plasma membrane or taken up by coated pits (cp) or plasmalemmal vesicles (pv) and delivered by receptor-mediated endocytosis or receptor-independent endocytosis, respectively, to lysosomes. Most LDL is transcytosed (TR) by plasmalemmal vesicles reaching the subendothelial space either as intact or possibly partially degraded particles. Incorporated within the interstitial fluid, it can be taken up by the peripheral cells by a combination of receptor-mediated (via LDL receptor) or receptor-independent processes (as described for the hepatocyte and endothelial cell). A certain amount of LDL is drained by the interstitial fluid into the lymph that brings these particles back in the blood circulation.

tein) is not recognized by the LDL receptor. Such modified LDL is taken up by a distinct binding site, the acetyl-LDL receptor (scavenger receptor; [11–13]; see Chapters 1 and 6). This receptor with a cleanup role is expressed (as is the ß-VLDL receptor) in cells of reticuloendothelial (mononuclear phagocytic) origin, including endothelial cells.

B. Metabolic Interactions with Endothelium

Endothelial Cell-Induced Modification of LDL

It has been reported that LDL can be modified by incubation with cultured endothelial cells (CECs). These modifications include an increased negative charge and buoyant density, a decrease in the relative content of cholesterol and phospholipids (associated with phospholipase A activity), and fragmentation of apoprotein B. This is characteristically associated with the formation of lipid peroxides. Similar changes could also be obtained in a cell-free system by incubating LDL with high concentration of Cu^{2+} or Fe^{3+}. Low density lipoprotein modifications were markedly inhibited by antioxidants [81,93–95,-160,161]. The biologically modified LDL is taken up by the acetyl-LDL receptor of macrophages up to ten times more rapidly than native LDL [146,147]. The oxidized LDL contains a form of degraded apo B that is recognized by the acetyl-LDL receptor of macrophage [95]. It has been also shown that cells other than EC, such as monocytes and neutrophils in vitro, can oxidize LDL (making it cytotoxic) [18,56]. Therefore, it has been suggested that such modified LDL might be involved in the lipid accumulation in atherogenesis [102,103]. However, since no modified LDL particles are detected in blood plasma, the physiopathological significance in vivo of this modification observed in vitro is still uncertain.

LDL-Induced Modifications of Endothelial Cells

There are some conflicting reports on the role of LDL in EC functions. Some findings revealed that both HDL and LDL from normolipidemic human subjects stimulate proliferation of ECs in culture. Both the protein and lipidic components of these lipoproteins contribute to this effect [20].

Experiments from other laboratories have shown that native LDLs injure cultured ECs from human umbilical vein [54,55] and enhance monocyte adhesion to ECs [3]. This cytotoxicity is inhibited by antioxidants [28]. Although the effect is not cell specific and it has been claimed that it is independent of the LDL receptor [54,170], recent reports claim that the cytotoxic compounds transported by LDL are introduced into ECs by receptor-mediated endocytosis using the LDL receptor [17,82].

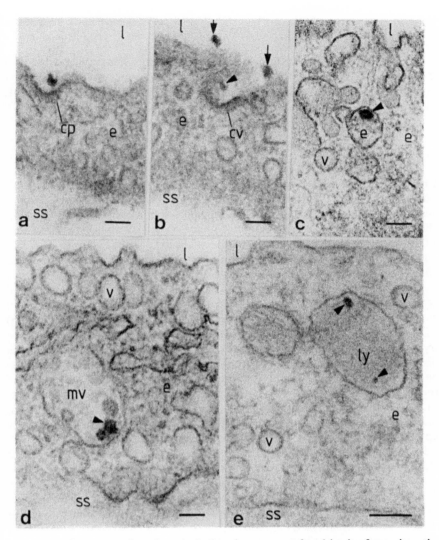

Figure 11 Segments of aortic endothelium from rats perfused in situ for various time intervals with human LDL (10 mg LDL cholesterol/dl) further detected within vascular lumen and tissues by immunocytochemistry. Gallery of electron micrographs depicting sequential steps of LDL endocytosis. Two minutes after perfusion, LDL particles mark some coated pits (cp) (A) or forming coated vesicles (cv), (B) while other particles are attached to the luminal plasma membrane (*arrow*) (B). At 5 min after perfusion, LDL labels some large uncoated vesicles tentatively identified as endosomes (e) (C) and multivesicular bodies (mv) (D). At 5–10 min perfusion, presumably degraded LDL particles (*arrowheads*) can be observed within lysosomes (ly) (e). e, endothelium; l lumen; v, plasmalemmal vesicles; ss, subendothelial space. Bars = 100 nm. (Reproduced by permission of the Rockefeller University Press from *J. Cell Biol.* 1983, *96*:1677.)

Oxidized LDL was found to be cytotoxic for cultured human or bovine ECs [5], to reduce EC uptake of [^{14}C]sucrose [7], and to suppress the production of PDGF-like protein by ECs; similar effects were obtained with acetyl-LDL [36].

C. Endocytosis

Endocytosis is to a large extent a high affinity process mediated by the LDL receptor pathway; it involves coated pits, coated vesicles, endosomes, multi-vesicular bodies, and lysosomes, where final degradation occurs (Fig. 11). Based on morphometric analysis of the time course distribution of perfused LDL particles (detected immunocytochemically), it was shown that an additional pathway may be represented by a receptor-independent process (fluid phase endocytosis) carried out by a fraction of plasmalemmal vesicles. In general, endocytosis of LDL by either microvascular or arterial endothelium is less extensive than is transcytosis [162,165]. Studies on the rat aorta and coronary artery in which untagged homologous (rat) or heterologous (human) LDL was perfused in situ and then identified in tissues by immunocytochemistry revealed that the receptor-mediated internalization of these particles was not significantly enhanced by increasing the tracer concentration but was decreased at low temperature. Endocytosis of homologous LDL was more pronounced than was that of heterologous LDL. Since in those experiments LDL was used at concentrations higher than those occurring in the normal rat plasma, it was assumed that at low LDL concentrations, saturable high affinity uptake could be augmented in relation to the nonsaturable pathway [162].

In spontaneously hyperlipoproteinemic old Spraque-Dawley rats in which plasma native lipoproteins were rendered visible by mordanting the tissue specimens with tannic acid (Fig. 12), particles of the size of LDL were detected within endothelial structures normally involved in endocytosis (Fig. 13). When cationic ferritin pI 8.4 was injected in vivo in some of these animals, the newly formed cationic ferritin–LDL complexes were internalized only to lysosomes and virtually failed to be transcytosed. This strongly polycationic ligand is endocytosed by coated pits/coated vesicles as well as by a fraction of plasmalemmal vesicles [122]. These findings suggest that cationic LDL may be more extensively delivered to the lysosomal compartment of a directly accessible cell such as an EC.

LDL Uptake by the Lung Capillary Endothelium

Since the entire heart output goes directly to the lungs, understanding of the nature and the extent of LDL interactions with pulmonary endothelium is of paramount importance [98,119]. This is particularly relevant in the design and the future of LDL–drug complexes.

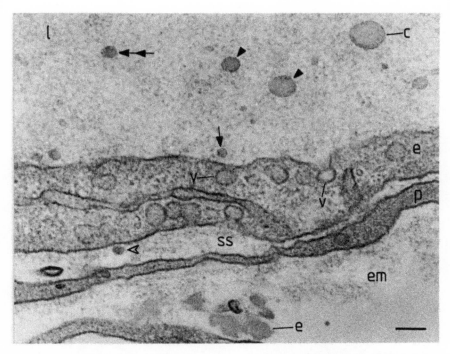

Figure 12 Segment of a blood vessel from a spontaneously hyperlipoproteinemic rat. Specimen preparation for electron microscopy included mordanting with tannic acid (see Ref. 122), which enhanced the contrast of circulating lipoproteins (LP). The vessel lumen (1) contains LP of the size of chylomicrons (C), chylomicron remnants (*arrowheads*), very low density lipoproteins (*double arrow*), and low density lipoprotein (*arrow*); some of the latter also can be seen in the subendothelial space (ss) (*empty arrowhead*). e, endothelial cell; v, plasmalemmal vesicles; p, pericyte; e, elastin; em, extracellular matrix. Bar = 100 nm.

Cholesterol is an important component of the lung surfactant being provided for by the circulating LDL (in humans and several animal species). There is evidence that both LDL and HDL can modulate cholesterol metabolism in the lung. It was demonstrated that the lipoproteins collected from sheep lung lymph show physical and chemical modification reflecting the result of their metabolic interactions with pulmonary cells (especially endothelia) and their associated lipid-hydrolyzing enzymes and lipid transfer proteins. In recent experiments we have used [³H]cholesterol–LDL and [¹²⁵I]LDL as well as LDL–gold complexes to determine the interaction of this lipoprotein with the endothelial cell surface [92,119]. The results indicated that radiolabeled LDL recirculated through the hamster lung in situ was taken up in a saturable manner and reached a plateau at ~20 min. Addition of heparin reduced the

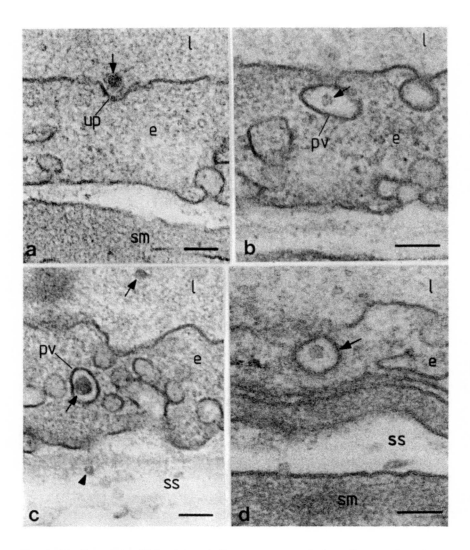

Figure 13 Segments of blood vessels from spontaneously hyperlipoproteinemic rats (specimens treated as in Fig. 12). Lipoprotein particles (*arrows*) of various sizes are seen in the lumen or associated with endothelial (e) structures potentially involved in endocytosis and transcytosis such as uncoated pits (up) of plasma membrane (a), plasmalemmal vesicles (pv) open to the luminal front (b) or apparently internalized (c); some particles are found in the subendothelial space (ss) (*arrowhead*). (d) LP particle within a coated vesicles. l, lumen; sm, smooth muscle cell. Bars = 100 nm.

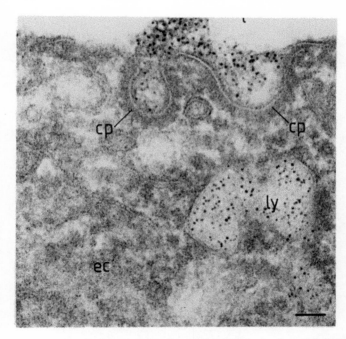

Figure 14 Hamster lung perfused for 60 min in situ with LDL–gold complex. In alveolar capillary endothelium (ec) the ligand labels heavily two coated pits (cp) and a lysosomelike structure (ly). l, capillary lumen. Bar = 100 nm. (Reproduced by permission from *Am. Rev. Respir. Dis.* 1986, *134*:1266.)

uptake by ~45%, whereas competition with unlabeled LDL inhibited the uptake by ~65%. These findings suggested that in the lung LDL is taken up by a dual process: (a) a specific mechanism that is saturable and time- and concentration-dependent, and (b) a receptor-independent uptake that is not saturable or dependent on time. Low density lipoprotein–gold complex bound to endothelial luminal plasma membrane labeling structures involved in receptor-mediated endocytosis (i.e., coated pits, coated vesicles, and endosomelike and lysosomelike features) (Fig. 14) and plasmalemmal vesicles potentially involved in transcytosis (Fig. 15). These observations suggest that lung capillary endothelial surface is provided with the machinery capable of monitoring the uptake and transport of LDL to the lung parenchyma. More detailed studies are required to determine the minute metabolic modifications of the circulating LDLs during their passage through the lung vasculature.

LDL Uptake by Cultured Endothelial Cells

Although the experience acquired using cultured endothelial cells has provided conflicting results, there are some lines of almost general agreement.

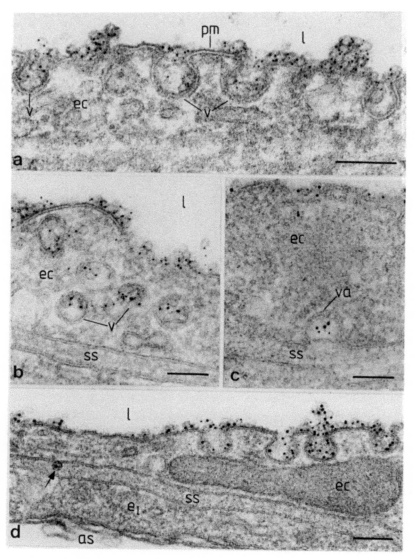

Figure 15 Segments of alveolar capillary endothelia from a hamster lung perfused in situ with LDL–gold complex. (a) Particles bound to plasma membrane (pm) and labeled plasmalemmal vesicles (v) open to the lumen (l). (b) At later time points, labeled vesicles (v) appear internalized, or (c) open to the abluminal front (va). (d) At 30 min of perfusion, rare particles (*arrow*) can be seen in the subendothelial space (ss). ec, endothelial cell; e_I, type I epithelial cell; a, alveolar space; l, capillary lumen. Bar = 100 nm. (Reproduced by permission from *Am. Rev. Respir. Dis.* 1986, *134*:1266.)

Endothelial cells and monocyte-macrophages appear thus far to be the only cells known to posses the LDL receptor, the ß-VLDL receptor, and the acetyl-LDL (or scavenger) receptor [4]. In rapidly dividing endothelial cells, LDLs play a regulatory role in cell cholesterol metabolism similar to that of other cells. However, in confluent (contact-inhibited) monolayers, LDL receptor activity and LDL degradation are very low [29–32,66,110,111,168] and the LDL receptors are dispersely distributed on ECS [111]. Reportedly, confluent human ECs are capable of catabolizing LDL when exposed to physiological concentration of this lipoprotein [21]. Early studies of LDL binding to cultured rabbit aortic endothelium suggested the existence of at least two classes of binding sites. The high affinity sites were fully saturated at very low concentration (~5 μg LDL protein/ml). Since the inhibition of EC sterol synthesis increased progressively with LDL concentrations, it has been assumed that an additional uptake by low affinity sites contributes to the control of EC cholesterol metabolism; such a process may be prevalent in vivo [104].

Low affinity LDL uptake and metabolism were less affected by the state of confluency. At physiological LDL concentrations in the culture medium (550 μg/ml) the fluid phase and low affinity adsorptive endocytosis of LDL accounted for 12% and 70%, respectively [160,161]. In cultured aortic endothelium, the fluid-imposed wall shear stress enhanced significantly the receptor-mediated internalization and degradation of [^{125}I]LDL. Nonspecific uptake of the same ligand was not influenced by shear stress levels [139]. Recent data revealed that intracytoplasmic delivery of monoclonal antibody directed against the 90 kilodalton coated vesicle membrane peptide inhibits receptor-mediated LDL catabolism in cultured ECs [23].

Quantification of the binding of LDL–gold complexes to the plasma membrane at 4°C showed no difference between arterial and venous ECs in culture. Preincubation with lipoprotein-depleted serum increased the number of LDL binding sites on these cells [79]. Low density lipoprotein uptake appears to be inversely correlated with the sialic acid content of ECS [44].

Cultured aortic ECs from diabetic minipigs have shown an LDL uptake about 40% higher than in ECs from normal animals [47]. An interesting avenue is the possibility of promoting the transfer of LDL cholesteryl esters from the aortic interstitium to a cellular catabolic compartment such as ECs, as suggested by the enhanced receptor-mediated endocytosis of LDL by ECs in culture under the influence of the calcium channel blocker verapamil [142].

Endothelial Uptake of Chemically Modified LDL

Biologically modified or damaged LDLs are cleared from the circulation and from the interstitial fluid by endothelial cells and macrophages, respectively, via the acetyl-LDL (scavenger) receptor [11]. Although modified LDLs have not been demonstrated in the plasma in vivo, it is possible that they are produced and rapidly cleared. In vitro studies have shown that LDLs chemically modified by acetylation, acetoacetylation, or treatment with malondialdehyde are avidly degraded by macrophages which become foam cells [73,74]. The acetyl-LDL receptor recognizes an epitope on the apoprotein moiety of the delipidated oxidized LDL [95]. After intravascular injection, the modified LDL is largely degraded in the liver I (6,73,74,84,85). In vivo and in vitro studies have demonstrated that EC, especially those lining the sinusoidal capillaries of liver, spleen, bone marrow, and adrenal gland, participate to various degrees in the degradation of acetyl-LDL [50,84,85,89,99,141,169]. Although acetyl-LDL receptor pathway is separate from the LDL pathway, the site of ligand binding for both is the coated pits [13,14,99]. Acetyl-LDL is not taken up by brain capillaries [39]. The uptake of reductively methylated LDL will be further discussed in Section V. D.

For drug targeting, the acetyl-LDL receptor has the potential advantage of residing on a more restricted number of cells than the LDL receptor, i.e., endothelial cells (especially those of sinusoid capillaries), monocytes, and macrophages.

When LDL was given a positive charge by reaction with N,N-dimethyl-1,3-propanediamine, the rate of lysosomal degradation of the lipoprotein by fibroblasts was increased about a hundredfold [13,14].

Endocytosis of LDL in Hypercholesterolemic Animals

In the advanced stage of plaque formation during diet-induced hyper-lipoproteinemia in hamsters, some EC overlying the atheroma may become loaded with lipid deposits to form foam cells [41,72,123,124]. Endothelial cell transformation in foam cells was observed both in hyperlipidemic rabbits (in which the main cholesterol carrier is ß–VLDL) and in hyperlipidemic hamsters (LDL major cholesterol carrier) [33,72,92,123,124]. Presumably the process is largely mediated by endocytosis via active ß–VLDL and LDL receptors, respectively. Such EC-derived foam cells progressively lose anionic sites and some glycoconjugates of their cell surface [41].

D. Transcytosis

To reach various cells, including those of the vessel wall, plasma LDL has to pass through the vascular endothelium. As other cells which need LDL as carrier of the cholesterol required for membrane synthesis, endothelial cell has to perform LDL uptake for its own needs and in addition has to transport LDL to the cells of the surrounding tissue. Inability of endothelium to monitor the transport of physiological amounts of LDL or an excessive accumulation of cholesterol within the artery walls largely underlies atherosclerosis.

The routes and mechanism of LDL transcytosis across various types of endothelium in different tissues have been incompletely explored in vivo [75,124].

In organs provided with sinusoid capillaries (e.g., liver, spleen, bone marrow, and to a certain extent adrenal cortex) it is assumed that particles of the size of LDL (20-25 nm) can readily pass through the large discontinuities of the endothelial cell. Actually the interactions of LDL and other lipoproteins with such endothelia are much more complex than a simple sieving process.

In the fenestrated endothelia of most visceral capillaries, LDL particles are not usually seen penetrating the diaphragmed fenestrae but rather associated with plasmalemmal vesicles. In these capillaries, as in those with continuous endothelium, LDL could not be observed penetrating the intercellular junctions [123,124,162,165].

In the continuous endothelium of large vessels and capillaries of heart, lung, and somatic tissues, particles of LDL size [122], LDL particles identified immunocytochemically [162], or LDL-gold complexes [92,123,124,163-166] (visualized by electron microscopy) have been observed in plasmalemmal vesicles in rare transendothelial channels, and in the subendothelial interstitia.

Transendothelial transport of LDL was investigated mostly in cultured endothelial cells and much less in vivo. In both cases the inquiry was addressed mainly to aortic endothelium, very few studies being conducted on capillary endothelia.

Information from In Vitro Studies

Cell cultures by their nature can tell little about the actual transport processes in vivo. Such studies are limited to the binding, internalization, and degradation of LDL by the cultured EC. Additional data may be obtained on EC monolayers grown in a twochamber system [48,53,87,156].

Early studies on LDL interaction with rabbit aortic endothelium in culture suggested the existence of two classes of binding sites: high affinity sites saturable at very low LDL concentrations (~ 5 μg LDL protein/ml) and low affinity sites whose uptake increases with increasing LDL concentration. It became conceivable that in vivo, where ECs are exposed to LDL concentrations far above those required to saturate the high affinity sites, the LDL uptake may be attributable to low affinity sites [104]. The latter may be instrumental in LDL transcytosis [123,124,162-166]. Cultured porcine aortic ECs at confluence were examined for the transport of rhodaminelabeled LDL. There was little transport at 4°C, but at 37°C transcytosis of the probe was significant, dose-dependent, and saturable at 0.4 mg protein/ml. The process was energy-dependent since its rate was affected by temperature and inhibited by 2-deoxyglucose and NaN_3. Reportedly, rhodamine-LDL was transported via transcytotic vesicles and not through intercellular junctions [48]. In a similar transport chamber system, in searching for the differential transport of LDL and malondialdehyde-modified LDL (MDA-LDL), it was found that the EC monolayers transport preferentially native LDL versus MDA-LDL. The biding and transport of the latter was inhibited by fucoidan (which competes for the scavenger receptor). It was suggested that transendothelial transport of LDL is regulated by a scavenger receptor and a nonspecific transport system [109]. It has been shown that exposure to fatty acids (especially albumin-bound oleic acid) increases human LDL transfer across cultured porcine pulmonary artery EC [53]. During their transendothelial passage, LDLs were not degraded in appreciable amounts [48,53]. By addition of ß-VLDL to the cultured EC monolayers, LDL transport rate was almost doubled without altering electrical resistance. Since no difference was observed in the LDL transport across EC obtained from normal or Watanabe heritable hyperlipidemic rabbits, it was assumed that LDL receptor was not involved in the transport of this lipoprotein [87]. Low density lipoprotein transport was also unidirectionally enhanced by monocyte chemotactic migration across cultured EC monolayers [156].

In Vivo and In Situ Experiments

Among the few experimental data obtained on LDL transport through capillary endothelium are those carried out on lung. In situ perfusion of the hamster lung with radiolabeled or gold-conjugated human LDL revealed that ~ 50% of LDL uptake is receptor-independent. Experiments with double-labeled LDL showed a preferential transport of [³H]cholesterol versus [¹²⁵I]apo B, which suggested that in part the LDL cholesterol is transferred to the lung parenchyma without the apoprotein moiety: this dissociation process appears to be initiated and monitored by the capillary endothelium. As shown by the distribution of LDL–Au particles, transcytosis of this complex is carried out by plasmalemmal vesicles [92].

Low density lipoprotein transport in mammalian aorta has been shown to take place from both the luminal and adventitial side (via vasa vasorum), the relative concentration levels and their spatial gradients being much higher near the intimal surface [8,144]. Local variation in the permeability of normal aortic wall to LDL [112] was identified as small punctate foci but their cellular nature is still unclear [83,113,150]. Through the use of an aorta–heart perfusion system it was detected by autoradiography that the [¹²⁵I]LDL occurs mostly in the intima and the innermost part of the media [140,144]. The cellular aspects of LDL transport across arterial endothelium were examined in rat, rabbit, and guinea pig by perfusion in situ of radioiodinated or gold-conjugated LDL followed by radioassay, immunocytochemistry, and electron microscopy [162,163]. Although such an approach does not allow quantitation of the process, some significant aspects were obvious. It was found that most perfused LDL is transported through aortic or coronary endothelium by a receptor-independent fluid phase uptake carried out by a fraction of plasmalemmal vesicles (Fig. 16). Unlike endocytosis, LDL transcytosis was less affected by lowering the temperature of the perfusates but was markedly enhanced at high concentrations of LDL. In rats, no evident difference in the

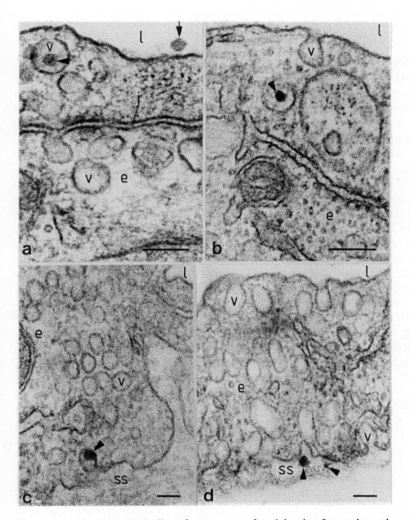

Figure 16 Aortic endothelium from rats perfused in situ for various time intervals with human LDL (10 mg LDL cholesterol/dl) and detected within vessel lumen and tissues by immunocytochemistry. Gallery of electron micrographs illustrating LDL transcytosis. (A) Two minutes after perfusion, LDL particles are encountered close to endothelial plasma membrane (*arrow*) or contained within plasmalemmal vesicles (v) near the luminal front. (B) Five minutes after perfusion, LDL particles mark plasmalemmal vesicles apparently internalized (*arrowhead*) or (C) in vesicles open to the abluminal front (*arrowhead*) or (D) in vesicles apparently discharging their content in the subendothelial space (ss) (*arrowheads*). (D) The latter particles appear usually smaller and more irregular than those detected on endothelium or within plasmalemmal vesicles. e, endothelial cell; l, lumen; v, plasmalemmal vesicle. Bars = 100 nm. (Reproduced by permission of the Rockefeller University Press from *J. Cell Biol.* 1983, *96*:1677.)

transport rate was noticed between homologous (rat) and heterologous (human) LDL [162]. Apo B–gold complexes perfused in situ were found both in features involved in endocytosis and in plasmalemmal vesicles. The latter were labeled by the apo B–Au complex at early stage (5 min) on the luminal front; at later stages (10–15 min) labeled vesicles opened on the abluminal front, some of them apparently discharging their content in the subendothelial space. The uptake and accumulation within the aortic wall were almost similar for native and methylated LDL. The methylated apo B–gold particles, perfused in the same conditions as native apo B–gold, were found located mainly in plasmalemmal vesicles. Morphometric analysis showed that for this ligand, transcytosis was about fivefold higher than endocytosis. Radiolabeled methyl-LDL perfused in rat in situ was recovered in the aortic wall in amounts similar to those measured for radiolabeled native LDL. Since the uptake of methyl-LDL largely reflects a receptor-independent process [15,16], the apparent discrepancy between these data and those obtained with LDL–Au and methyl-LDL–Au complexes can be explained at least in part by the very low endocytic rate of methyl-LDL by the cells of the aortic wall, its accumulation virtually representing the transcytosed fraction [163–166].

Using a different approach, Wiklund et al. [171] assessed the role of the LDL receptor in the LDL influx in the rabbit aortic wall by injecting in vivo simultaneously [^{125}I]LDL and [^{125}I]methyl-LDL (the latter is not recognized by the LDL receptor); no significant difference between the two probes was found. It was concluded that transport of LDL across EC is not dependent on a receptor-mediated mechanism. However, ECs themselves contribute with ~40% to LDL degradation in the intima, the process involving the high affinity LDL receptor normally expressed in vivo [145,171]. Spatial distribution and quantification of LDL in the aortic wall can be determined by an immunotransfer procedure [60]. The receptor-dependent LDL transport appears to be the prevalent mechanism for the removal of plasma LDL by the liver [37]. The receptor-independent transport activity is widely distributed in several organs [117,171]; in WHHL rabbits the absence of LDL receptors leaves the independent pathway the only one in operation [136].

LDL Transcytosis in Arterial Endothelium of Hyperlipidemic Animals

There is an almost general agreement that in the atherosclerotic arterial plaques of hyperlipidemic animals, there is an increased endothelial permeability to plasma macromolecules including LDL [149,167]. Direct quantitative evidence and identification of pathways and mechanisms involved in such a process, however, are very scarce, most of the evidence being only circumstantial and based mostly on LDL accumulation in the atherosclerotic lesions [59,60,128–133,172]. A little better understood is the process involving the ß-VLDL (see Section IV. D). In atherosclerotic rabbits, LDL assimilation (apparently by both receptor-mediated and receptor-independent processes) increases at sites of arterial atheromatous lesions [125], the insudation of LDL into the intima appearing as a result of enhanced vesicular transport [69,149]. As revealed by immunocytochemistry, LDL was deposited in the subendothelial space [69]. By using double-labeled cholesterol ([³H] and [¹⁴C]), it was shown that esterified cholesterol enters the arterial tissue mainly as part of the lipoprotein by transendothelial transport, whereas part of the influx of free cholesterol is by direct exchange between LDL and luminal endothelial plasma membrane [148]. High density lipoprotein decreases aortic permeability to LDL in hypercholesterolemic rabbits [68]; a similar effect was ascribed to diazepam [76]. Local experimental removal of endothelium does not favor deposition of lipids, which accumulate preferentially in areas of intimal thickening covered by regenerated endothelium [78]. Other authors claimed that damaged arterial wall exhibits a preferential accumulation of LDL [107].

LDL Transport in Human Aorta

The presence of LDL in the normal arterial intima has been demonstrated by immunofluorescence, immunohistochemistry, and electrophoretic and radioisotope techniques [1,128–133,173]. The amount of lipoprotein influx within the intima is correlated with plasma cholesterol concentration, intimal accumulation of LDL being almost double its concentration in the serum, while intimal deposition of albumin is only a quarter of the plasma concentration [128]. Transport of LDL into the normal human aortic tissue shows some similarities with that seen in the aorta of hypercholesterolemic rabbits [148].

Lipoproteins containing apo B and resembling plasma LDL have been demonstrated immunocytochemically in the grossly normal intima and in fatty-fibrous plaques of human aorta [128–132], where they complex specifically with the matrix glycosaminoglycans. Although apparently similar to plasma LDL, the apo B–containing lipoproteins extracted from human aortas contain some particles larger than LDL, are more electronegative than LDL [57–59,61], and their apo B is highly degraded. These particles are avidly ingested by macrophages, in which they stimulate cholesterol esterification and accumulation with subsequent transformation in foam cells [58,59]. In aliquots of interstitial fluid collected from normal human aortic intima, an LDL concentration more than twice the concentration in the plasma was found; no measurable amounts of LDL were detected in the inner media [130,131].

Human atherosclerotic lesions have been shown to accumulate intracellularly and extracellularly apo B. As detected immunohistologically, coronary arteries display mostly extracellular apo B. Intracellular apo B would correspond to arteries that usually have earlier atherosclerotic involvement [173]. Human atheroma contain modified LDL similar to that extracted from normal aorta. In atherosclerotic subjects, a considerable enhancement of the flux of free and esterified cholesterol from plasma into the inner layers of the vessel wall was detected. However, no significant correlation between cholesterol influx and cholesterol content of the arterial wall in these patients could be established [9]. In two patients with homozygous familial hypercholesterolemia, autopsic material examined histologically suggested a correlation between apo B deposits and proliferation of sulfated glycosaminoglycans [174].

From the preceding data we can assume that LDL access, transit, and metabolism within the vessel wall is a complex, multifactorial processing of which the transendothelial transport is only a component.

E. Cholesterol Efflux from the Artery Wall

Cholesterol accumulation within the artery wall can be expressed as an imbalance between the inflow and outflow of lipoproteins or their components. By extrapolation to other cell systems [97] it may be assumed that efflux of cholesterol could occur as (a) free cholesterol (after intracellular deesterification of endocytosed cholesteryl esters), (b) lipoprotein particles, or (c) cholesteryl ester efflux mediated by cholesteryl ester transfer protein (CETP). Results obtained in a model system of cultured smooth muscle cells and macrophages showed that CETP may promote cholesteryl ester efflux from aortic interstitium or desintegrated cells [143].

VI. HIGH DENSITY LIPOPROTEINS

High density lipoprotein represents a surface remnant of the chylomicrons and VLDL, being generated during the transport and lipolysis of these triglyceride-rich lipoproteins.

High density lipoprotein particles, 7-10 nm in diameter, constitute the predominant plasma lipoprotein in many animal species and in young humans. In adult subjects, the number of circulating HDLs is about 10- to 20-fold higher than the number of all other lipoprotein particles. The plasma concentration of HDL (200-500 mg/dl) is almost equal to that of LDL, but since more HDL than LDL occurs in the interstitial fluid, the total body content of HDL probably exceeds that of LDL [27].

High density lipoprotein, with its flexible subclasses, appears as a very dynamic system, the site of cholesterol esterification and main carrier for the transfer of cholesterol from the peripheral tissues to the liver (reverse cholesterol transfer). During this process, HDL behaves as a temporary station for some plasma lipids and proteins, each component turning over periods of time ranging from 1 h to 1 day.

By close association with the plasma membrane of various cells, HDL can exchange with these cholesterol and cholesteryl ester molecules [97]. It promotes release of cholesterol from the arterial wall cells in culture, but a specific binding seems not to be a prerequisite for cellular cholesterol efflux [43]. Although some data have been reported suggesting that HDL interacts with cells, including ECs via membrane lipids rather than specific receptors [151], evidence is being accumulated that most tissues express specific binding sites for HDL. High density lipoprotein receptor has not been purified as yet; it expresses a high affinity ($K_d = \sim 10^{-8}$ M), seems to be different from the apo B and apo E receptors, and is regulated in a direction opposed to that of LDL receptor.

A. Metabolic Interactions with Endothelium

High density lipoprotein inhibits the uptake of LDL by humans ECs in culture and has a protective action against the injurious effect of LDL on such cells. These early observations promoted and extended the clinical evidence that low concentration of HDL is a risk factor for coronary heart disease. It has been also shown that HDL stimulates proliferation of bovine ECs in culture [152,153,155[and the addition of 25-hydroxysterol results in a fivefold to tenfold increase in the number of saturable, high affinity EC binding sites for HDL [154]. Competition experiments suggested that HDL binding sites on ECs also interact with other lipoproteins. Apo E-free HDL binds to cultured human EC with high affinity, indicating that HDL interaction with its cognate receptor is not mediated by the recognition of one specific apoprotein [49].

Since ECs lack enzymes for synthesis of arachidonic acid, they have to acquire arachidonate from exogenous sources such as plasma lipoproteins. Incubation of cultured arterial ECs with HDL stimulates prostacyclin synthesis by supplying the cells with arachidonate, whereas LDL has no such effect. Stimulation by HDL of the EC production of a potent platelet antiaggregant and vasodilator such as prostaglandin may constitute another beneficial effect of HDL in cardiovascular disease [34,100].

In cultured aortic ECs, HDL binding sites are upregulated when acetyl-LDL is delivered to cells via the scavenger receptor pathway. Thus the HDL receptor of ECs (as well as of other cells of the vessel wall) may facilitate the reverse cholesterol transport from the cells that have internalized excess cholesterol by the acetyl-LDL receptor [10]. Moreover, it was demonstrated that exposure of foam cells to HDL depletes them of excess cholesterol [11,62], suggesting a potential role of this lipoprotein in the prevention of foam cell formation during atherogenesis. In dog lymph the presence of discoid HDL and its increase upon cholesterol feeding was taken as strong circumstantial evidence for the peripheral formation of this form of HDL as an early event in reverse cholesterol transport [26]. Despite several lingering uncertainties, the two main mechanisms currently ascribed to the antiatherogenic activity of HDL are (a) the inhibition of LDL uptake by the cells of the artery wall, and (b) the reverse cholesterol transport from these cells [143]. Cholesterol is then esterified by LCAT and transported to the liver, where cholesteryl ester is hydrolyzed and the resulting cholesterol is used for synthesis of lipoproteins or bile acids or excreted as such in the bile.

B. Endocytosis and Transcytosis

In capillaries isolated from rat fat tissue, it was reported that fluorescent-labeled HDL was ingested at more than twice the efficiency of LDL uptake [108]. Internalization and degradation of HDL by the bovine aortic EC in culture was 10– to 20–fold lower than LDL. In both sparse and confluent endothelial monolayers the HDL uptake was assumed (yet not directly demonstrated) to occur by fluid and nonspecific adsorptive endocytosis via a pathway distinct from that of LDL internalization [155].

In rabbit with diet-induced hypercholesterolemia, leading to overproduction of ß–VLDL, the fractional catabolic rate of [^{125}I]HDL was augmented and its accumulation within the aorta was more than ninefold lower than in control animals. It was considered that the low plasma level and reduced aortic accumulation of HDL may accentuate cholesterol loading induced by ß–VLDL [22].

When human grossly normal aorta and atherosclerotic plaques were examined for their content in HDL and apo A-I, it was found that while in normal intima most apo A-I is in the HDL density fraction, in the plaque, ~50% of this apoprotein is in the $d > 1.21$ fraction. The 1.063–1.21 density fraction isolated from plaques contains particles similar to HDL (6–12 nm) but enriched in free cholesterol. This was interpreted as the in vivo counterpart of the cholesterol transfer from cell membranes to the HDL infiltrated within the artery wall [52].

Incorporation into HDL of a triantennary galactose-terminated cholesterol derivative (Tris-Gal-Chol) led to a specific interaction of this compound with the asialoglycoprotein receptor on hepatocyte. Conversely, a Tris-Gal-Chol–LDL compound interacted mainly with the galactose receptor in the Kupffer cells. The endocytosed Tris-Gal-Chol–HDL was degraded in lysosomes. This was taken as a suggestion to use this derivative for targeting drugs to either Kupffer or parenchymal liver cells, as discussed by Van Berkel and colleagues in Chapter 7 [158,159].

Endocytosis and transcytosis of HDL across various endothelia, in normal and pathologic conditions, constitute at present a largely open field. Further investigation is fully justified by the notorious antiatherogenic effect of HDL and its potential usefulness for site-specific drug delivery.

VII. LIPOPROTEINS IN THE INTERSTITIAL FLUID AND LYMPH

Once passed beyond endothelial barrier, lipoproteins enter the extracellular compartment where their circulation within the interstitial fluid may interact with various components of the extracellular matrix before reaching the cells of the host tissue. Upon direct contact with the cell membranes, LP may be taken up by the cells, predominantly by a receptor-mediated process, or may exchange cholesterol bidirectionally with the cell membranes. Part of the interstitial fluid is drained into the initial lymphatics, which progressively merge into larger lymph vessels; eventually they are collected into the thoracic duct, which brings lymph into the venous blood system. There is almost general agreement that the prenodal peripheral lymph represents an adequate sample of the interstitial fluid [126,127]. All classes of LP are present in lymph: the lymph/plasma concentration varies from 0.03 for VLDL to 0.2 for HDL. In some organs such as heart and lung the LP lymph/plasma ratio may amount to 0.6

For reasons that are not completely understood, the lymph LPs are more heterogeneous in size (e.g., discoidal HDL particles) than are their plasma counterparts, from which they also may differ chemically (e.g., lymph LDL and HDL are enriched in cholesterol).

Studies on humans who received intravenous injection of [^{14}C]cholesterol showed that the free cholesterol/esterified cholesterol ratio in the lymph LDL and HDL was high, suggesting that both lipoproteins can act as acceptors for tissue-free cholesterol, but the predominant acceptor was HDL [106]. Investigations on nonhuman primates demonstrated that the primary effects of increased dietary cholesterol was the augmentation of the cholesteryl esters of all lymph LP, whereas the unesterified cholesterol was unchanged [67]. Similar studies in men indicated that both LDL and HDL undergo modifications during their passage into peripheral lymph [105]. The ratio of LDL to HDL cholesterol in the lymph differs among species and among organs. There is a particularly high LDL concentration in the cardiac lymph both in dogs and in pigs [64]. In cholesterol-fed dogs the lymph/plasma ratios of LDL were reduced as compared to normal dogs, while HDL was unaffected, suggesting a selective interstitial distribution of lipoproteins, which may be modified to various degrees in the interstitial fluid. It seems that each tissue has a characteristic mixture of lipoproteins that reacts differently to a high cholesterol diet [65]. Comparison of interstitial fluid collected from human aortic intima and serum revealed the LDL from proliferative gelatinous lesions had a decreased negative charge [130,131]. In the interstitial fluid collected from human aortic fatty streaks, LDL was reduced to one-third of the concentration in normal intima, while a$_2$-macroglobulin and albumin concentrations were unchanged [130–133]. In diet-induced hyperlipoproteinemic rabbits and hamsters, in early prelesional stages of atherogenesis most of the lipoproteinemic material transcytosed into the subendothelial spaces occurred as extracellular phospholipid liposomes rich in unesterified cholesterol and displaying on their contour apo B [33,80,92,123,124]. This was taken as a suggestion that in these conditions, the extravasated LPs are modified either during their transendothelial transport or while in the extracellular matrix. It is believed that (as part of a local protective reaction) the LP-derived components are subsequently reassembled and taken up by the local macrophages and smooth muscle cells to become foam cells [123].

VIII. CONCLUDING REMARKS

Interactions of plasma lipoproteins with the vascular endothelium are multiple and complex. The luminal endothelial surface expresses a broad spectrum of binding sites, which indirectly or directly secure a close association between LP and endothelium. Some of them participate in the intravascular metabolism of LP; these include heparan sulfate proteoglycans of the glycocalyx, which immobilize the chylomicrons and VLDL to be exposed to the hydrolytic activity of membrane-associated lipoprotein lipase. Other binding sites or receptors mediate the specific and regulated uptake and transport of LP.

Each class of LP can be taken up by ECs via endocytosis (predominantly receptor-mediated) and internalized to lysosomes where degradation occurs. Each class of LP can also be transported across the ECs via transcytosis (mostly receptor-independent) to be delivered to the surrounding tissues. During transcytosis LP particles behave as hydrophilic macromolecules being, sorted, gated, and transported across EC according to their size, charge, and chemistry. The fractional contribution of endocytosis and transcytosis varies with the lipoprotein, animal species, location and structure of endothelium, and physio(patho)logical conditions involved. Endothelial cell is among the few cells which express a full set of receptors for the endocytosis of ß–VLDL, LDL, acetyl-LDL (scavenger), and HDL. It is still unclear whether transcytosis operates with the same type of receptors but of much lower affinity or occurs totally independent of receptors.

In their interaction both EC and LP participate with the whole spectrum of physical and chemical properties. Upon a transient close apposition between LP and EC luminal plasma membrane, the two partners can exchange cholesterol by direct transfer. Endothelial cells can acquire cholesterol for their own metabolic needs either this way or, more extensively, through receptor-mediated endocytosis.

For site-specific drug delivery using LP as carriers, designing a successful targeting depends essentially on a rigorous knowledge of LP binding sites on the endothelium of the desired tissue and the capability to monitor adequately the transcytotic process of the LP–drug complex and its delivery to the target cells.

ACKNOWLEDGMENTS

We gratefully acknowledge the word processing done by D. Neacsu.

The work carried out in our laboratory and reported in this review was supported by the Ministry of Education, Romania, and by the National Institutes of Health Grant HL-26343.

REFERENCES

1.   Adams CM. Tissue changes and lipid entry in developing atheroma. In: Atherogenesis: initiating factors, Ciba Found. Symp. 12 (new series). Amster dam:Elsevier, 1973: 5–37.

2.   Alderson LM, Endemann G, Lindsey S, Pronzcuk A, Hoover RL, Hayes KC. LDL enhances monocyte adhesion to endothelial cells in vitro. Am J Pathol 1986; 123:334–342.

3.   Attie AD, Pittman RC, Steinberg D. Hepatic catabolism of low density lipoprotein: mechanisms and metabolic consequences. Hepatology 1982; 2:269–281.

4.   Baker DP, Van Lenten BJ, Fogelman AM, Edwards PA, Kean C, Berliner JA. LDL, scavenger and ß–VLDL receptors on aortic endothelial cells. Arteriosclerosis, 1984; 4:248–255.

5.   Bernheimer AW, Robinson WG, Linder R, Mullins D, Jip YK, Cooper NS, Seidman Y, Uwajima T.   Toxicity of enzymatically oxidized low density lipoprotein. Biochem Biophys Res Commun 1987; 148:260–266.

6.   Blomhoff R, Drevon CA, Eskild W, Helgerund P, Norum KR, Berg T. Clearance of acetyl low density lipoprotein by rat liver endothelial cells. Implications for hepatic cholesterol metabolism. J Biol Chem 1984; 259:8898-8903.

7.   Borsum T, Henriksen T, Reisvaag A. Oxidized low density lipoprotein can reduce the pinocytic activity in cultured human endothelial cells as measured by cellular uptake of [^{14}C]sucrose. Atherosclerosis 1985; 58:81–96.

8.   Bratzler RL, Chisolm GM, Colton CK, Smith KA, Lees RS. The distribution of labeled low-density lipoproteins across the rabbit thoracic aorta   in   vivo. Atherosclerosis 1977; 28:289–307.

9.   Bremmelgaard A, Stender SJ, Lorentzen J, Kjeldsen K. In vivo flux of plasma cholesterol into human abdominal aorta with advanced atherosclerosis. Arteriosclerosis 1986; 6:442–452.

10.   Brinton EA, Kenagy RD, Oram JF, Bierman EL. Regulation of high density lipoprotein binding activity of aortic endothelial cells by treatment with acetylated low density lipoprotein. Arteriosclerosis, 1985; 5:329–335.

11.   Brown MS, Ho YK, Goldstein JL. The cholesterol ester cycle in macrophage foam cells. Continual hydrolysis and reesterification of cytoplasmic cholesteryl esters. J Biol Chem 1980; 255:9344–9352.

12.   Brown MS, Goldstein JL. Lipoprotein metabolism in the macrophage: implications for cholesterol deposition in atherosclerosis. Ann Rev Biochem 1983; 52:223–261.

13.   Brown MS, Goldstein JL. How LDL receptors influence cholesterol and atherosclerosis. Sci Am 1984; 251:58–66.

14.   Brown MS, Goldstein JL. The receptor model for transport of cholesterol in plasma. Ann NY Acad Sci 1985; 454:178–182.

15.   Carew TE, Pittman RC, Steinberg D. Tissue sites of degradation of native and reductively methylated [^{14}C]sucrose-labeled low density lipoprotein rats. Contribution of receptor-dependent and receptor-independent pathways. J Biol Chem 1982; 257:8001–8008.

16.   Carew TE, Pittman RC, Marchand ER, Steinberg D. Measurement in vivo of irreversible degradation of low density lipoprotein in the rabbit aorta. Predominance of intimal degradation. Arteriosclerosis, 1984; 4:214–224.

17.   Catapano AL. Transport of cytotoxic compounds to cells via the LDL receptor pathway. Med Sci Res 1987; 15:411–413.

18.   Cathcart MK, Morel DW, Chisolm GM. Monocytes and neutrophils oxidize low density lipoprotein making it cytotoxic. J Leukocyte Biol 1985; 38:341–350.

19.   Chajek-Shaul T, Friedman G, Stein O, Olivecrona T, Stein Y. Binding of lipoprotein lipase to the cell surface is essential for the transmembrane transport of chylomicron cholesteryl ester. Biochim Biophys Acta 1982; 712:200–210.

20.   Chen J-K, Hoshi H, McClure DB, McKeehan WL. Role of lipoproteins in growth of human adult arterial endothelial and smooth muscle cells in low density lipoprotein-deficient serum. J Cell Physiol 1986; 129:207–214.

21.   Coetzee GA, Stein O, Stein Y. Uptake and degradation of low density lipoproteins (LDL) by confluent contact-inhibited bovine and human endothelial cells exposed to physiological concentrations of LDL. Atherosclerosis 1979; 33:425–431.

22.   Daugherty A, Lange LG, Sobel BE, Schonfeld G. Aortic accumulation and plasma clearance of β–VLDL and HDL: effects of diet-induced hyper-cholesterolemia in rabbits. J Lipid Res 1985; 26:955–963.

23.   Dennis PA, Mendrick DL, Rennke HG, Diehl TS, Davies PF. Intracytoplasmic delivery of monoclonal antibody V5/1 directed against 90 KD coated vesicle membrane polypeptide inhibits receptor-mediated LDL metabolism in cultured vascular endothelium. Fed Proc 1987; 46:3838. (abstr)

24.   Dessay KS, Gotlieb AI, Steiner G. Very low density lipoprotein binding to cultured aortic endothelium. Can J Physiol Pharmacol 1985; 63:809–815.

25.   De Zanger R, Wisse E. The filtration effect of the rat liver fenestrated sinusoidal endothelium on the passage of (remnant) chylomicrons to the space of Disse. In: Knook DL, Wisse E, eds. Sinusoidal liver cells. Amsterdam: 1982: 69–76.

26.   Dory L, Boquet LM, Hamilton RL, Sloop CH, Roheim PS. Heterogeneity of dog interstitial fluid (peripheral lymph) high density lipoproteins: implications for a role in reverse cholesterol transport. J Lipid Res 1985; 26:519–527.

27.   Eisenberg S. High density lipoprotein metabolism. J Lipid Res 1987; 25:1017–1058.

28.   Evensen SA, Galdal KS, Nilsen E. LDL-induced cytotoxicity and its inhibition by antioxidant treatment in cultured human endothelial cells and fibroblasts. Atherosclerosis 1983; 49:23–30.

29.   Fielding PE, Vlodavsky I, Gospodarowicz D, Fielding CJ. Effect of contact inhibition on the regulation of cholesterol metabolism in cultured vascular endothelial cells. J Biol Chem 1979; 254:749–755.

30.   Fielding CJ, Vlodavsky I, Fielding PE, Gospodarowicz D. Characteristics of chylomicron binding and lipid uptake by endothelial cells in culture. J Biol Chem 1979; 254:8861–8868.

31.   Fielding CJ. The endothelium triglyceride-rich lipoproteins, and atherosclerosis: insights from cell biology and lipid metabolism. Diabetes 1981; 30 (suppl. 2):19–23.

32.   Fielding PE, Fielding CJ. Role of lipoproteins in the regulation of cultured endothelial cell cholesterol metabolism. In Jaffe EA, ed. Biology of endothelial cells. Boston:Martinus Nijhoff, 1984: 356–364.

33.   Filip DA, Nistor A, Bulla A, Radu A, Lupu F, Simionescu M. Cellular events in the development of valvular atherosclerotic lesions induced by experimental hypercholesterolemia. Atherosclerosis 1987; 67:199–214.

34.   Fleisher LN, Tall AR, Witte LD, Miller RW, Cannon PJ. Stimulation of arterial endothelial cell prostacyclin synthesis by high density lipoproteins. J Biol Chem 1982; 257:6653–6655.

35.   Fournier N. Mechanism of lipid transport in the heart. In: Horisberger M, Bracco V, eds. Lipids in modern nutrition. New York:Raven Press, 1987:29–42.

36.   Fox PL, DiCorleto PE. Lipid containing w-3-fatty acids specifically inhibit production of platelet-derived growth factor-like protein by endothelial cells. J Cell Biol 1987; 105:22a.

37.   Franke H, Durer V, Schlog B, Dargel R. In vivo binding and uptake of low density lipoprotein–gold and albumin–gold conjugates by parenchymal and sinusoidal cells of the fetal rat liver. Cell Tissue Res 1987; 249:221–226.

38.   Friend DR, Pangburn S. Site-specific drug delivery. Med Res. Rev 1987; 7:53–106.

39.   Gaffney J, West D, Arnold F, Sattar a, Kumar S. Differences in the uptake of modified low density lipoproteins by tissue cultured endothelial cells. J Cell Sci 1985; 79:317–325.

40.   Garfield AS, Schotz MC. Lipoprotein lipase. In: Gotto AM Jr, ed Plasma lipoproteins. Amsterdam:Elsevier, 335–358.

41.   Ghinea N, Leabu M, Hasu M, Muresan V, Colceag J, Simionescu N. Prelesional events in atherogenesis. Changes induced by hypercholesterolemia in the cell surface chemistry of arterial endothelium and blood monocytes, in rabbit. J Submicrosc Cytol 1987; 19:209–227.

42.   Ghitescu L, Fixman A, Simionescu M, Simionescu N. Specific binding sites for albumin restricted to plasmalemmal vesicles of continuous capillary endothelium: receptor-mediated transcytosis J Cell Biol 1986; 102:1304–1311.

43.   Gianturco SH, Eskin SG, Navarro LT, Labart CJ, Smith LC, Gotto AM Jr. Abnormal effects of hypertriacylglycerolemic very low density lipoproteins on 3-hydroxy-3-methylglutaryl-Co. A reductase activity and viability of cultured bovine aortic endothelial cells. Biochim Biophys Acta 1980; 681:143–152.

44.   Gorog P, Pearson JD. Surface determinants of low density lipoprotein uptake by endothelial cells. Atherosclerosis 1984; 53:21–29.

45.   Gotto AM Jr, Pawnall HJ, Havel RJ. Introduction to the plasma lipoproteins. In: Segrest JP, Albers JJ, eds. Methods in enzymology, vol. 128. Plasma lipoproteins. New York:Academic Press, 1986: 3–41.

46.   Gustafson S, Ostlund-Lindqvist AM, Vessby B. Uptake and degradation of human very-low-density lipoproteins by rat liver endothelial cells in culture. Biochim Biophys Acta 1985; 843:308–315.

47. Grunwald J, Hesz A, Robenek H, Brucker J, Buddecke A. Proliferation, morphology and low density lipoprotein metabolism of arterial endothelial cells cultured from normal and diabetic minipigs. Exp Mol Pathol 1985; 42:60–70.

48. Hashida R, Anamizu C, Kumura J, Okuma S, Yoshida Y, Takano T. Transcellular transport of lipoprotein through arterial endothelial cells in monolayer culture. Cell Struct Funct 1986; 11:31–42.

49. Havekes L, Schonten D, van Hinsbergh VW, De Wit E. Characterization of the biding of apoprotein E-free high density lipoprotein to cultured human endothelial cells. Biochem Biophys Res Commun 1984; 122:785–790.

50. Havekes L, Mommaas-Kienhuis Am, Schonten D, De Wit E, Scheffer M, van Hinsbergh VW. High affinity uptake and degradation of acetylated low density lipoprotein by confluent human vascular endothelial cells. Atherosclerosis 1985; 56:81–92.

51. Havel RJ. Lipid transport function of lipoproteins in blood plasma. Am J Physiol 1987; 253 (Endocrinol. Metab. 16):E1–E5.

52. Heideman CL, Hoff HF. Lipoproteins containing apolipoprotein A-I extracted from human aortas. Biochim Biophys Acta 1982; 711:431–444.

53. Hennig B, Shasby DM, Spector AA. Exposure to fatty acid increases human low density lipoprotein transfer across cultured endothelial monolayers. Circ Res 1985; 57:776–780.

54. Henriksen T, Evensen SA, Carlander B. Injury to human endothelial cells in culture induced by low density lipoproteins. Scand J Clin Lab Invest 1979; 39:361–368.

55. Henriksen T, Evensen SA, Carlander B. Injury to cultured endothelial cells induced by low density lipoproteins; protection by high density lipoproteins; protection by high density lipoproteins. Scand J Clin Lab Invest 1979; 39:369–375.

56. Hiramatsu K, Rosen H, Heinecke JW, Wolfbauer G, Chait A. Superoxide initiates oxidation of low density lipoprotein by human monocytes. Arteriosclerosis 1987; 7:55–60.

57. Hoff HF, Gaubatz JW. Isolation, purification and characterization of a lipoprotein containing apo B from the human aorta. Atherosclerosis 1982; 42:273–292.

58. Hoff HF, Morton RE. Lipoproteins containing apo B extracted from human aortas. Structure and function. Ann NY Acad Sci 1985; 454:183–194.

59. Hoff HF, Yamauchi Y, Bond MG. Reduction in tissue LDL accumulation during coronary artery regression in cynomolgus macaques. Atherosclerosis 1985; 56:51–60.

60. Hoff HF, Dusek DM, Lynn MP. Spatial distribution and accumulation of low density lipoproteins in the abdominal aorta of swine: determination by a novel electrotransfer procedure. Lab Invest 1986; 55337–386.

61. Hong JL, Pflug J, Reichl D. Comparison of apoprotein B of low density lipoproteins of human interstitial fluid and plasma. Biochem J 1984 222:49–55.

62. Innerarity TL, Pitas RE, Mahley RW. Modulating effects of canine high density lipoproteins on cholesteryl ester synthesis induced by ß–very low density lipoproteins in macrophages: possible in vitro correlates with atherosclerosis. Arteriosclerosis 1982; 2:114–124.

63. Jonasson L, Bondjers G, Hansson GK. Lipoprotein lipase in atherosclerosis: its presence in smooth muscle cells and absence from macrophages. J Lipid Res 1987; 28:437–445.

64. Julien P, Downar E, Angel A. Lipoprotein composition and transport in the pig and dog cardiac lymphatic system. Circ Res 1981; 49:248–254.

65. Julien P, Fong B, Angel A. Cardiac and peripheral lymph lipoproteins in dogs fed cholesterol and saturated fat. Arteriosclerosis 1984; 4:435–442.

66. Kenagy R, Bierman EL, Schwartz S, Albers JJ. Metabolism of low density lipoprotein by bovine endothelial cells as a function f cell density. Arteriosclerosis 1984; 4:365–367.

68. Klimov AN, Popov AV, Nagornev VA, Pleskov VM. Effect of high density lipoproteins on permeability of rabbit aorta to low density lipoproteins. Atherosclerosis 1985; 55:217–223.

69. Kurozumi T, Imamura T, Tanaka K, Yae Y, Koga S. Effects of hypertension and hypercholesterolemia on the permeability of fibrinogen and low density lipoprotein in the coronary artery of rabbits. Immunoelectron-microscopic study. Atherosclerosis 1983; 49:267–276.

70. Leabu M, Ghinea N, Muresan V, Colceag J, Hasu M, Simionescu N. Cell surface chemistry of arterial endothelium and blood monocytes in the normolipidemic rabbit. J Submicrosc Cytol 1987; 19:193–208.

71. Lindsey S, Hayes KC. Modified β–VLDL alters monocyte adherence to endothelial cells. Arteriosclerosis 1986; 6:525. (abstr)

72. Lupu F, Danaricu I, Simionescu N. Development of intracellular lipid deposits in the lipid-laden cells of atherosclerotic lesions. A cytochemical and ultrastructural study. Atherosclerosis 1987; 67:127–142.

73. Mahley RW, Weisgraber KA, Innerarity TL, Windmueller HG. Accelerated clearance of low-density and high-density lipoproteins and retarded clearance of E apoprotein-containing lipoproteins from the plasma of rats after modification of lysine residues. Proc Natl Acad Sci USA 1979; 76:1746–1750.

74. Mahley RW, Innerarity TL. Lipoprotein receptors and cholesterol homeostasis. Biochim Biophys Acta 1983; 737:197–222.

75. Majno G, Joris I, Zand T. Atherosclerosis: new horizons. Hum Pathol 1985; 16:3–5.

76. Manttari M, Malkonen M, Manninen V. Effect of diazepam on endothelial permeability, plasma lipids and lipoproteins in cholesterol -fed rabbits. Acta Med Scand (suppl) 1982; 660:109–113.

77. Mazzone T, Jensen M, Chait A. Human arterial wall cells secrete factors that are chemotactic for monocytes. Proc Natl Acad Sci USA 1983; 80:5094–5097.

78. Minick CR, Stemerman MB, Insull W Jr. Effect of regenerated endothelium on lipid accumulation in the artery wall. Proc Natl Acad Sci USA 1977; 74:1724–1728.

79. Mommaas-Kienhuis AM, Krijbolder LH. Visualization of binding and receptor-mediated uptake of low density lipoproteins by human endothelial cells. Eur J Cell Biol 1985; 36:201–208.

80. Mora R, Lupu F, Simionescu N. Prelesional events in atherogenesis. Colocalization of apolipoprotein B, unesterified cholesterol and extracellular phospholipid liposomes in the aorta of hyperlipidemic rabbit. Atherosclerosis 1987; 67:143–154.

81.   Morel DW, DiCorleto PE, Chisolm GM. Endothelial and smooth muscle cells alter low density lipoprotein in vitro by free radical oxidation. Arteriosclerosis 1984; 4:3573–3364.

82.   Morel DW, DiCorleto PE, Chisolm GM. Modulation of endotoxin-induced endothelial cell toxicity by low density lipoprotein. Lab Invest 1986; 55:419–426.

83.   Morel EM, Colton CK, Stemerman MB, Smith KA. Transport of ^{125}I-low density lipoprotein (LDL) through punctate regions of enhanced endothelial permeability in rabbit aorta, in vivo. In: Third Intl Symp Biol Vasc Endoth Cell. Cambridge, 1984:35. (abstr vol)

84.   Nagelkerke JF, Barto KP, van Berkel TJC. In vivo and in vitro uptake and degradation of acetylated low density lipoprotein by rat liver endothelial, Kupffer and parenchymal cells. J Biol Chem 1983; 258:12221–12227.

85.   Nagelkerke JF, van Berkel TJC. Rapid transport of fatty acids from rat liver endothelial to parenchymal cells after uptake of cholesteryl ester-labeled acetylated LDL. Biochim Biophys Acta 1986; 875:593–598.

86.   Naito M, Wisse E. Filtration effect on endothelial fenestrations on chylomicron transport in neonatal rat liver sinusoids. Cell Tissue Res 1978; 190:371–382.

87.   Navab M, Hough GP, Berliner JA, Frank JA, Fogelman AM, Haberland MA, Edwards PA. Rabbit beta-migrating very low density lipoprotein increases endothelial macromolecular transport without altering electrical resistance. J Clin Invest 1986; 78:389–397.

88.   Nestel PJ. The regulation of lipoprotein metabolism. In: Gotto AM Jr, ed. Plasma lipoproteins. Amsterdam:Elsevier, 1987: 153–182.

89.   Netland PA, Zetter BR, Via DP, Voyta JC. In situ labeling of vascular endothelium with fluorescent acetylated low density lipoprotein. Histochem J 1985; 17:1309–1320.

90.   Nicoll A, Duffield R, Lewis B. Flux of plasma lipoproteins into human arterial intima. Comparison between grossly normal and atheromatous intima. Atherosclerosis 1981; 39:229–242.

91.   Nilson-Ehle P. Lipolytic enzymes and plasma lipoprotein metabolism. Ann Rev Biochem 1980; 49:667–693.

92.   Nistor A, Simionescu M. Uptake of low density lipoproteins by the hamster lung. Interactions with capillary endothelium. Am Rev Respir Dis 1986; 134:1266–1272.

93.   Parthasarathy S, Steinbrecker VP, Barnett J, Witztum JL, Steinberg D. Essential role of phospholipase A_2 activity in endothelial cell-induced modification of low density lipoprotein. Proc Natl Acad Sci USA 1985; 82:3000–3004.

94.   Parthasarathy S, Printz, Boyd DJ, Joy L, Steinberg D. Macrophage oxidation of low density lipoprotein generates modified form recognized by the scavenger receptor. Arteriosclerosis 1986; 6:505–511.

95.   Parthasarathy S, Fong LG, Otero D, Steinberg D. Recognition of solubilized apoproteins from delipidated, oxidized low density lipoprotein (LDL) by the acetyl-LDL receptor. Proc Natl Acad Sci USA 1987; 84:537–540.

96.   Pedersen ME, Cohen M, Schotz MC. Immunocytochemical localization of the functional fraction of lipoprotein lipase in the perfused heart. J Lipid Res 1983; 24:512–521.

97.   Phillips MC, Rothblat GH. Cholesterol flux between high density lipoproteins

and cells. In: Catapano AL et al, eds. Atheroscl review , vol. 16. New York:Raven Press, 1987:57–86.

98. Pietra GG, Spagnoli LG, Capuzzi DM, Sparks CE, Fishman AP, Marsh JB. Metabolism of ^{125}I-labeled lipoproteins by the isolated rat lung. J Cell Biol 1976; 70:30–46.

99. Pitas RE, Boyles J, Mahley RW, Bissell DM. Uptake of chemically-modified low density lipoproteins in vivo is mediated by specific endothelial cells J Cell Biol 1985; 100:103–117.

100. Pomerantz KB, Fleisher LN, Tall AR, Cannon PJ. Enrichment of endothelial cell arachidonate by lipid transfer from high density lipoproteins: relationship to prostaglandin I synthesis. J Lipid Res 1985; 26:1269–1276.

101. Poznansky MJ, Juliano RL. Biological approaches to the controlled delivery of drugs: a critical review. Pharmacol Rev 1984; 36:227–336.

102. Quin MT, Parthasarathy S, Steinberg D. Oxidized low density lipoprotein is chemotactic for human monocytes. Arteriosclerosis 1986; 6:563a.

103. Quin MT, Parthasarathy S, Fong LG, Steinberg D. Oxidatively modified low density lipoproteins: a potential role in recruitment and retention of monocyte/macrophages during atherogenesis. Proc Natl Acad Sci USA 1987; 84:2995–2998.

104. Reckless JPD, Weinstein DB, Steinberg D. Lipoprotein and cholesterol metabolism in rabbit arterial endothelial cells in culture. Biochim Biophys Acta 1978; 529:475–487.

105. Riechl D, Postiglione A, Myant NB, Pflug JJ, Press M. Observations on the passage of apoproteins from plasma lipoproteins into peripheral lymph in two men. Clin Sci Mol Med 1975; 49:419–426.

106. Reichl D, Myant NB, Rudra DN, Pflug JJ. Evidence for the presence of tissue-free cholesterol in low density and high density lipoproteins of human peripheral lymph. Atherosclerosis 1980; 37:489–495.

107. Roberts AB, Lees AM, Lees RS, Strauss HW, Fallon JT, Taveras J, Kopiwoda S. Selective accumulation of low density lipoproteins in damaged arterial wall. J Lipid Res 1983; 24:1160–1167.

108. Robinson CS, Wagner RC. Differential endocytosis of lipoproteins by capillary endothelial vesicles. Microcirc Endoth Lymph 1985; 2:313–329.

109. Rose DC, Williams SK. Differential transport of LDL and modified LDL across endothelial cell monolayers. J Cell Biol 1987; 105:328a.

110. Sanan DA, Strumpfer AE, Van der Westhuyzen DR, Coetzee GA. Native and acetylated low density lipoprotein metabolism in proliferating and quiescent bovine endothelial cells in culture. Eur J Cell Biol 1985; 36:81–90.

111. Sanan DA, Van der Westhuyzen DR, Gevers W, Coetzee GA. The surface distribution of low density lipoprotein receptors on cultured fibroblasts and endothelial cells. Ultrastructural evidence for dispersed receptors. Histochemistry 1987; 86:517–523.

112. Santillan GG, Schuh J, Chan SI, Bing RJ. Binding and internalization of low density lipoproteins by perfused arteries. Biochem Biophys Res Commun 1980; 95:1410–1415.

113. Schnitzer JJ, Tompkins RG, Colton CK, Smith KA, Stemerman MB. Distribution of ^{125}I-low density lipoprotein in selected tissues of the squirrel monkey. In:Third Intl Symp Biol Vasc Endoth Cell. Cambridge, 1984:127. (abstr. vol.)

114.   Scow RO, Blanchette-Mackie EJ, SMith LC. Role of capillary endothelium in
       the clearance of chylomicrons. A model for lipid transport from blood by
       lateral diffusion in cell membranes. Circ Res 1976; 39:149–162.
115.   Scow RO, Blanchette-Mackie EJ, Smith LC. Transport of lipid across capillary
       endothelium. Fed Proc 1980; 39:2610–2617.
116.   Shaw JM, Shaw KV, Yanovich S, Iwanik M, Futch WS, Rosowsky A, Schook
       LB. Delivery of lipophilic drugs using lipoproteins. (in press)
117.   Shepherd J, Packard CJ. Receptor-independent low-density lipoprotein
       catabolism. In:Methods in enzymology, vol. 129. New York:Academic Press,
       1986:566–590.
118.   Simionescu M, Ghitescu L, Fixman A, Simionescu N. How plasma macro-
       molecules cross the endothelium. News Physiol Sci 1987; 2:97–100.
119.   Simionescu M. Receptor-mediated transcytosis of plasma molecules by vascular
       endothelium. In:Simionescu N, Simionescu M, eds Endothelial cell biology in
       health and disease. New York:Plenum, 1988:69–104.
120.   Simionescu N. Transcytosis and traffic of membranes in the endothelial cell.
       In:Schweiger HG, ed. International cell biology 1980–1981. Berlin:Springer-
       Verlag, 1981:657–672.
121.   Simionescu N. Cellular aspects of transcapillary exchange. Physiol Rev 1983;
       63:1536–1579.
122.   Simionescu N, Simionescu M. Interactions of endogenous lipoproteins with
       capillary endothelium in spontaneously hyperlipoproteinemic rats. Microvasc
       Res 1985; 30:314–332.
123.   Simionescu N, Vasile E, Lupu F, Popescu G, Simionescu M. Prelesional events
       in atherogenesis. Accumulation of extracellular cholesterol-rich liposomes in the
       arterial intima and cardiac valves of the hyperlipidemic rabbit. Am J Pathol
       1986; 123:109–125.
124.   Simionescu N. Prelesional changes of arterial endothelium in the hyperlipopro-
       teinemic atherogenesis. In:Simionescu N, Simionescu M, eds. Endothelial cell
       biology in health and disease. New York:Plenum, 1988:385–429.
125.   Slater HR, Shephered J, Packard CJ. Receptor-mediated catabolism and tissue
       uptake of human low density lipoprotein in the cholesterol-fed atherosclerotic
       rabbit. Biochim Biophys Acta 1982; 713:435–445.
126.   Sloop CH, Dory L, Hamilton RL, Krause BR, Roheim PS. Characterization
       of dog peripheral lymph lipoproteins: the presence of a disc-shaped "nascent"
       high density lipoprotein. J Lipid Res 1983; 24:1429–1440.
127.   Sloop CH, Dory L, Roheim PS. Interstitial fluid lipoproteins. J Lipid Res
       1987; 28:225–237.
128.   Smith EB, Staples EM. Intimal and medial plasma protein concentrations and
       endothelial function. Atherosclerosis 1982; 41:295–308.
129.   Smith EB. Endothelium and lipoprotein permeability. In:Woolf N, ed. Biology
       and pathology of the vessel wall. Bern:Karger, 1983:279–293.
130.   Smith EB, Ashall C. Low density lipoprotein concentration in interstitial fluid
       from human atherosclerotic lesions. Relation to theories of endothelial damage
       and lipoprotein binding. Biochim Biophys Acta 1983; 754:249–257.
131.   Smith EB, Ashall C. Variability of the electrophoretic mobility of low density
       lipoprotein. Comparison of interstitial fluid from human aortic intima and
       serum. Atherosclerosis 1983; 49:89–98.

132. Smith EB, Ashall C. Compartmentalization of water in human atherosclerotic lesions. Changes in distribution and exclusion volumes for plasma macromolecules. Arteriosclerosis 1984; 4:21–27.

133. Smith EB. Plasma macromolecules in interstitial fluid from normal and atherosclerotic human aorta. Monogr Atheroscler 1986; 14:179–183.

134. Spady DK, Bilheimer DW, Dietschy JM. Rates of receptor-dependent and independent low density lipoprotein uptake in the hamster. Proc Natl Acad Sci USA 1983; 80:3499–3503.

135. Spady DK, Turley SD, Dietschy JM. Receptor-independent low-density lipoprotein transport in the rat in vivo. J Clin Invest 1985; 76:1113–1122.

136. Spady DK, Huettinger M, Bilheimer DW, Dietschy JM. Role of receptor-independent low density lipoprotein transport in the maintenance of tissue cholesterol balance in the normal and WHHL rabbit. J Lipid Res 1987; 28:32–41.

137. Sparks CE, Deboff JL, Capuzzi DM, Pietra GG, Marsh JB. Proteolysis of very low density lipoprotein in perfused lung. Biochim Biophys Acta 1978; 529:123–130.

138. Spector AA. Plasma albumin as a lipoprotein. In:Scann AM, Spector AA, eds. Biochemistry and biology of plasma lipoproteins. New York:Marcel Dekker, 1986:247–279.

139. Spraque EA, Steinbach BL, Nerem RM, Schwartz CJ. Influence of a laminar steady-state fluid-imposed wall shear stress on the binding, internalization and degradation of low density lipoproteins by cultured arterial endothelium. Circulation 1987; 76:648–656.

140. Stein O, Stein Y, Eisenberg S. A radioautographic study of the transport of ^{125}I-labeled serum lipoproteins in the rat aorta. Z Zellforsch 1973; 138:223–232.

141. Stein O, Stein Y. Bovine aortic endothelial cells display macrophage-like properties towards acetylated ^{125}I-labeled low density lipoprotein. Biochim Biophys Acta 1980; 620:631–635.

142. Stein O, Leitersdorf E, Stein Y. Verapamil enhances receptor-mediated endocytosis of low density lipoproteins by aortic cells in culture. Arteriosclerosis 1985; 5:35–44.

143. Stein O, Halperin G, Stein Y. Cholesteryl ester efflux from extracellular and cellular elements of the arterial wall. Model systems in culture with cholesteryl linoleyl ether. Arteriosclerosis 1986; 6:70–78.

144. Stein Y, Stein O. Lipid synthesis and degradation and lipoprotein transport in mammalian aorta. In: Atherogenesis: initiating factors, Ciba Found. Symp. 12. Amsterdam:Elsevier, 1973:165–183.

145. Steinberg D, Pittman RC, Carew TE. Mechanisms involved in the uptake and degradation of low density lipoprotein by the artery wall in vivo. Ann NY Acad Sci 1985; 454:195–206.

146. Steinbrecher UP, Parthasarathy S, Leake DS, Witztum JL, Steinberg D. Modification of low density lipoprotein by endothelial cells involves lipid peroxidation and degradation of low density lipoprotein phospholipids. Proc Natl Acad Sci USA 1984; 81:3883–3887.

147. Steinbrecher UP. Oxidation of human low density lipoprotein results in derivatization of lysine residues of apolipoprotein B by lipid peroxide decomposition products. J Biol Chem 1987; 262:3603–3608.

148. Stender S, Zilversmith DB. Comparison of cholesteryl ester transfer from chylomicrons and other plasma lipoproteins to aortic intima-media of cholesterol-fed rabbits. Arteriosclerosis 1982; 2:493–499.

149. Stemerman MB. Effects of moderate hypercholesterolemia on rabbit endothelium. Arteriosclerosis 1981; 1:25–32.

150. Stemerman MB, Morel EM, Burke KR, Colton CK, Smith KA, Lees RS. Local variation in arterial wall permeability to low density lipoprotein in normal rabbit aorta. Arteriosclerosis 1986; 6:64–69.

151. Tabas I, Tall AR. Mechanism of the association of HDL with endothelial cells, smooth muscle cells and fibroblasts. J Biol Chem 1984; 259:13897–13905.

152. Tauber J-P, Cheng J, Gospodarowicz D. Effect of high and low density lipoproteins on proliferation of cultured bovine vascular endothelial cells. J Clin Invest 1980; 66:696–708.

153. Tauber J-P, Cheng J, Massoglia S, Gospodarowicz D. High density lipoproteins and the growth of vascular endothelial cells in serum-free medium. In Vitro 1981; 17:519–530.

154. Tauber J-P, Goldminz D, Gospodarowicz D. Up-regulation in vascular endothelial cells of binding sites of high density lipoprotein induced by 25-hydroxycholesterol. Eur J Biochem 1981; 119:327–339.

155. Tauber J-P, Goldminz D, Vlodavsky I, Gospodarowicz D. The interaction of the high-density lipoprotein with cultured cells of bovine vascular endothelium. Eur J Biochem 1981; 119:317–325.

156. Territo M, Berliner JA, Fogelman AM. Effect of monocyte migration on low density lipoprotein transport across aortic endothelial cell monolayers. J Clin Invest 1984; 74:2279–2284.

157. Tomlinson E. (Patho)physiology and the temporal and spatial aspects of drug delivery. In Tomlinson E, Davis SS, eds. Site-specific drug delivery. Chichester:Wiley, 1986:1–26.

158. Van Berkel TJC, Kruijt JK, Kempel HJ. Specific targeting of high density lipoproteins to liver hepatocytes by incorporation of a tris-galactoside-terminated cholesterol derivative. J Biol Chem 1985; 260:12203–12207.

159. Van Berkel TJC, Kruijt JK, Harkes L, Nagelkerke JF, Spanjer H, Kempen H-JM. Receptor-dependent targeting of native and modified lipoproteins to liver cells. In:Tomlinson E, Davis SS, eds. Site-specific drug delivery. Chichester: Wiley 1986:49–68.

160. van Hinsbergh VW, Havekes L, Emeis JJ, van Corven E, Scheffer M. Low density lipoprotein metabolism by endothelial cells from human umbilical cord arteries and veins. Arteriosclerosis 1983; 3:547–559.

161. van Hinsberg VW, Scheffer M, Havekes L, Kempen HIM. Role of endothelial cells and their products in the modification of low density lipoprotein. Biochim Biophys Acta 1986; 878:49–64.

162. Vasile E, Simionescu M, Simionescu N. Visualization of the binding, endocytosis and transcytosis of low density lipoprotein in the arterial endothelium in situ. J Cell Biol 1983; 96:1677–1689.

163. Vasile E, Simionescu N. Transcytosis of low density lipoprotein through vascular endothelium. In:Seno S, Copley AL, Venkatachalam MA, Hamashima Y, Tsujii T, eds. Glomerular dysfunction and biopathology of vascular wall. Tokyo:Academic Press, 1985:87–101.

164. Vasile E, Popescu G, Simionescu N. Enhanced transcytosis and accumulation of ß–very low density lipoproteins in the aorta of rabbits with experimental hyperlipidemia. In:XVIth International Congress of the International Academy of Pathology. Vienna, 1986:68. (abstr vol)

165. Vasile E, Popescu G, Simionescu M, Simionescu N. Interaction of low density lipoprotein and ß–very low density lipoprotein with the arterial endothelium in normal and hypercholesterolemic animals. In 4th Intl. Symp. Biol. Vasc. Endoth. Cell. Noordwijkerhout, 1986:123. (abstr)

166. Vasile E, Simionescu M, Simionescu N. Transport pathways of ß-VLDL in aortic endothelium of normal and hypercholesterolemic rabbits. Atherosclerosis 1988; 75:195–210.

167. Verlangieri AJ, Cardin BA, Bush M. The interaction of aortic glycosaminoglycans and ^{3}H-insulin endothelial permeability in cholesterol-induced rabbit atherogenesis. Res Commun Chem Pathol Pharmacol 1985; 47:85–96.

168. Vlodavsky I, Fielding PE, Fielding CJ, Gospodarowicz D. Role of contact inhibition in the regulation of receptor-mediated uptake of low density lipoprotein in cultured vascular endothelial cells. Proc Natl Acad Sci USA 1978; 75:356–360.

169. Voyta JC, Via DP, Butterfield CE, Zetter Br. Identification and isolation of endothelial cells based on their increased uptake of acetylated-low density lipoprotein. J Cell Biol 1984; 99:2034–204.

170. Warren JS, Ward PA. Review: oxidative injury to the vascular endothelium. Am J Med Sci 1986; 29:97–103.

171. Wiklund O, Carew TE, Steinberg D. Role of the low density lipoprotein receptor in penetration of low density lipoprotein into rabbit aortic wall. Arteriosclerosis 1985; 5:135–141.

172. Yamauchi Y, Hoff HF. Apolipoprotein B accumulation and development of foam cell lesions in coronary arteries of hypercholesterolemic swine. Lab Invest 1984; 51:325–332.

173. Yomantas S, Elner VM, Schaffner T, Wissler RW. Immunohistochemical localization of apolipoprotein B in human atherosclerotic lesions. Arch Pathol Lab Med 1984; 108:374–378.

174. Yutani C, Go S, Imakita M, Ishibashi-Neda H, Hatanaka K, Yamamoto A. Autopsy findings in two patients with homozygous familial hypercholesterolemia. Special references to apolipoprotein B localization and internalization defect of low density lipoprotein. Acta Pathol Jpn 1987; 37:1489–1504.

3

Techniques for Complexing Pharmacological Agents to Lipoproteins and Lipid Microemulsions

Bo Lundberg
Åbo Akademi University, Åbo, Finland

I. INTRODUCTION

A. The Drug Targeting Concept

The idea of targeting macromolecular drug conjugates to a specific cell population and thereby improving the therapeutic index is a fascinating concept in the field of drug administration. The need for selective drug delivery systems is obvious. The therapeutic efficacy of drugs is often diminished by their inability to reach the site of action at an accurate dosage. The need for drug targeting is most acute in the chemotherapy of cancer. The delivery of cytotoxic agents specifically to neoplastic target cells would improve the therapeutic index by increasing tumor cell death and decreasing toxic effects on normal cells. Recent advances in cell biology and pharmacology have formed novel rationales for targeted carrier systems dependent on, e.g., cellular receptors or immunological recognition (for review see Ref. 1).

A basic prerequisite for a functioning targeting system is an adequate carrier and methods for complexing drugs to it. Systems for targeted drug delivery have been suggested based on the use of antibodies [2]., hormones [3], DNA [4], proteins [5], and liposomes [6] as carriers. The most exploited system has been that of the liposomes and several reviews have been written on their feasibility for drug targeting (see, e.g. Ref. 7). Despite the ambitious proposals, the liposomal system still has only marginal applications in chemotherapy. However, the know-how gathered from the work done with liposomes has much to teach us about the general prerequisites of targeted drug delivery systems.

From general considerations it can be deduced that a good drug carrier for in vivo use should meet special requirements:

1. The drug–carrier complex must be stable during storage and against biological components in the circulation or in another possible application site.
2. The carrier should be biocompatible and biodegradable and the drug–carrier conjugate must not produce unacceptable levels of toxicity or immunological reactions.
3. The carrier should be suitable for targeting.
4. The drug–carrier complex must allow release of the drug at the target site.
5. The carrier must not cause unspecific uptake by nontarget cells.
6. For large-scale clinical use the carrier system must be pharmaceutically acceptable in regard to formulation homogeneity, cost of manufacture, and ease of handling and administration.

B. Lipoproteins as Potential Drug Carriers

Plasma lipoproteins and largely also lipid microemulsions meet most of the requirements for a satisfactory drug carrier just summarized. From a general point of view it is mainly the nature of endogenous particles that makes the lipoproteins such appealing drug carriers. In that respect they should be well tolerated by the body, not cause immunological reactions, be completely biodegradable, and not be rapidly cleared in the circulation by the reticuloendothelial system (for exceptions refer to Chapter 1, 2, or 5). Furthermore, the lipoproteins and especially low density lipoproteins (LDLs) have an inherent target system in their receptor-mediated uptake. By the LDL receptor pathway an LDL–drug complex can reach the lysosomes, where the carrier LDL is then enzymatically degraded. De Duve et al. [8] defined a "lysosomotropic" carrier as a drug–carrier complex that can be endocytozed by the target cells and processed in their lysosomes with release of the drug in free and active form.

Special attention has been paid to the possible use of LDL as a carrier for targeting antitumoral drugs to neoplastic cells [9]. The rationale for this is that certain types of cancer cells have higher LDL receptor activity than the corresponding normal cells [10]. Furthermore, the LDL particle can be modified in such ways that it is recognized by receptors other than the classical LDL receptor; e.g., acetylated LDL is rapidly taken up by macrophages [11] and lactosylated LDL is effectively catabolized by hepatocytes [12]. In fact, the targeting of LDL-drug complexes to macrophages might be the issue that comes closest to the range of practical applications. Besides LDL, high density lipoproteins (HDLs) have also been shown to bind to cells at specific sites and to be taken up by receptor-mediated endocytosis [13]. High density lipoprotein has a complex metabolism and the longest half-life in the circulation (4–5 days). Intracellularly, HDL is transported within endosomes, which do not fuse with lysosomes [14]. A drug administered as a complex with HDL will thus remain in the circulation system for a longer time and have intracellular route different from that of a drug bound to LDL.

C. Lipid Microemulsions as Potential Drug Carriers

The second topic of this review, lipid microemulsions, is in spite of the structural similarity more problematic than lipoproteins as a drug–carrier system. There is an obvious risk that, like liposomes, they will be recognized by the immunological system as foreign particles and rapidly removed from the circulation by reticuloendothelial cells in the liver, spleen, and elsewhere. Another uncertain factor is the possibility of microemulsion–drug complexes escaping the bloodstream. The continuous endothelium acts as a barrier to the passage of liposomes [15] and will probably do so regarding lipid microemulsion particles. On the other hand, the advantages of lipid microemulsions as drug carriers are clear; e.g., the use of appropriate lipid constituents guarantees a nontoxic and biodegradable drug vehicle, and the prerequisites for drug incorporation are excellent. A fascinating dimension for the pharmacological use of lipid microemulsions is the possibility of binding ligand for various cell surface receptors to the particle surface. The targeting of lipid emulsion particles can in principle be performed with all the methods that have been applied on liposomes, notably the use of monoclonal antibodies as a homing device [7,16]. This targeting approach has already been fulfilled by the reassembly of biologically active model LDL, through the binding of apolipoprotein B (apo B) to lipid microemulsions [17,18]. When a lipid microemulsion particle without any surface ligand is under consideration, the term "passive targeting" can be used to refer to the natural localization patterns of such a particle when introduced into the body. An apparent advantage of lipid microemulsion particles over vesicles as drug carriers is their high physical stability due to the compact neutral lipid core. A major drawback of vesicles with the entrapped water core is their tendency to be disrupted by serum components [19] (Fig. 1 and Table 1).

Besides the cellular uptake of whole carrier particles, drug delivery can also take place throug surface transfer. This phenomenon is a passive transport mechanism, which is thought to involve diffusion of molecules through the aqueous phase [20]. The transfer process has been shown to take place with similar rates from both lipoproteins [21] and lipid microemulsions [22]. If drug delivery solely by cellular uptake of whole drug carrier particles

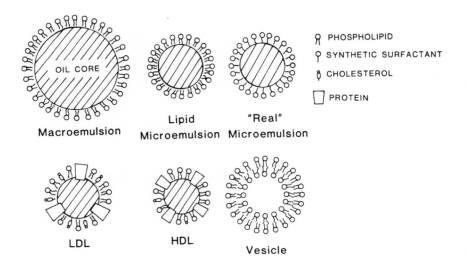

Figure 1 Schematic description of the structure of emulsions, lipoproteins, and vesicles.   The figures describe the similarity among lipid microemulsion, "real" microemulsion, and lipoproteins and the difference between lipid microemulsions and vesicles regarding the particle core, consisting of oil or an entrapped water pool, respectively.

is wanted, the transfer process can be a disturbing factor, which causes unspecific drug uptake.

Figure 2 depicts different routes for cellular uptake of lipoproteins and lipid microemulsions with incorporated drugs. Native LDL and HDL are bound to specific receptors and taken up by endocytosis, while modified LDL is internalized by different routes, e.g., via scavenger- receptors for acetyl-LDL on macrophages. Lipid microemulsions are taken up by "passive" targeting

Table 1 Comparison of Lipid Microemulsion, "Real" Microemulsion, and Vesicles

Properties	Lipid Microemulsion	"Real" Microemulsion	Vesicles
Diameter, nm	20–100	10–250	30–500
Thermodynamically stable	No	Yes	No
Stability in circulation	Good	–	Poor
Dilution by H_2O	Stable	Altered	Stable

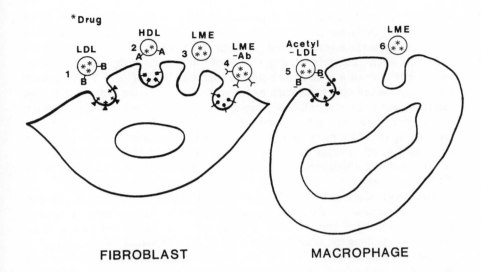

FIBROBLAST   MACROPHAGE

Figure 2 Schematic representation of possible mechanisms for cellular uptake of drug complexed to lipoproteins and lipid microemulsions. (1, and 2) Internalization of LDL–drug and HDL–drug complexes via receptor–mediated endocytosis. (3) Unspecific uptake of lipid microemulsion (LME)–drug complex by pinocytosis. (4) Internalization of an LME–drug complex with covalently attached antibody (Ab), which binds to a target antigen followed by endocytosis. (5) Macrophages can take up modified LDL, acetyl-LDL–drug complexes, via the scavenger receptor mechanism. (6) Phagocytosis of LME–drug complexes by macrophages.

through pinocytosis and such particles with attached ligands (e.g., antibodies) are bound to target antigens and internalized by endocytosis. Lipid microemulsions can also be internalized by macrophages through phagocytosis.

II. COMPLEXING OF DRUGS TO LIPID MICROEMULSIONS

A. Basic Definitions

Emulsions are heterogeneous systems of one liquid dispersed in another in the the form of droplets. Unstabilized emulsions of two immiscible liquids such as triacylglycerol (oil) and water separate rapidly when shanken together. However, if a surface-active emulsifier (surfactant) is present, the high interfacial tension (γ) between the liquids is reduced and a stable suspension can be obtained. The surfactants are amphiphilic molecules with both hydrophilic and lipophilic parts and tend to accumulate at the interface with the polar part in water and the apolar part extended into the oil phase.

Emulsifiers may be lipids such as phospholipids, monoacylglycerols, and soaps or synthetic surfactants. When the surface-active agent absorbs onto an interface the surface tension changes from γ_0 for the pure liquid to γ for the absorbed layer. The relative magnitudes of γ and γ_0 have significant bearings on the behavior of a lipid in an aqueous solution. Thus if $\gamma_0 > \gamma$, the system forms aggregated structures such as bilayers. If $\gamma < 0$, the net effect on the interface would be a spontaneous emulsification of the two phases. In order to be effective an emulsifier should thus generate a low interfacial γ value. Depending on which of the immiscible phases is continuous, the emulsions can be termed as oil-in-water (O/W) or water in oil (W/O). Modifications of these are the multiple emulsions, which are either water-in-oil-in-water (W/O/W) or (O/W/O) systems.

Emulsions stabilized by surfactants traditionally have wide use in pharmaceutics (for review see Ref. 23). However, the traditional surfactant chemistry has little bearing if the emulsion particle is to be introduced into a living system as a vehicle for drug delivery. The special demands on such a system involve the choice of surfactants, oil phase, and emulsification technology. When it comes to the incorporation of drugs into microemulsion particles, this is generally made as an integral part of the emulsification process by dissolving the drug in surfactant or oil phase. In this chapter, great importance is attached to the problem of the preparation of lipid microemulsions, which is not altogether easy. To avoid confusion I want to point out that the term microemulsion in this context is used in a restricted manner. In conventional surface chemistry, the term microemulsion (or swollen micellar system) refers to single thermodynamically stable solution phases, which form spontaneously by mixing oil and water in the presence of surfactant and cosurfactant [24]. Lipid microemulsions have droplet diameters within the limits of "real" microemulsions (10–250 nm), but they are not thermodynamically stable (Table 1).

B. Emulsion Stability

The physical stability of emulsions is one of the foremost problems affecting their pharmaceutical use. Instability of emulsions is manifested in changes in physical properties such as droplet size distribution, which is a consequence of the flocculation or coalescense of globules. The stability of emulsions depends on low interfacial free energy; the electrostatic charge of dispersed particles; the droplet size; ordered hydration layers; and surface barriers, involving steric, viscous, and elastic properties.

Although a low interfacial tension greatly enhances the emulsification process, its importance for emulsion stability should not be overstressed. Emulsions having the same interfacial free energy may have widely different stabilities.

The theory of stability of lyophobic colloids (the DLVO theory) takes into account two kinds of long-range forces which determine the closeness of contact of two particles and hence their tendency to aggregate: the London–van der Waals forces of attraction, and the electrostatic repulsion between electric double layers. Although the attractive forces cannot be attenuated, the net charge of emulsion particles is an important factor for emulsion stability, which can be quite easily manipulated. The decrease in stability against coalescence when salt is added–e.g., when NaCl is added to an oil–phospholipid emulsion [25]–indicates that the electrostatic repulsion forces originating from charged lipids in the emulsifier play an important role for the emulsion's stability. To avoid the adverse effect of NaCl the solution can be made isotonic with blood by the addition of 2.5% glycerol. It is also possible to use glucose.

It is generally accepted that optimal emulsions are formed when the surfactant/dispersed phase ratio is such that it generates globule size at a minimum [26]. The excellent stability of many lipid microemulsions is obviously to a large extent dependent on their small particle size.

A factor that can add considerably to the stability of emulsions is the solvation of adsorbed sufactant layers. The existence of ordered water layers around O/W emulsion particles is especially important if they do not have any net charge. The hydration property of phospholipids is favorable for emulsion stabilization [27]. It can also be conceivable that the protein part of lipoproteins effectively contributes to the water solvation of these particles.

The surface barrier factors are short-range forces and can be directly related to coalescence behavior. Steric or entropic repulsion arises because long chains of a surfactant suffer loss of entropy when restricted on contact. The formation of aggregates or coalescense at collisions between emulsion particles is further prevented if the dispersed droplets exhibit a degree of elasticity on contact.

Emulsion stability can be judged experimentally by means of visual observation combined with droplet size determination. An extensive flocculation or coalescence, possibly followed by creaming or settling, is easily recognized by the naked eye. To detect slowly proceeding aggregation or coalescence of globules, which is common for lipid microemulsions, methods for size and size distribution determinations are needed. Such methods are described in Section III. F. However, a considerable problem is to assess, within a reasonable time, the stability of an emulsion that ought to be stable for months or even years. No ideal method exists to solve this problem, but centrifugation or addition of salt can give a fairly good approximation. To test the stability of a drug-bearing emulsion it is essential to include the drug, since this may alter the stability properties considerably.

C. Choice of Emulsifier

The choice of a proper emulsifier for drug-bearing emulsion systems is a delicate problem, especially if the drug is intended for parenteral use. In addition to the demand for good emulsifying properties, the emulsifier should be nontoxic and biodegradable. Unfortunately, surfactants with excellent emulsifying properties often have pronounced toxic effects. The need for biological acceptability restricts the selection to certain natural surface-active lipids and some artificial surfactants with low toxicity.

For the stabilization of oil globules in water it is essential that the surfactnat possesses a degree of hydrophilicity to confer an enthalpic stabilizing force and a degree of hydrophobicity to secure adsorption at the interface. In the practical approach to the emulsion formation the hydrophile–lipophile balance (HLB) system can help to select the most effective emulsifier for a given oil [23]. At the higher end of the scale the surfactants are hydrophilic O/W emulsifiers. However, the usefulness of HLB values for the selection of surface-active agents for lipid microemulsions is limited and they cannot explain why mixed surfactants have greater stabilizing power than single ones.

Another method for selection of emulsifiers uses the phase inversion temperature (PIT) value [28]. This experimental value takes into account the nature of the oil, the interaction between the surfactants and oil, and the nature of the water phase. The use of PIT values has not yet been employed for the preparation of lipid microemulsions.

Lipid emulsions have an established use for intravenous (I.V.) nutrition [29]. Such emulsions are stabilized by phospholipids isolated from natural sources. The safe use of parenteral emulsions stabilized by phospholipids proves that these can be chosen as the basic surface-active components for lipid microemulsion preparation. However, all phospholipids are not alike from the toxicological point of view, since soya phosphatides [30], were found to induce a fall in blood pressure, whereas egg yolk phosphatides were not [31].

The composition of commercial phospholipid preparations (lecithins) varies according to the supplier (Table 2). The formulas of the main phospholipids (Fig. 3) indicate a balance between hydrophilic and lipophilic parts in the molecules resulting in emulsifying properties. A major component in most commercial lecithins is phosphatidylcholine (PC). It has O/W emulsion-promoting characteristics and can be regarded as a neutral phospholipid with zero net charge (pH 7). Phosphatidylethanolamine (PE) and, to a lesser extent, phosphatidylinositol (PI) have W/O emulsifying properties. PE is neutral while PI, like phosphatidylserine (PS) and phosphatidic acid (PA), is negatively charged. Phospholipids with a net negative charge have a key role in lecithins of commercial origin, since they increase both the degree of dispersion and the emulsion stability [32]. It is further conceivable that the stability of nutritional fat emulsions is enhanced by the presence of a lamellar liquid crystalline phase in the form of phospholipid liposomes [33]. In a lipid microemulsion for drug delivery, the presence of such structures may be considered inappropriate.

Table 2 Phospholipid Composition of the Commercially Available Soybean Lecithins

Producer	Trade name	PC (mol%)	PE (mol%)	PI (mol%)	PS (mol%)	PA (mol%)	Lyso (mol%)	Anionic lipids (mol%)	P_{tot}[a] (wt%)
Ross & Rover, New York, NY	YELKIN	41	34	19	–	6	–	25	49
Staley Manufacturing Co., USA	STASOL 3318	34	10	29	–	12	15	41	65
Unimills, Hamburg, Germany	Bolec Z	33	29	24	–	14	–	38	51
Lucas Meyer, Hamburg, Germany	AZOL	45	19	11	25	–	–	36	41
Aarhus Oleifabrik, Denmark	OKH	11	19	41	–	29	–	70	68
Unknown, Netherlands	SBP–BZ	33	32	21	–	14	–	35	78
Lucas Meyer, Hamburg, Germany	Epikuron–200, PC	100	–	–	–	–	–	–	98

[a] Total phospholipids in products calc. mol. wt. = 750.
Key: PC = phosphatidyl choline: PE = phosphatidyl ethanolamine; PI = phosphatidyl inositol; PS = phosphatidyl serine; PA = phosphatidic acid; Lyso = lysolecithin.
Source: Ref. 32.

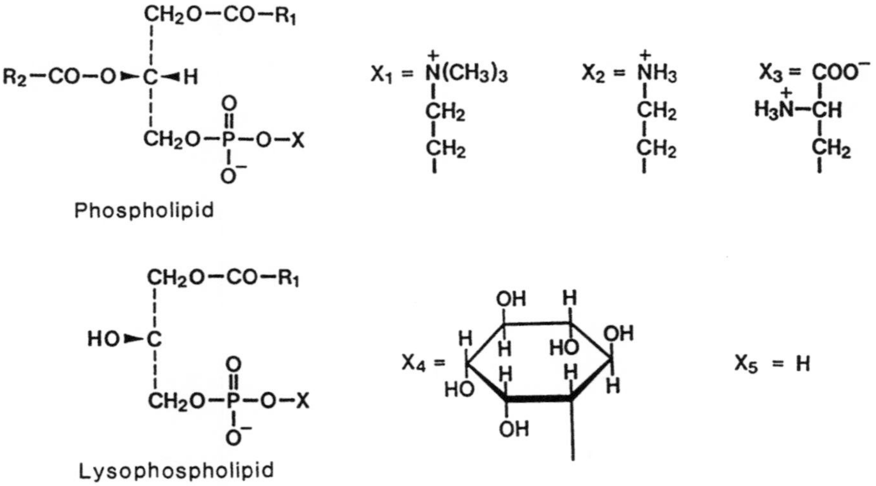

Phospholipid

Lysophospholipid

Figure 3 Structure of glycerophospholipids. R_1 = saturated fatty acid, R_2 = unsaturated fatty acid, X = head alcohol.

Head alcohol	Phospholipid
X_1 = choline	phosphatidylcholine (lecithin)
X_2 = ethanolamine	phosphatidylethanolamine (cephalin)
X_3 = serine	phosphatidylserine
X_4 = myoinsoitol	phosphatidylinositol
X_5 = H	phosphatidic acid

Although fat emulsions like Intralipid [34] stabilized with phospholipids can be stored for several years at $0-4°C$ without any significant changes of physical and chemical properties, the stability is not unproblematic. It has been stated that the emulsions must be stored at low temperature and without addition of supplements as antibiotics, vitamins, or potassium, because they may break the emulsion [35]. The faulty properties of phospholipids with two fatty acid chains can in part be assigned to their molecular structure. They belong to the class of insoluble, swelling amphiphilic lipids, which tend to form planar aggregates in water [36]. This aggregation behavior is related to the cylindrical shape of the monomer molecule itself, which best fits into a planar aggregate. Furthermore, the capacity of phospholipids to decrease the interfacial tension between oil and water is only moderate [37]. These limitations in the emulsifying properties of phospholipids mean that for the preparation of proper lipid microemulsions a great deal of energy (e.g., sonication, homogenization) must be applied to the system, with a risk for decomposition of lipids and incorporated drugs.

The emulsifying properties of phospholipids can be improved by addition of lipids or artificial surfactants, which tend to form spherical micelles. Lysophosphatidylcholine (lysPC) is a typical representative of this class of soluble amphiphilic compound [36]. The conical shape of the lysPC molecule fits well into the curved microemulsion surface and the interfacial tension-reducing effect is significant [37]. However, the strong surface activity of compounds like lysPC also leads to toxic biological effects like disruption of membranes (manifested in hemolysis) [38].

Even if the strong cytolytic property of lysPC-like surfactants prevents their use as major emulsifiers, they may be used as additives to the swelling phospholipids. Dicetylphosphate (DCP) has often been used to give liposomes a negative charge and improve their stability, although it has been shown to cause tissue necrosis in mice [39]. Fatty acid soaps are suitable emulsifiers for O/W emulsions, but they cannot be employed in emulsions for internal use. Other potential additives are short-chain PCs (< 10 carbons), a group of micelle-forming lipids, which spontaneously form stable unilamellar vesicles with long-chain phospholipids [40]. Acetylated monoglycerides (AMGs) represent modified natural lipids with low toxicity, which have been used to improve the stability of emulsions stabilized by phospholipids [41]. However, the addition of micelle-forming surfactants does not always facilitate emulsification, as demonstrated by the adverse effect of bile salts [42]. The explanation is that only the fraction of PC that is not present in bile salt–PC mixed micelles contributes to emulsification.

The most promising alternatives for emulsification-enhancing additives to phospholipids may be found among the nonionic surfactants. In contrast to most ionic surfactants, several nonionic ones have good biological tolerance. Triton X-100 (Fig. 4) is a moderately hemolytic surfactant widely used in experimental biology, but its potential emulsifying properties have been neglected. The same holds for N-octyl-ß-D-glucopyranoside (n-octylglucoside), which is unique among nonionic surfactants because it has a high CMC and is easily dialyzed. However, the most promising alternatives seem to be found in the Pluronic series (F38, F68, F77, F108, F127). These oxyethylene–oxypropylene copolymers (Fig. 4) have been reported to express low toxicity toward biological membranes; the hydrophilic congener F68 does not induce erythrocyte hemolysis even at saturating concentrations [43]. No side effects were observed upon injection of Pluronic F68 [31], whereas a Pluronic F108 emulsion resulted in side effects such as vomiting and minor edema in the paws of rabbits [41]. Together with soya phospholipids the F68 constitutes the surfactant system of the emulsion Lipomul-I.V., intended for intravenous use [44]. For other congeners in the Pluronics series the toxicity increases with increasing lipophilicity.

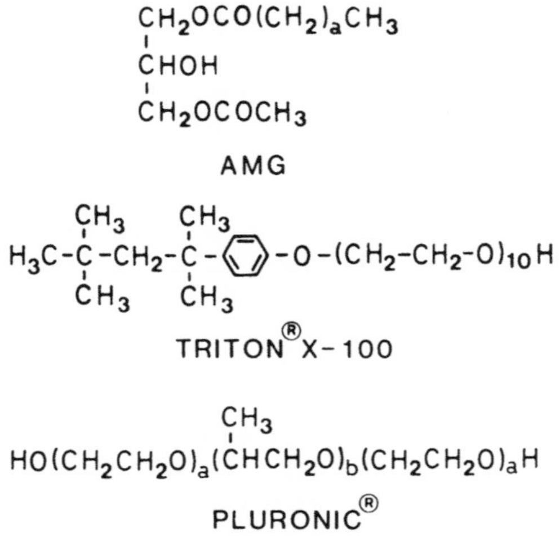

$$CH_2OCO(CH_2)_aCH_3$$
$$|$$
$$CHOH$$
$$|$$
$$CH_2OCOCH_3$$

AMG

$$\begin{array}{cc} CH_3 & CH_3 \\ | & | \\ H_3C-C-CH_2-C-\langle\bigcirc\rangle-O-(CH_2-CH_2-O)_{10}H \\ | & | \\ CH_3 & CH_3 \end{array}$$

TRITON® X-100

$$CH_3$$
$$|$$
$$HO(CH_2CH_2O)_a(CHCH_2O)_b(CH_2CH_2O)_aH$$

PLURONIC®

Figure 4 Structure of synthetic surfactants with potential use as components in surfactant mixtures for preparation of drug-carrying lipid microemulsions. AMG = acetylated monoglycerides; Triton X-100 = octylphenolpoly(ethyleneglycolether)$_{10}$; Pluronic = oxyethylene–oxypropylene copolymer.

The rational choice of surfactant mixtures implies simple and fast methods for assessment of emulsifying efficacy and emulsion stability. A routine method, which has been employed in this laboratory, uses optical density measurements at 450 nm as an estimation of turbidity and hence mean particle size of an emulsion system [45]. Series of emulsions are made with a systematic variation of the composition of the emulsifier, using a reproducible emulsification method, and then the optical density is recorded at different times. Preprarations that are not optically clear or weakly opaque by visual inspection can be discarded at once. Turbidity measurements of this kind are obviously more informative than interfacial tension values. The addition of a series of surfactants (including Triton X-100, n-octylglucoside, sodium oleate, short-chain PCs, Pluronic F68, monooleoylglycerol, lysPC, PA, DCP, AMG) to an egg yolk phosphatidylcholine (EYPC) film between water and trioleoylglycerol caused only minor changes of the interfacial tension (B. Lundberg, unpublished results). On the other hand, the effect of some of these additives on the turbidity and hence the particle size of a trioleoylglycerol/EYPC (2:1) emulsion was quite dramatic (Fig. 5). The data for monooleoylglycerol and lysPC are not shown since the first had a poor effect and the second had low efficacy within physiologically acceptable concentrations. The effect of Pluronic F68 was only moderate, but owing to the low toxicity, it was considered suitable as the basic additive to the PC system. Dicaproyl PC and n-octylglucoside had only moderate effects of their own and were totally incom-

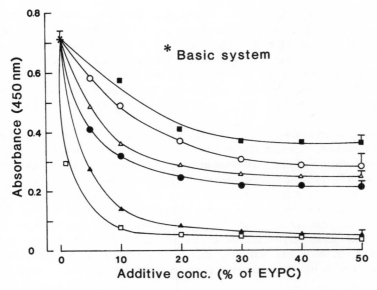

Figure 5 The effects of the addition of different synthetic surfactants on the turbidity of a basic emulsion system consisting of 1 mg trioleoylglycerol and 0.5 mg egg yolk phosphatidylcholine (EYPC) per milliliter. The turbidity is inversely related to the emulsion particle size. Additives: (O) Pluronic F68, (□) Triton X-100, (△) acetylated monoglycerides, (●) octadecylamine, (■) dicaproylphosphatidylcholine, (▲) Pluronic F68 + Triton X-100 (2:1). Error bars represent standard deviations for at least four experiments. (B. Lundberg, unpublished results.)

patible with Pluronic F68. Of all the surfactants tested, Triton X-100 had the most outstanding effect as an additive, even at low proportions compared to PC. However, because of the membrane-disrupting properties, it was chosen only as a minor additive. With trioleoylglycerol as oil phase a surfactant mixture of PC with 10% Pluronic F68 and 5% Triton X-100 was found to give excellent emulsification even with a low energy input. Compounds that add a net negative charge to the emulsion particles, PA, sodium oleate, and DCP had small, similar emulsification-promoting effects in combination with the PC–Pluronic F68–Triton X-100 surfactant system. Octadecylamine was found to be a suitable additive to the same system for obtaining a positive surface charge. From the toxicological point of view PA is preferable to DCP and octadecylamine [39].

D. Choice of Oil Phase

The high demands on biological acceptability and biodegradability that are placed on the surfactant part of the lipid microemulsion must also be applied to the oil phase. Considering these factors, vegetable oils seem to be most suitable. Among oils tested for intravenous use are soya bean oil, olive oil, cottonseed oil, sesame oil, and cod liver oil. Of these the soya bean oil has been found to be the most appropriate and has been used extensively in an emulsion (Intralipid) for parenteral nutrition [46]. The major components of vegetable oils are triacylglycerols, with varying fatty acid composition. The cholesteryl esters represent another neutral lipid group that should be well tolerated.

It is essential to realize that the oil phase is by no means simply an inert carrier or reservoir. The interfacial adsorption of surfactant and the mutual interactions between oil and surfactant are important factors in the emulsification process [47]. A basic prerequisite regarding the physical properties of the oil phase is a liquid (or liquid crystalline) state at body temperature; further, the emulsification process must be performed above the melting point of the oil [45]. These factors are seen in Fig. 6, which shows the turbidity of emulsions with cholesteryl linoleate or trioleoylglycerol as oil phase. The cholesteryl ester, with less favorable oil–surfactant interactions and higher microviscosity than the triacylglycerol, is more difficult to disperse. Addition

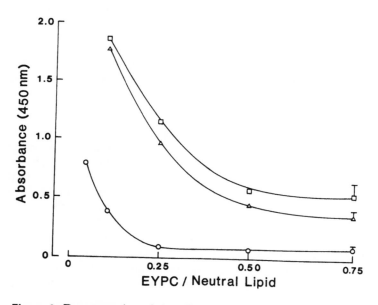

Figure 6 Demonstration of the effects of the oil phase properties on the emulsification process. Trioleoylglycerol (○) is much easier to emulsify than cholesteryl linoleate (□), while a trioleoylglycerol–cholesteryl linoleate (1:4) mixture takes up a middle position (△). The surfactant mixture consisted of egg yolk phosphatidylcholine (EYPC) + 10% Pluronic F68 + 5% Triton X-100. Error bars represent standard deviations for at least three experiments. (B. Lundberg, unpublished results.)

of 20% trioleoylglycerol to the cholesteryl ester results in an appreciable improvement in this respect. The possibility of improving the emulsification properties of an oil phase by addition of relatively small proportions of another oil should be noted when planning a drug–oil phase.

The task of finding a biocompatible and easily emulsified oil is, however, only part of the choice of an oil phase for the incorporation of drugs into lipid microemulsions. If the drug is a lipophilic one, it may be dissolved in a suitable oil and emulsified together with that oil. Such a procedure has been used for the incorporation of, e.g., cyclandelate [48], diazepam [49], and etonidate [50] into fat emulsions for intravenous use. However, hydrophobicity and low water solubility of a compound do not automatically imply that it is unlimitedly oil soluble. A drug in the liquid state will be completely soluble in an oil if their solubility parameters do not differ by more than three units [51]. However, most drugs (>95%) are solids, which behave in a quite different way. The octanol–water partition coefficient (K_{ow}) is widely used among pharmacologists as a measure of lipophilicity of organic molecules. For compounds below their melting point, as log K_{ow} goes up (i.e., lipophilicity increases) oil solubility goes down. Thus for solids, the use of the octanol–water partition coefficient as a measure of lipophilicity is incorrect and the oil solubility is determined solely by the strength of their crystal lattice forces [51]. It has also been pointed out that specific solute–solvent interactions are more important than the bulk properties of the pure components in determining drug solubilites in nonaqueous systems [52]. This means that even a small change in the composition of the oil phase can cause a marked change in the solubility of a compound in it. A demonstration of this is the fact that an addition of 2% fatty acids to an oil caused an increase of about 40% in cholesterol solubility [53]. The addition of acetylated monoglycerieds to soya bean oil favors the dissolution of barbituric acids in the oil phase [48]. Unfortunately, systematic studies of drug solubility in oils suitable for liquid microemulsion preparation are very scarce.

If a drug is too polar for incorporation into an oil phase, it can be made more lipophilic by derivatization. Occasionally the drug derivative itself has physicochemical properties suitable for a microemulsion oil phase, exemplified by a steroid mustard carbamate [54]. A more common situation is one where the apolar prodrug has to be dissolved in an oil to obtain the proper physical state, as for dexamethasone palmitate [55]. The synthesis of lipophilic prodrugs from structurally simple drugs with few polar groups, like the attachment of indomethacin to acylglycerols [56], is quite easily accomplished. However, more complicated molecules, with several polar groups in scattered positions, like the anthracycline antitumor antibiotics, are not easily made lipophilic. One at least partially successful attempt has been made by esterifying doxorubicin with retinoic and linoleic acids [57]. The design of drug molecules that can be carried by lipoproteins and microemulsions is treated in more detail in Chapter 4 of this volume.

E. Drug-Emulsion Conjugation

There are two principal alternatives for drug complexing to lipid microemulsions (Fig. 7): (a) incorporation of lipophilic drugs or prodrugs into the oil phase of the emulsion particle and (b) intercalation or anchoring of amphiphilic drugs or prodrugs into the surface monolayer of the emulsion particle. Regarding lipoproteins there is the further possibility of covalent linkage of drugs to the apolipoprotein moiety (for details see Chapter 4).

It should be noted that an apolar drug molecule may partition into the surface monolayer in parallel with the distribution of a small fraction of such hydrophobic compounds as triacylglycerols and cholesteryl esters among the surface lipids [58]. Moreover, it must be stressed that even though an amphiphilic drug molecule is properly anchored in the surface monolayer (e.g., through a hydrocarbon chain), the drug–emulsion complex may not be entirely stable. The surface lipid cholesterol, with a low water solubility, is rapidly exchanged between emulsion droplets and cells [22].

A

Lipophilic drug (Comp. 25)

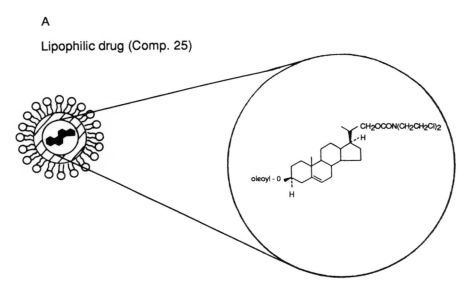

Figure 7 Schematic description of three different modes of complexing drugs to lipid microemulsions and lipoproteins. (A) Incorporation of a lipophilic drug [e.g., comp. 25 = N-(4)-3ß-(oleoyloxy)androst-5-en-17ß-yl(pentyl)oxy)carbonyl)-N,N-bis(2-chloroethyl)amine (see Ref. 69)] into the oil core of the carrier particle. (B) Anchoring or intercalation of an amphiphilic drug [e.g., ara-CMPH = arabinofuranosyl-cytosine-5'-n-hexadecyl phosphate (see Ref. 90)] into the surface monolayer of the carrier particle. (C) Covalent linkage of a drug [e.g., DNR = daunorubicin (see Ref. 89)] to an apolipoprotein or lipid component of the carrier particle.

B

Amphiphilic drug (ara-CMPH)

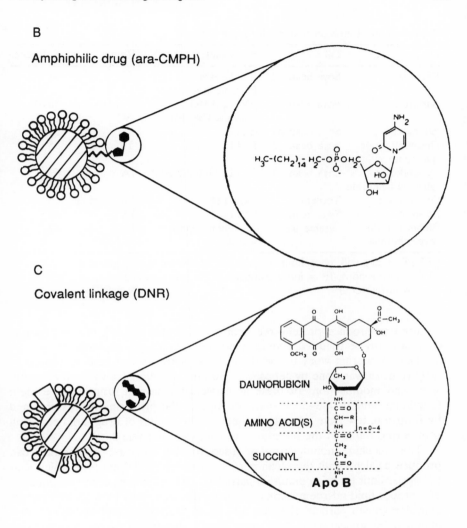

C

Covalent linkage (DNR)

The incorporation of drugs into emulsion core or surface monolayer is usually completed in connection with the emulsification process by including the drug in the oil phase or with the surfactant. The emulsification process itself is thus the focus of drug–emulsion conjugation techniques. We earlier stated that surfactants can decrease the mechanical energy needed to break up an oil phase into emulsion droplets. Since the efficacy of nontoxic surfactants in reducing the interfacial tension is limited and many drugs are sensitive to decomposition, the importance of mild emulsification techniques is stressed. Most methods currently used for lipid microemulsion preparation have been

Table 3 Lipid Emulsion Systems for Drug Delivery

Drug	Oil	Surfactant[a]	Route[b]	Reference
Diazepam	Soya bean oil	EYPL, AMG, Pluronic 108	IV	41
Barbituric acid	Soya bean oil	EYPL, AMG, Pluronic F68, F108	IV, IP SC	48,62
Etomidat	Soya bean oil	EYPL	IV	50
Dexamethasone palmitate	Soya bean oil	EYPL	IV	55
Cyclandelate glyceryl trimitrate	Soya bean oil	EYPL, Pluronic F108	IA	63
Griseofulvin	Triolein	Tween 80	Intraluminal	65
Bisantrene	Soya bean oil	EYPL	IA	66
Bleomycin plus hyperthermia	Sesame oil	Aluminum mono-stearate	IA	67

[a] EYPE = egg yolk phospholipids; AMG = acetylated monoglycerides.
[b] IV = intravenous; IP = intraperitoneal; SC = subcutaneous.
   IA = intraarterial.

adjusted to produce protein-free lipoprotein models [59,60], but in principle they can also be used to produce microemulsion–drug complexes.

The pioneering work in the field of parenteral use of lipid emulsions as drug vehicles was done by Jeppsson. He used nutritional soya bean oil emulsions and modifications of these as carriers for several drugs, with evident functional benefits [48–50,61–64]. In spite of the promising results obtained by Jeppsson, few have taken up his ideas. The studies of drug–lipid emulsion complexing found in the literature are compiled in Table 3.

The "machine homogenization" used for fat emulsion preparation can produce particles with the mean particle diameter of 200 nm [49], which is at the upper limit for microemulsion particles. The first report of the preparation of a proper lipid microemulsion came from this laboratory [45]. A transparent emulsion (mean particle diameter 50 nm) of trioleoylglycerol and EYPC was obtained by sonication in two steps: first a suspension of oil in water was prepared and then the phospholipid emulsifier was added at a second sonication step. When the proportion of EYPC to oil was gradually increased, the optical density of the preparations decreased to a limiting value, at a weight ratio of 2.2:1 between oil and emulsifier. Calculations indicated that at this volue the oil globules were covered by a monomolecular surface film of EYPC. Later the same emulsification technique was used to produce emulsions of EYPC, cholesterol, and trioleoylglycerol as protein-free lipoprotein models with a mean particle diameter of about 30 nm, which is near that of native LDL [59]. The emulsions proved to be quite homogeneous with respect to particle size and lipid ratios, as shown by gel filtration and electron

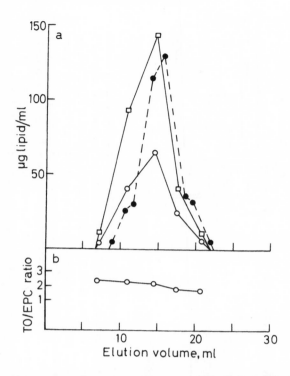

Figure 8 Elution pattern of a 2.2:1 trioleoylglycerol (TO)/egg yolk phosphatidylcholine (EPC) emulsion with concentration of TO (□) and EPC (○) in the fractions [the elution pattern of LDL (●) is shown as comparison], (b) The TO/EPC ratios of the fractions. (From Ref. 59.)

microscopy (Fig. 8). To obtain proper microemulsions, with cholesteryl esters as oil phase, the original sonication steps were replaced by a combined injection–sonication procedure [17]. In this technique a specially designed apparatus is used for the simultaneous sonication by a probe sonicator and injection of lipids dissolved in organic solvents (Fig. 9A). This procedure has further been successfully used for the preparation of protein-free lipoprotein models [68] and microemulsion of a potential cytotoxic drug for reconstitution with apo B [69].

Lipid microemulsions of good quality of cholesteryl esters surface-stabilized by PCs have been prepared by prolonged sonication (5 h), but this procedure can hardly be applied to drug systems, since the risk for chemical degradation during sonication is obvious. More promising from this point of view is the injection technique described in Refs. 70 and 71. According to this procedure a dried lipid mixture was dissolved in anhydrous isopropanol and then injected

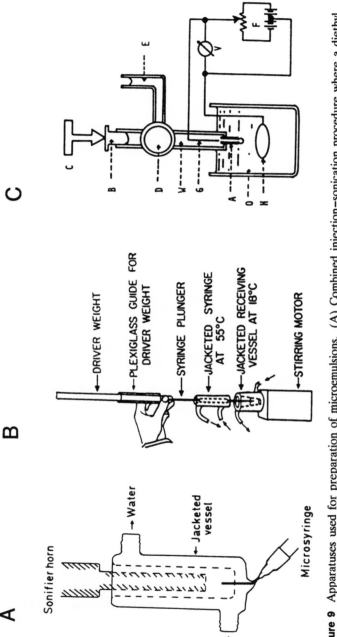

Figure 9 Apparatuses used for preparation of microemulsions. (A) Combined injection–sonication procedure where a diethyl ether–chloroform 2:1 solution containg the lipid (or drug) was slowly injected through the bottom of the jacketed sonication vessel. Simultaneously, the sonication was performed with the horn dipped about 1 cm into the buffer solution at 52°C. (From Ref. 17; see also Refs. 22 and 69.) (B) The lipids dissolved in 2–propanol were injected with a water-jacketed Hamilton syringe at 55°C just below the surface of a rapidly mixing buffer maintained at 18°C. (From Ref. 71; see also Ref. 70.) (C) Schematic diagram of the apparatus for electrocapillary emulsification. A, Teflon tube; B, syringe; C, driving motor; D, three-way cork; E, water phase reservoir; F, DC power supply; G and H, platinum electrodes; O, oil phase; V, voltmeter; W, water phase. In the case of O/W-type emulsification, the oil phase is introduced from the syringe into the aqueous phase. (From Ref. 72.)

into buffer (maintained at 18°C) with a thermostatted Hamilton syringe at 55°C (Fig. 9B). The isopropanol was subsequently removed from the solution by centrifugation through a buffer-depleted Sephadex G-50 column. The resulting microemulsions were essentially homogeneous with respect to particle size (~40 nm) and chemical composition, provided that trioleoylglycerol was present in the core and the PC component was either a disaturated or monosaturated–monosaturated one. Factors of crucial importance were the temperature of the jacketed syringe and the water content of the 2-propanol.

An interesting emulsification technique, with potential use for preparation of drug–emulsion systems, is the electrocapillary method [72]. This is based on the decrease in interfacial tension to almost zero by applying a potential difference to the oil/water interface (Fig. 9C). The emulsification is spontaneous without the surface tension lowering effect of a surfactant, and the use of large amounts of mechanical energy is avoided.

This survey does not deal with multiple emulsions, which have promising biopharmaceutical applications [73] but cannot be regarded as typical lipid microemulsions. Their formulation and stability have been reviewed recently [74].

LIPOPROTEINS AS DRUG CARRIERS

A. Overview

The initiating factor behind the idea of using plasma lipoproteins as drug carriers was the elucidation of receptor-mediated uptake system by Goldstein and Brown [75]. This specific mechanism enables intracellular uptake of an LDL–drug complex with subsequent lysosomal release of the drug [9]. A further stimulus was the finding that certain cancer cells show higher LDL receptor activity than the corresponding normal cells [76]. However, this potential targeting effect can only be partial, since normal cells also take up LDL, although at a lower rate. An alternative approach to the delivery of drugs via the receptor-mediated uptake of whole lipoprotein–drug complexes is the use of lipoproteins as vehicles for sustained drug release. In this respect HDL should be especially favorable because of its long residence time in the circulation.

Even if drug targeting via the LDL pathway has its shortcomings, the lipoprotein particle has many advantages as a drug carrier:

1. Being an endogenous component, it can avoid such typical carrier problems as immunological reactions and rapid plasma clearance due to uptake by reticuloendothelial cells.
2. Receptor-mediated endocytosis enables intracellular uptake of the drug.
3. The lipoprotein particle is totally biodegradable.

4. The small particle size avoids "size" problems like lung emboli and promotes diffusion to extravascular compartments.
5. It provides a biocompatible vehicle for lipophilic drugs, which is of special importance if the drug is sensitive to decomposition.

On the other hand, lipoproteins have disadvantages as drug carriers:

1. Their complex and unstable nature.
2. Potential drug cytotoxicity to normal cells through defective targeting.

The prospect of using lipoproteins as drug carriers is widened by possible modification of native lipoproteins. Such measures can radically change the cellular uptake and the body distribution of the lipoproteins [77], which offer special opportunities for targeting purposes. The HDL class of plasma lipoproteins represents a largely disregarded potential drug carrier system. Recent evidence indicates that HDL particles, too, are taken up by receptor-mediated endocytosis and intracellularly transported within endosomes, which do not fuse with lysosomes [14]. Such a pathway might enable a different intracellular route of drug administration than the "lysosomotropic" one for LDL–drug complexes.

From a structural and colloid-chemical point of view the lipoproteins are very similar to lipid microemulsions. However, the apolipoprotein part of the lipoproteins, which certainly adds to their colloidal stability, also makes them prone to denaturing agents. The preservaton of the native structure is essential for the in vivo behavior of lipoprotein–drug complexes, since altered lipoproteins are recognized as foreign material and removed from the circulation by cells of the reticuloendothelial system. Such behavior would dramatically decrease the specificity of drug delivery via the lipoprotein receptor pathways.

The complexing of drugs to lipoproteins can be performed in corresponding ways as for lipid microemulsions (Fig. 7). Lipophilic drugs can be incorporated into the oil core and amphiphilic drugs, into the surface monolayer of lipoproteins. An additional possibility regarding lipoproteins is the covalent linkage of drugs to their protein parts.

The incorporation of membrane-penetrating amphiphilic drugs into lipoproteins in principle offers no problem. A simple incubation of the drug with the lipoproten preparation leads to uptake by surface transfer [78]. However, the incorporation of lipophilic drugs into the oil core of lipoproteins is much more difficult since an uncatalyzed surface transfer of such drugs will not proceed at an appreciable rate. The known methods involve one of two principles: (a) the lipoprotein core is extracted by solvent or detergent and then the particle is reassembled with a drug containing core, or (b) the physical transfer of the apolar drug to lipoprotein is enhanced by, e.g., a lipid transfer protein. A special case is the possibility of covalent linkage of drugs to lipoproteins, which in principle can be applied to amphiphilic, lipophilic, and water-soluble drugs. In this section the methods with potential use for lipoprotein–drug conjugation found in the literature are outlined, and suggestions on their possible applicability in practical use are offered.

B. Lipoprotein Reassembly Methods

Lipoprotein Isolation

The initial step in the use of lipoproteins as drug carriers is to find a suitable method for preparation of lipoproteins. There is no problem for basic research, since there are many good routine methods available. When it comes to large scale separation of lipoproteins for clinical use as drug carriers, however, more emphasis must be placed on simple and cheap, but still reproducible and pharmaceutically acceptable methods. One alternative is centrifugation, preferably single spin gradient [79] or zonal [80] ultracentrifugation. The precipitation methods in common clinical use [81] could be another possibility. The methods based on use of polyethylene glycol 6000 or heparin should be gentle ones. A simple method, which, however, yields the lipoproteins in a diluted form, is gel filtration on agarose [82] or corresponding gels. The use of plasmapharesis for the isolation of LDL and the preparation of drug–lipoprotein complex from the patients' own LDL has also been proposed [83]. Such a procedure might have the advantage of reducing the endogenous LDL level and hence the competition between LDL–drug complex and native LDL. However, judging from today's knowledge it is impossible to point out any generally superior isolation method. The method must depend on the actual problem: how much lipoprotein is needed, the purity demanded, and the lipoprotein class to be isolated.

Solvent Extraction

The first method for the reconstitution of biologically active LDL was presented by Krieger et al. [84]. This so-called Krieger method used potatoe starch to stabilize an apo B–phospholipid complex, while the neutral lipid core was extracted with heptane and replaced with an exogenous cholesteryl ester. The reconstituted LDL retained its β mobility on agarose gel electrophoresis and its ability to be precipitated by an antibody to native LDL and by heparin–manganese. The Krieger method has been used to incorporate a variety of hydrophobic compounds into the core of LDL, including dioleyl methotrexate [85], 25-hydroxycholesteryl oleate [86], cholesteryl nitrogen mustard (phenesterine) [87], and pyrene coupled to a derivative of cholesteryl oleate [88]. Although the Krieger method produces recombinants with receptor-specific uptake in vivo an N-trifluoroacetyladriamycin-14-valerate (AD-32)–LDL complex prepared according to this method was more rapidly removed from the plasma after intravenous injection than native LDL was [89]. This rapid plasma clearance was found to be due to the freeze-drying of LDL in the presence of insoluble starch. When LDL was frozen and freeze-dried

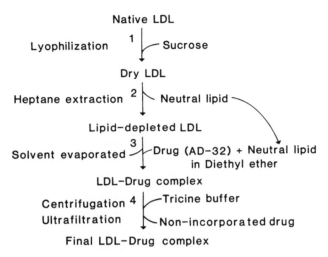

Figure 10 Schema of the modified Krieger method for the reconstitution of the LDL–drug complex. (1) LDL was lyophilized in the presence of potato starch. (2) The dry LDL was extracted with heptane at 4°C and the heptane phase was evaporated, leaving neutral lipids. (3) A drug–neutral lipid mixture in diethylether was then added to the lipid-depleted LDL and the solvent was evaporated. (4) The drug–LDL complex was solubilized in Tricine buffer; insoluble, nonincorporated drug was separated by centrifugation; finally, the preparation was passed through a 0.45 μm sterile filter. (From Ref. 89.)

in the presence of sucrose instead of starch, a plasma disappearance rate equal to that of native LDL was obtained [89].

It was also found that when the evaporated heptane extracts were solubilized together with the drug a normal in vivo fate was obtained. Obviously the drug AD-32 itself does not possess physicochemical properties suitable for a core oil phase, but it does so in combination with the native neutral lipids of LDL. The amount of AD-32 incorporated in LDL was lower with this modified Krieger method (100 molecules/LDL particle) than with the original one (400 molecules/LDL particle). The complex remained stable during dialysis, density gradient centrifugation, and gel filtration. The cellular drug accumulation was found to be dependent on the LDL receptor activity of the cells. The procedures of the modified Krieger method are summarized in Fig. 10. They include lyophilization and heptane extraction of LDL, mixing of the resulting apo B-phospholipid preparation with drug and extracted LDL neutral lipids, and finally solubilization of reconstituted LDL–drug complexes by addition of buffer followed by purification by centrifugation and filtration. The Krieger method and its modified version have been adopted as useful reconstitution methods by several groups, but a bad point noted in this laboratory [69] and also by others [90] is the low recovery of drug in the final preparation.

Solvent extraction has long been a routine method for the delipidation of HDL [91]. The resulting apolipoproteins, mainly apo A-I and apo A-II, have an outstanding capacity to solubilize lipids, especially PCs at their transition temperatures. Reassembled HDL particles in the same size range as native HDL have been prepared by the Krieger method [92] and by association of delipidated apo HDL with sonicated microemulsions of phospholipids and neutral lipids [93]. The in vivo fate of reassembled HDL has been studied and the plasma decay kinetics of the core lipids were not distinguishable from those of biologically labeled lipids in native HDL [92]. However, the possibility for using reassembled HDL as a drug carrier seems to have been overlooked.

Detergent Solubilization

Detergents can be used for the delipidation and solubilization of the water-insoluble apo B. The isolated apoprotein can then be used for reassembly of LDL with lipid(–drug) microemulsions. Such a method has been applied to the incorporation of cholesteryl oleate [17] and a cytotoxic steroid mustard carbamate [69] into the core of the reconstituted LDL particle. The procedures, schematically shown in Fig. 11, included solubilization of apo B from LDL with sodium deoxycholate (SDOC), separation of apo B–SDOC complexes by gel filtration and then completion of the LDL reconstitution by association of the solubilized apo B with a neutral lipid (or drug)–EYPC microemulsion prepared by the injection-sonication method. By these procedures an optically clear preparation of reassembled LDL was obtained, which was stable as determined by density gradient ultracentrifugation and gel filtration. The LDL particles had a mean diameter of 21 nm and exhibited ß migration at agarose electrophoresis. The in vivo cellular uptake and metabolization were found to be similar to that of native LDL.

A reconstitution method similar to that described above has been developed by another group [18]. The major differences are that they used very long sonication times (5 h) for preparation of the microemulsion and a slow addition of SDOC-solubilized apo B to the microemulsion in a dialysis bag where the removal of SDOC took place. The properties of the reassembled LDL obtained by these procedures were similar to those for the method described previously.

Detergent solubilization methods are rather tedious for routine drug–LDL conjugation. A simple and fast detergent dialysis method has been tested but with unsatisfactory result [90]. The lipophilic drug AD-32 was solubilized by octylglucoside, LDL was added to the solution, and finally the complexes formed were fractionated by extensive dialysis, gel filtration, and filtration through 0.45 μm filters. The method was not considered useful since the stabil-

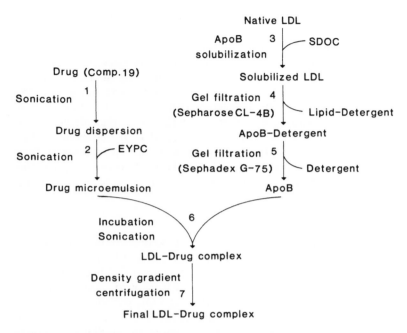

Figure 11 Schema of the detergent method for the reconstitution of the LDL–drug complex. (1) The drug was dispersed in buffer by the combined injection–sonication method (see Fig 9A). (2) The drug dispersion was stabilized by bath-type sonication with egg yolk phosphatidylcholine (EYPC). (3) LDL was solubilized by addition of sodium deoxycholate (SDOC). (4) Solubilized apolipoprotein B (apo B) and lipid–detergent mixed micelles were separated by Sepharose CL-4B gel filtration. (5) To remove detergent from apo B the solubilized apo B was passed through a Sephadex G-75 column. (6) The reconstitution of the LDL–drug complex was accomplished by incubation and sonication of detergent-delipidated apo B with the drug microemulsion. (7) The LDL–drug complexes were purified and isolated by density gradient centrifugation. (From Ref. 69.)

ity of the drug–LDL complexes was poor and the removal of detergent was not complete.

The in vivo biological activity of the recombinant LDL obtained by dtergent solubilization methods was similar to that of native LDL [17,18]. However, the in vivo behavior is more doubtful. Preliminary studies (S. Vitols and B. Lundberg, unpublished results) showed that a considerable portion of a reconstituted LDL preparation was cleared rapidly from the plasma (in mice), indicating uptake by the reticuloendothelial system. This might be a critical point regarding the practical use of drug–LDL complexes prepared by detergent solubilization methods for drug delivery.

Enzymatic Digestion

Delipidation methods involving organic solvents or detergents involve a risk for changes in the protein structure. A presumably more gentle method uses enzymatic delipidation of LDL [69]. The cholesteryl ester core was hydrolyzed with sterol ester hydrolase (EC 3.1.1.13) in the presence of EYPC vesicles and albumin in order to bind the reaction products; free cholesterol and free fatty acids. The resulting apo B–polar lipid complexes were associated with a suspension of the lipophilic drug in buffer. After purification by centrifugation and filtration the cytotoxic activity of the LDL–drug complex was tested on cells in culture. It was demonstrated that the preparation was able to kill all the cells by uptake via the LDL receptor pathway. In vivo studies have not been performed.

C. Transfer Methods

Dry-Film Stir

The dry-film stir method for incorporation of drugs into lipoproteins is technically simple and reasonably mild. The principle behind the method is physicochemical transfer of drug molecules from a dried film on the walls of the incubation tube [94] or on the surface of an inert carrier such as Celite 545 [78]. The procedures involve evaporation to dryness of the drug in organic solvent as a thin film on the support and then incubation with a lipoprotein preparation in the presence of necessary preservants such as antioxidant, antimicrobial agent, dark room, and inert gas. After completed incubation (usually at 37°C) the drug–lipoprotein complex can be purified by Sephadex G-120-15 column chromatography [94] or by centrifugation (3000 rpm for 10 min) if Celite 545 is employed [78]. For sterilization and removal of aggregated complex a 0.45 μm filter can be used. The successful partitioning of a number of different structured drugs into LDL or modified LDL has been reported [90], they include hexadecylmethotrexate (HMTX) (55 molecules/LDL particle), AD-32 (100 molecules/LDL particle), muramyltripeptide phophatidyl-ethanolamine (MTP-PE) (140 molecules/acetylated LDL particle), and arabino-furanosylcytosine-5'-(n-hexadecylphosphate) (ara-CMPH) (90 molecules/LDL particle) (Fig. 12). On the other hand, only 0.5% of the added daunomycin was incorporated into LDL [94] and it was not possible to associate adriamycin (adr) with LDL, since aggregation and flocculation occurred, while HDL–adr was a surprisingly stable complex [9]. The colloidal stability of the lipoprotein–drug complexes will thus depend on the characteristics of both components and stability problems seem hard to foresee. In general, the drug incorporation efficiency of LDL was much higher than that of HDL [90], which was also noted for the incorporation of ß-sitosteryl-ß-D-glucopyranoside into lipoproteins; LDL gave values about ten times higher than HDL calculated on lipoprotein protein contents [78].

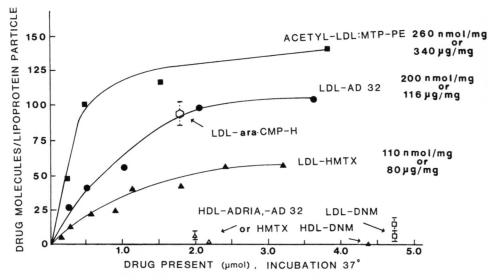

Figure 12 Stoichiometry of lipoprotein–drug complexes. Lipoprotein–drug complexes were prepared using drug and constant amounts of lipoprotein by a dry-film stir technique at 37°C. HDL-adria and LDL-dnm (upper point) were prepared by the aqueous addition and detergent dialysis methods, respectively. HDL-HMTX and HDL-AD-32 were prepared by the solvent extraction method. The drug levels were determined following the fractionation of each complex using Sephadex G15-120 chromatography and 0.45 μm filtration. The number of drug molecules per particle were estimated assuming molecular weights of 2.5×10^6 for LDL and acetyl-LDL and 2.5×10^5 for HDL species, with protein representing 25 wt% for LDL and acetyl-LDL and 50 wt% for HDL. (From Ref. 90.)

The conjugation of drugs to lipoproteins by the dry-film stir method meets the requirements for simplicity and rapidity that must be placed on any practically useful method. The biological behavior of the complexes seems to be acceptable both in vitro [94] and in vivo [78], although nonreceptor interactions of the lipoprotein–drug complexes with cells might be a disturbing factor [95]. The largest question mark that remains is whether it is possible to incorporate a lipophilic drug into the core of a lipoprotein particle, with sufficient efficacy, without any promoting factor.

Aqueous Addition

The aqueous addition method is, of course, adaptable only for complexing of water-soluble drugs with lipoproteins. It has been used for preparation of aclacinomycin A–LDL complex [96], conjugation of adriamycin and daunomycin with HDL [90], and complexing the photoactive anticancer porphyrin derivative photofrin II with LDL and HDL [97]. Conjugation of drug and lipoprotein was accomplished simply by addition of an aqueous solution of the drug to the lipoprotein preparation, followed by isolation of the complex by, e.g., dialysis, gel filtration, and ultrafiltration. Such measures gave incorporation of 15–450 molecules of aclacinomycin A per LDL particle [96], 3–5 adriamycin per HDL [90], and ~130 photofrin II per LDL [97].

From surface transfer studies one can deduce that drugs conjugated with lipoproteins by aqueous addition will undergo a fast transfer to other lipoproteins and cells [20]. Such specific physical transfer will of course reduce the specificity of the drug delivery. For water-soluble drugs the use of lipoproteins as sustained drug release systems is the most realistic alternative, although the delivery of aclacinomycin A [96] and photofrin II [97] to cells in vitro by LDL seems to be achieved mainly by the LDL receptor pathway.

Facilitated Transfer

The exchange of strongly lipophilic lipids, like triacylglycerols and cholesteryl esters, from lipoproteins to cultured cells in serum-free medium is extremely slow [22]. However, the addition of transfer mediators like solvents [98] or lipid transfer proteins [99] can appreciably increase the transfer rate. A corresponding approach may be applicable also to lipophilic drugs.

A solvent that is well tolerated in biological systems is dimethyl sulfoxide (DMS)). This solvent has been used for incorporation of labeled cholesteryl esters into lipoproteins and has also been tested for incorporation of AD-32 into LDL, but with inferior results compared to that obtained by the modified Krieger method (S. Vitols, personal communication). Although DMSO is useful for incorporation of trace amounts of neutral lipids into lipoproteins, transfer factors present in plasma seem to be more promising. Cholesteryl esters have been transferred from microemulsions to LDL, using lipoprotein-free plasma, with good efficacy (40% recovery) [100]. It was established that the transferred cholesteryl esters were located in the core of a structurally intact LDL. The labeling procedure did not alter either the in vitro metabolism of LDL by fibroblasts or the rate of plasma clearance in the rat. In a recent study purified lipid transfer protein was used for incorporation of cholesteryl esters into acetyl-LDL [101]. Thus far the transfer protein approach has not been used to incorporate drugs into lipoproteins, but at least one group has started such a project (T.J.C. van Berkel, personal communication). However, the structural requirements on the compounds which can be transferred by transfer proteins must be elucidated before the usefulness of this method for drug–lipoprotein conjugation can be judged.

Apolipoprotein Transfer

It is now believed that during lipoprotein metabolism many apoproteins undergo transfer among the lipoproteins. High density lipoprotein apoproteins (apo A, C, E groups) are also readily transferred from lipoproteins onto lipid emulsions [102]. The most interesting species of the readily exchangeable apoproteins might be apo E. Lipid emulsions with apo E were shown to be more rapidly metabolized by the hepatocyte than were unsupplemented emulsions [103]. However, although the transfer of apoproteins to drug-containing microemulsions surely is a feasible measure, no such study seems to have been published.

Apo B is usually termed an unexchangeable protein [104]. However in the light of recent studies this might not be altogether true, since LDL and lipid microemulsions have been shown to express strong interactions that were prevented by apo A-I [105]. The inability of lipid emulsions to bind any significant amount of apo B from a mixture of lipoproteins might thus partially be explained by the presence of apo A-I.

Studies in this laboratory (B. Lundberg, unpublished results) have demonstrated fast and complex interactons between lipid microemulsions and LDL, when incubated together. These interactions also involved transfer of apo B from LDL to microemulsions and the resulting complexes exhibited a good biological activity in vitro. The possible utilization of direct LDL–lipid microemulsion interactions for drug–lipoprotein complexation is, however, still at the speculative stage.

D. Covalent Binding

The complexation of drugs to lipoproteins by covalent linkage is discussed in Chapter 4 of this volume and thus is only briefly outlined here. The lipoprotein surface will in principle offer good possibilities for covalent conjugation (e.g., lysine residues). However, the complex formed must meet some important requirements, the most central being good physical stability, cleavability in lysosomes but not in the circulation, and noninterference with receptor recognition. Covalent linkage of anthracyclines to lysine residues of LDL resulted in a progressive decline in the affinity of the conjugate for the receptor in vitro with increasing substitution [89]. The in vivo fate of the conjugates, on the other hand, was similar to that of native LDL. The attachment of ten methotrexate molecules to the surface of each LDL particle resulted in decreased physical stability of the lipoprotein and a reduced activity in vitro of the complex against leukemic cells compared to free drug [106].

E. The Isolation of Drug-Carrier Complexes

Generally the same methods can be used for separation of lipid microemulsion-drug and lipoprotein–drug complexes. The critical step for the successful production of such complexes is to find a suitable conjugation method; their isolation should generally not present difficulties. The most frequently used methods are gel filtration chromatography, centrifugation, and microfiltration or combinations of these methods. The choice of method(s) is strongly dependent upon the conjugation procedure employed. If the solid-film stir method is used, a simple filtration may be sufficient; but if detergent or solvent is included, more rigorous separation methods are necessary.

Gel Filtration

Gel filtration is a convenient and gentle method for isolation of a drug–carrier complex. If unbound drug is to be separated from the complex, small pore gels (e.g., Sephadex G-120-15) can be used, whereby the complex is eluted with the void volume and the drug is retained n the column [94]. When large aggregates like lipid microemulsions and reconstituted lipoproteins are to be separated, large pore gels (e.g., Sepharose CL-4B, BioGel A-5m) are the appropriate choice [17]]. With a column of suitable length, sufficient separation usually can be achieved, but the metod is time-consuming unless a high pressure apparatus is used [107]. Another drawback is the dilution of the sample, which often necessitates a subsequent concentration step. Chromatographic methods have also proven to be efficient when used in affinity modes. Ligands such as heparin or antibodies to apoproteins, immobilized onto a solid support, have created efficient systems for isolation or purification of lipoprotein fractions [108].

Centrifugation

Isopycnic or gradient density centrifugation methods are especially useful when free apolipoproteins are to be separated from drug–carrier complexes. In some recombination methods the particle sizes of the precursor drug–microemulsion and the final reassembled lipoprotein are so alike that they cannot be completely separated by gel filtration. With a linear NaBr or sucrose gradient centrifugation this problem is easily solved [17]. The fractions can be collected from the bottom or the top of the centrifuge tube or by tube slicing. Punctuation of the bottom of the tube has been used with satisfactory results. The tedious dialysis step, after a gradient centrifugation, is a disadvantage of this technique.

Ultrafiltration

The filtration of the lipoprotein- or microemulsion–drug complexes through a 0.45 μm microfilter is usually the final step to assure sterility before test of biological activity in vitro or in vivo. At the same time possible aggregated material is removed. In combination with the dry-film stir method, filtration may be used as sole purification method, perhaps in combination with a short conventional centrifugation [90]. An annoying drawback with filters is their tendency to stack.

F. Characterization of Drug-Carrier Complexes

Physical Properties

Gel filtration and density gradient centrifugation, combined with quantitative chemical analyses, can be used for characterization of the drug–carrier complex principally in the same way as for the isolation (Fig. 13). The characteristics obtained by these methods are structural integrity, approximate particle size, particle density, and chemical composition and homogeneity of the particles. The particle size and size distribution can also be estimated from negative staining electron micrographs, although results sometimes diverge significantly from those obtained by gel filtration [71]. Useful information about particle size and charge can be obtained by agarose and polyacrylamide gradient gel electrophoresis [109]. The particle size of emulsions, intended for intravenous administration, has been determined by a laser centrifugal photosedimento-meter, which is capable of detecting particles down to 20 nm [49]. In my experience, a very convenient particle sizing method is photon correlation spectroscopy (PCS) (or time-dependent light scattering). This technique is a fast and reliable one, which provides absolute size distributions, with a small quantity of sample, in the size range from about 0.02 μm to several micro-meters (Table 4). The instrumentation is commercially available with automation including data analyses.

More detailed information of the structure and interactions on the molecular level can be obtained by physical techniques such as differential scanning calorimetry (DSC), nuclear magnetic resonance (NMR), fluorescence, and circular dichroism (CD) spectroscopy. Thermal characterization by DSC has shown that LDL undergoes two distinct thermal transitions: the first, at 20–40°C, is associated with an order–disorder transition of the core–located cholesteryl esters; the second, at ~85°C, is associated with disruption of the LDL particle [110]. The significance of the LDL core transition below physio-logical temperature is demonstrated by the fact that the preparation of stable lipid microemulsions and model lipoproteins requires a liquid (or liquid cry-stalline) oil core [17]. Thermal analysis of bulk systems can be employed for the design of neutral lipid–drug mixtures with suitable oil phase properties [111].

The mobility and position of lipids, and potentially also drugs, in carrier particles can be studied by NMR spectroscopy [112]. Yet there is a practical obstacle because an optimal use of NMR implies deuterium– or carbon–13–en-riched lipids and drugs, whose availability is limited today. The fluidity of the oil core can also be assessed by measuring the mobility of an electron spin resonance (ESR) probe [113].

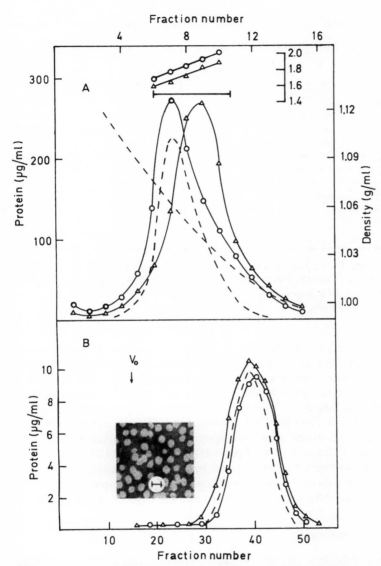

Figure 13 Characterization of compound 25–*m*-LDL complexes prepared by the detergent (○) and hydrolase (△) method, respectively.   (A) Density gradient centrifugation in a 0–40% sucrose gradient.   Bar = fractions taken for incubations with cells and further characterization; the insert shows ratios between compound 25 and protein (apo B) in the fractions; the dotted line represents the position of native LDL in the density gradient.   (B) Sepharose CL-4B column chromatography.   The dotted line shows the elution pattern of native LDL as a comparison; the arrow indicates the void volume (V_0) of the column; the insert is an electron micrograph of cytotoxic *m*-LDL particles; bar = 30 nm. (From Ref. 69.)

Table 4 Size Distribution from Photon Correlation Spectroscopy for a Lipid Microemulsion[a]

Particle size distribution (nm)[b]	Graph of distribution of mass	Percent by mass
21.8–26.4 |+		16.5
26.4–31.9 |————————————— +		27.8
31.9–38.7 |——————————————————— +		27.8
38.7–46.9 |————————————————— +		25.3
46.9–56.8 |————————————— +		14.2
56.8–68.8 |————————— +		6.7
68.8–83.3 |——————— +		4.3
83.3–101.0 |————— +		3.0
101.0–122.3 |——— +		1.5
122.3–148.2 |— +		0.6
148.2–179.5 |+		0.2

[a] Oil phase, 1 mg trioleyglycerol; emulsifier mixture, 0.5 mg egg yolk phosphatidyl choline + 0.1 mg Pluronic F68 + 0.1 mg Triton X-100. The sample was sonicated for 3 x 10 s with 10 s intervals by a probe sonicator.

[b] Average mean, 48.5 nm.

More within the practical scope than NMR are fluorescence measurements, which can give similar information with high sensitivity [114]. The supply of fluorescence-labeled lipids and drugs is already considerable. The location of a fluorescent probe in a particle can be verified by fluorescence energy transfer [115]. By the method of fluorescence polarization, using fluorescence probes such as perylene or 1,6–diphenyl1-1,3-hexatriene (DPH), the microviscosity within lipid domains in emulsions and lipoproteins can be estimated [116].

The secondary structure of apolipoproteins, and eventual changes in them after drug incorporation, can be assessed by circular dichroism spectra [117]. The circular dichroism of chromophores (e.g., cholesteryl *cis*-parinarate) in the core of LDL is sensitive to the physical stae of the core oil phase [115].

Biological Characterization

The biological activity is in fact the most important feature of drug–carrier complexes and should be tested at an early stage of an investigation. Antibody precipitation experiments can give a reliable conception of the biological feasibility of lipoproteins [118]. However, such studies are only preludes to tests on living cells. The validity of this statement is demonstrated by the fact that acetyl-LDL loses its ability to bind to the classic LDL receptor of nonmacrophage cells, while it remains precipitable by antibodies to native LDL [119].

Cell culture experiments are easily performed and can give in vitro values of binding, uptake, and degradation of LDL–drug complexes [69]. The application of fluorescence-labeled drugs has obvious advantages above the use of the classical ^{125}I-labeling of lipoproteins for the study of the cellular uptake of lipoprotein–drug complexes. The technique of digital imaging fluorescence microscopy combines the use of fluorescent lipid analogs or drugs with fluorescence microscopy. By this method the binding and uptake of drugs by individual cells can be assessed and their intracellular fate can be traced [120]. However, even if the in vitro characteristics of a drug–carrier complex are satisfactory, it does not a priori imply that this also applies to the in vivo behavior. Low density lipoprotein reconstituted by the Krieger method is reported to bind to the LDL receptor in cultured fibroblasts with the same affinity as native LDL and the uptake and degradation are similar [84], yet the reconstituted LDL showed a much more rapid plasma clearance compared to native LDL, indicating an extensive uptake by cells in the reticuloendothelial system [89]. It is thus essential to recognize the need for in vivo tests of a drug–carrier complex before its pharmacological applicability can be assessed.

IV. PROBLEMS AND PROSPECTS

The tremendous development of the fields of molecular biology and molecular immunology during the last decade has opened new areas for the controlled, selective delivery of drugs. Although the use of the body's own transport and recognition systems should be a sound approach to the drug targeting issue, many severe obstacles must be faced. A basic problem is to find a proper carrier and methods for complexing drugs to it. Thus far, surprisingly little attention has been given to the use of lipid microemulsions as drug vehicles, in spite of the fact that positive results have been obtained, even with "passive" targeting [41,66]. The "active" targeting of lipid microemulsions has hitherto been confined to the attachment of apolipoproteins to them, but the prospect should at least include all those alternatives applied to liposomes [7].

The preparation of lipid microemulsions with particle sizes small enough to avoid thrombophlebitis is well within the practical range by use of the emulsifiers and emusification techniques available. The effects of the drug on the oil phase physicochemical properties and the surfactant–oil phase interactions are factors that have to be carefully considered. Surfactant mixtures, with naturally occuring phospholipids as major components, are biodegradable, essentially nontoxic, and relatively nonimmunogenic and can, for example, be administered in large amounts over long periods as components of lipid emulsions for intravenous feeding.

The most critical part of the complexing of drugs to lipid microemulsions is the design of a proper drug–oil phase. It is essential to realize that even if a drug is lipophilic, its successful incorporation into an oil phase is far from certain and derivatization will often be needed. Ideally, a carrier-associated drug should be specifically selected or designed for the purpose. There is a challenge for pharmaceutical chemists! A further inducement to improve the lipophilicity of a drug is the possibility of increasing its ability to penetrate membranes and reach its site of action. The alternative to anchor the drug into the surface monolayer by, say, fatty acid chains seems to be applicable to the use of the carrier as a sustained release system, since surface transfer of drug will occur at a considerable rate.

The main topic of this book is, however, the use of lipoproteins for the administraton of drugs. There is no doubt that they are nearly ideal as drug carriers and fulfill most evaluation parameters that have to be considered regarding drug delivery systems. The lipoproteins possess favorable properties concerning load factors, immunogenicity, toxicity, and applicability to disease processes. Less ideal are the potential selectivity and the pharmaceutical feasibility.

The complexing of drugs to lipoproteins offers a more intricate problem than does complexing to lipid microemulsions. The reason for this is the need to preserve the native conformation of the apolipoprotein part of the lipoprotein particle in order to avoid a fast clearance by the reticuloendothelial system. Several of the methods taken into consideration in this survey were developed for the preparation of lipoprotein models, but they also have obvious applicability for drug complexing to lipoproteins. Some of these methods, however, seem to be rather too complicated and tedious for large scale clinical use and there is a need to simplify them. Future development may favor the use of different facilitated-transfer methods, which avoid the breaking up of the native lipoprotein structure. Still largely overlooked opportunities are offered by HDL, with its long half-life in circulation, and apo E, which is less complicated from the preparative point of view than is apo B.

The most critical evaluation parameter regarding lipoprotein–drug complexes is the plasma clearance rate. Before this parameter has been evaluated, the utility of a complexing technique cannot be established. Among the methods tested, a normal in vivo clearance rate has been reported for lipoprotein–drug complexes prepared by the modified Krieger method [89] and for lipoproteins with inorporated exogenous cholesteryl esters prepared by the lipid transfer method [100]. However, it is still far too early for a comparable evaluation of the different complexing methods and their potential practical applicability.

The field of drug delivery by lipoproteins and lipid microemulsions is still at an early stage of development. The scientific information available from cell biology, immunology, biochemistry, and physical chemistry should now form an adequate base for further progress and practical applications.

REFERENCES

1. Poznansky MJ, Juliano RL. Biological approaches to the controlled delivery of drugs: a critical review. Pharmacol Rev 1984; 36:277–325.
2. Lee FH, Hwang KM. Antibodies as specific carriers for chemotherapeutic agents. Cancer Chemother Pharmacol 1979; 3:17–24.
3. Wall ME, Abernathy GS, Carroll FI, Taylor DJ. The effect of some steroidal alkylating agents on experimental animal mammary tumor and leukemia systems. J Med Chem 1968; 12:810–818.
4. Trouet A, Deprez-de Campeneere D, de Duve C. Chemotherapy through lysosomes with DNA–daunorubicin complex. Nature 1972; 239:110–112.
5. Szeherke M, Wade R, Whisson ME. The use of macromolecules as carrier of cytotoxic groups. II. Nitrogen mustard–protein complexes. Neoplasma 1972; 19:211–215.
6. Gregoriadis G, Wills EJ, Swain CP, Tavill AS. Drug carrier potential of liposomes in cancer chemotherapy. Lancet 1974; 1:1313–1316.
7. Gregoriadis G, ed. Liposome technology, vol. 3. Boca Raton, FL:CRC Press, 1984.
8. De Duve C, de Barsy T, Poole B, Trouet A, Tulkens P, van Hoof F. Lysosomotropic agents. Biochem Biopharm 1974; 23:2495–2531.
9. Gal D, Ohashi M, MacDonald PC, Buchsbaum HJ, Simpson ER. Low density lipoprotein as a potential vehicle for chemotherapeutic agents and radionucleotides in the management of gynecologic neoplasms. Am J Obstet Gynecol 1981; 139:877–885.
10. Ho YK, Smith GS, Brown MS, Goldstein JL. Low density lipoprotein (LDL) receptor activity in human myelogenous leukemia cells. Blood 1978; 52:1099-1114.
11. Goldstein JL, Ho YK, Basu SK, Brown MS. Binding site on macrophages that mediates uptake and degradation of acetylated low-density lipoprotein producing massive cholesterol deposition. Proc Natl Acad Sci USA 1979; 76:333–337.
12. Attie AD, Pittman RC, Steinberg DS. Metabolism of native and lactosylated human low density lipoprotein: evidence for two pathways for catabolism of exogenous proteins in rat hepatocytes. Proc Natl Acad Sci USA 1980; 77:5923–5927.
13. Wu J-D, Butler J, Bailey JM. Lipid metabolism in cultured cells. XVIII. Comparative uptake of low density and high density lipoproteins by normal, hypercholesterolemic and tumor virus-transformed human fibroblasts. J Lipid Res 1979; 20:472–480.
14. Schmitz G, Robenek H, Lohmann V, Assmann G. Interaction of high density lipoproteins with cholesteryl ester-laden macrophages: biochemical and morphological characterization of cell surface receptor binding, endocytosis and resecretion of high density lipoproteins by macrophages. EMBO J 1985; 4:613–622.
15. Poste G, Bucana C, Raz A, Bugelski P, Kirsh R, Fidler IJ. Analysis of the fate of systemically administrated liposomes and implications for use in drug delivery. Cancer Res 1982; 42:1412–1422.
16. Shaw JM, Shaw KV, Schook LB. Drug delivery particles and monoclonal antibodies. In:Schook LB, ed. Monoclonal antibodies: Production, techniques and applications. New York:Marcel Dekker, 1987:285–310.

17. Lundberg B, Suominen L. Preparation of biologically active analogs of serum low density lipoprotein. J Lipid Res 1984; 25:550-558.

18. Ginsburg GS, Walsh MT, Small DM, Atkinson D. Reassembled plasma low density lipoproteins: phospholipid-cholesterol ester-apoprotein B complexes. J Biol Chem 1984; 259:6667-6673.

19. Senior J, Gregoriadis G. Stability of small unilamellar liposomes in serum and clearance from the circulation: the effect of the phospholipid and cholesterol components. Life Sci 1982; 30:2123-2136.

20. Phillips MC, Johnson WJ, Rothblat GH. Mechanisms and consequences of cellular cholesterol exchange and transfer. Biochim Biphys Acta 1987; 906:223-276.

21. Lundberg BB, Suominen LA. Physicochemical transfer of [^{3}H]cholesterol from plasma lipoproteins to cultured human fibroblasts. Biochem J 1985; 228:219-225.

22. Ekman S, Lundberg B. Transfer of lipids from microemulsions of cholesteryl ester, triglyceride, phosphatidylcholine and cholesterol to human fibroblasts in culture. Biochim Biophys Acta 1987; 921:347-355.

23. Attwood D, Florence AT. Surfactant systems: their chemistry, pharmacy and biology. New York:Chapman and Hall, 1983.

24. Prince LM, ed. Microemulsions: theory and practice. New York:Academic Press, 1977.

25. Rydhag L, Wilton J. The function of phospholipids of soybean lecithin in emulsions. JAOCS, 1981; 58:830-837.

26. Depraetere P, Florence AT, Puisieux F, Seiller M. Some properties of oil-in-water emulsions stabilized with mixed non-ionic surfactants (Brij 92 and Brij 96). Int J Pharm 1980; 5:291-304.

27. Lundberg B, Svens E, Ekman S. The hydration of phospholipids and phospholipid-cholesterol complexes. Chem Phys Lipids 1978; 22:285-292.

28. Shinoda K. The correlation between the dissolution state of nonionic surfactant and the type of dispersion stabilized with the surfactant. J Colloid Interf Sci 1967; 38:198-208.

29. Pelham LD. Rational use of intravenous fat emulsions. Am J Hosp Pharm 1981; 38:198-208.

30. Meyer CE, Francher JA, Schurr PE, Webster HD. Composition, preparation and testing of an intravenous fat emulsion. Metabolism 1957; 6:591-596.

31. Schuberth O, Wretlind O. Intravenous infusion of fat emulsions, phosphatides and emulsifying agents. Acta Chir Scand (Suppl), 1961; 278:3.

32. Rydhag L. The importance of the phase behaviour of phospholipids for emulsion stability. Fette Seifen Anstrichm 1979; 81:168-174.

33. Friberg S, Jansson P-O, Cederberg E. Surfactant association structure and emulsion stability. J Colloid Interf Sci 1976; 55:614-623.

34. Boberg J, Hakansson I. Physical and biological changes in an artificial fat emulsion during storage. J Pharm Pharmacol 1964; 16:641-646.

35. Todays Drugs, Br Med 1970; 352-353

36. Small DM. A classification of biologic lipids based upon their interaction in aqueous systems. JAOCS 1968; 45:108-119

37. Davis SS, Hansrani P. The coalescence behaviour of oil droplets stabilized by phospholipid emulsifiers. J Colloid Interf Sci 1985; 108:285-287.

38. Weltzien HU. Cytolytic and membrane-perturbing properties of lysophosphatidylcholine. Biochim Biophys Acta 1979; 559:259-287.

39. Adams DH, Joyce G, Richardson VJ, Ryman BE, Wisniewski HM. Liposome toxicity in the mouse central nervous system. J Neurol Sci 1977; 31:173–179.
40. Gabriel NE, Roberts MF. Interaction of short-chain lecithin with longchain phospholipids: characterization of vesicles that form spontaneously. Biochemistry 1986; 25:2812–2821.
41. Jeppsson RI. Plasma levels of diazepam in the dog and the rabbit after two different injection formulations, emulsion and solution. J Clin Pharm 1976; 1:181–187.
42. Linthorst JM, Clark SB, Holt PR. Triglyceride emulsification by amphipaths present in the intestinal lumen during digestion of fat. J Colloid Interf Sci 1977; 60:1–10.
43. McPherson JR Jr, McPherson JR. Experimental hyperlipemic agents: nontoxic alternative agents to Triton WR-1339. Proc Soc Exp Biol Med 1983; 172:133–134.
44. Waddell WR, Geyer RP, Olsen FR, Stare FJ. Clinical observations on the use of nonphosphatide (Pluronic) fat emulsions. Metabolism 1957; 6:815–821.
45. Lundberg B. Preparations of stable, optically clear emulsions of triolein and cholesteryl oleate by ultrasonication with egg lecithin. Chem Phys Lipids 1975; 14:260–262.
46. Schuberth O, Wretlind A. Erfahrungen über intravenöse Fettemulsionen. Arch klin Chir 1957; 287:486–489.
47. Becher P. Nonionic surface-active compounds. VII. Interfacial tensions of solutions of nonionic surface-active agents. J Colloid Interf Sci 1963; 68:665–673.
48. Jeppsson R, Schoefl GI. The ultrastructure of lipid particles in emulsions prepared with various emulsifiers. Aust J Exp Biol Medc Sci 1974; 52:697–702.
49. Jeppsson R, Groves JM, Yalabik HS. The particle size distribution of emulsions containing diazepam for intravenous use. J Clin Pharm 1976; 1:123–127.
50. Gran L, Bleie H, Jeppsson R, Maartmann-Moe H. Etomidate in Intralipid: a solution for painfree injection. Anaesthesist, 1983; 32:475–477.
51. Patton JS, Stone B, Papa C, Abramowitz R, Yalkowsky SH. Solubility of fatty acids and other hydrophobic molecules in liquid trioleoylglycerol. J Lipid Res 1984; 25:189–197.
52. Anderson BD, Rytting JH, Higuchi T. Solubility of polar organic solutes in nonaqueous systems: role of specific interaction. J Pharm Sci 1980, 69:676–680.
53. Kritchevsky D, Tepper SA. Solubility of cholesterol in various fats and oils. Proc Soc Exp Biol Med 1964; 116:104–107.
54. Firestone RA, Pisano JM, Falck JR, McPhaul MM, Krieger M. Selective delivery of cytotoxic compounds to cells by the LDL pathways. J Med Chem 1984; 27:1037–1043.
55. Mizushima Y, Hamano T, Yokoyama K. Use of a lipid emulsion as a novel carrier for corticosteroids, J Pharm Pharmacol 1982; 34:49–50.
56. Paris GY, Garmaise DL, Cimon DG, Swett L, Carter GW, Young P. Glycerides as prodrugs. 3. Synthesis and anti-inflammatory activity of (1-(p-chlorobenzoyl)-5-methoxy-2-methyl-indole-3-acetyl) glycerides (indomethacin glycerides) J Med Chem 1979; 23:9–12.
57. Vitols SG, Masquelier M, Peterson CO. Selective uptake of a toxic lipophilic anthracycline derivative by the low-density lipoprotein receptor pathway in cultured fibroblasts. J Med Chem 1985; 28:451–454.

58.   Miller KW, Small DM. Surface-to-core and interparticle equilibrium distribution of triglyceride-rich lipoprotein lipids. J Biol Chem 1983; 258:13772–13784.
59.   Lundberg B, Alajääski M. Lipoprotein models: ultrasonicated emulsions of phosphatidylcholine, cholesterol, and triolein. Acta Chem Scand 1979; 33:86–92.
60.   Ginsburg GS, Small DM, Atkinson D. Microemulsions of phospholipids and cholesteryl esters: protein-free models of low density lipoprotein. J Biol Chem 1982; 257:8216–8227.
61.   Jeppsson R. Effect of barbituric acids using an emulsion form intravenously. Acta Pharm Suecica 1972; 9:81–90.
62.   Jeppsson R. Effects of barbituric acids using an emulsion form intraperitoneally and subcutaneously. Acta Pharm Suecica 1972; 9:199–206.
63.   Jeppsson R, Ljungberg S. Intra-arterial administration of emulsion formulations containing cyclandelate and nitroglycerin. Acta Pharm Suecica 1973; 10:129–140.
64.   Winsnes M, Jeppsson R, Sjöberg B. Diazepam adsorption to infusion sets and plastic syringes. Acta Anaesth Scand 1981; 25:93–96.
65.   Noguchi T, Taniguchi K, Yoshifuji T, Muramishi S, Sezaki H. Lymphatic transport of griseofulvin in the rat and the possible factors determining the extent of lymphatic absorption. Chem Pharm Bull 1977; 25:2231–2238.
66.   Buck M, Tsukamoto T, Kvols LK, Kovach JS. Pharmacology studies of bisantrene, a poorly soluble anticancer drug, formulated in a lipid emulsion. In:Ishigami J, ed. Recent Adv Chemother, Proc Int Congr Chemother, 14th ed Anticancer Sect. 1. Tokyo:University of Tokyo Press, 1985:540–541.
67.   Kitamura M, Sugimachi K, Maekawa S, Kai H, Kuwano H, Matsuda H. Destruction of VX2 tumor in rabbits by hyperthermia plus bleomycin suspended in sesame oil. Cancer Treat Rep 1986; 70:1263–1269.
68.   Ekman S. [³H]cholesterol transfer from microemulsion particles of different sizes to human fibroblasts. Lipids 1987; 22:657–663.
69.   Lundberg B. Preparation of drug–low density lipoprotein complexes for delivery of antitumoral drugs via the low density lipoprotein pathway. Cancer Res 1987; 47:4105–4108.
70.   Sklar LA, Craig IF, Pownall HJ. Induced circular dichroism of incorporated fluorescent cholesteryl esters and polar lipids as a probe of low density lipoprotein structure and melting. J Biol Chem 1981; 256:4286–4292.
71.   Via DP, Craig IF, Jacobs GW, Van Vinkle WB, Charlton SC, Gotto AM Jr, Smith LC. Cholesteryl ester-rich microemulsions: stable protein-free analogs of low density lipoproteins. J Lipid Res 1982; 23:570–576.
72.   Watanabe A, Higashitsuji K, Nishizawa K. Studies on electrocapillary emulsification. J Colloid Interfac Sci 1978; 64:278–289.
73.   Whitehill D. Multiple emulsions and their future uses. Chem Drug 1980; 213:130–135.
74.   Florence AT, Whitehill DW. The formulation and stability of multiple emulsions. Int J Pharm 1982; 11:277–308.
75.   Goldstein JL, Brown MS. The low-density lipoprotein pathway and its relation to atherosclerosis. Ann Rev Biochem 1977; 46:897–930.
76.   Vitols S, Gahrton G, öst Å, Peterson C. Elevated low density lipoprotein receptor activity in leukemic cells with monocytic differentiation. Blood 1984; 63:1186–1193.

77. Shepherd J, Packard CJ. Receptor-independent low-density lipoprotein catabolism. In:Albers JJ, Segrest JP, eds. Methods of enzymology vol. 129. New York:Academic Press, 1986:566–599

78. Seki J, Okita A, Watanabe M, Nakagawa T, Honda K, Tatewaki N, Sugiyama M. Plasma lipoproteins as drug carriers: pharmacological activity and disposition of the complex of ß-sitosteryl-ß-D-glucopyranoside with plasma lipoproteins. J Pharm Sci 1985; 74:1259–1264.

79. Poumay Y, Ronveaux-Dupal M-F. Rapid preparative isolation of concentrated low density lipoproteins and of lipoprotein-deficient serum using vertical rotor gradient ultracentrifugation. J Lipid Res 1985; 26:1476–1480.

80. Patsch JK, Sailer S, Kostner G, Sandhofer F, Holasek A, Braunsteiner H. Separation of the main lipoprotein density classes from human plasma by rate-zonal ultracentrifugation. J Lipid Res 1974; 15:356–366.

81. Kannel WB, Castelli WP, Gordon T. Cholesterol in the predicition of aterosclerotic disease. Ann Intern Med 1979; 90:85–91.

82. Rudel LL, Marzetta CA, Johnson FL. Separation and analysis of lipoproteins by gel filtration. In:Albers JJ, Segrest JP, eds. Methods in enzymology, vol. 129. New York:Academic Press, 1986:45–47.

83. Vitols S. Low density lipoprotein receptors in leukemia: diagnostic and therapeutic implications. Ph.D. Thesis, Karolinska Institutet, Stockholm, 1985.

84. Krieger M, Brown, MS, Faust JR, Goldstein JL. Replacement of exogenous cholesteryl esters of low density lipoprotein with endogenous cholesteryl linoleate: reconstitution of a biologically active lipoprotein particle. J Bio Chem 1978; 253:4093–4101.

85. Krieger M, Smith LC, Anderson RGW, Goldstein JL, Kao YJ, Pownall HJ, Gotto AM Jr, Brown MS. Reconstituted low density lipoprotein: a vehicle for the delivery of hydrophobic fluorescent probes to cells. J Supramol Struct 1979; 10:467–478.

86. Krieger M, Goldstein JL, Brown MS. Receptor-mediated uptake of low density lipoprotein reconstituted with 25-hydroxy-cholesteryl oleate suppresses 3-hydroxy-3-methylglutaryl–coenzyme A reductase and inhibits growth of human fibroblasts. Proc Natl Acad Sc USA 1978; 75:5052–5056.

87. Firestone RA, Pisano JM, Falck JR, McPhaul MM, Krieger M. Selective delivery of cytotoxic compounds to cells by the LDL pathway. J Med Chem 1984; 27:1037–1043.

88. Mosley ST, Goldstein JL, Brown MS, Falck JR, Anderson RGW. Targeted killing of cultured cells by receptor-dependent photosensitization. Proc Natl Acad Sci USA 1981; 78:5717–5721.

89. Masquelier M, Vitols S, Peterson C. Low density lipoprotein (LDL) as a carrier of antitumoral drugs: in vivo fate of drug-LDL complexes in mice. Cancer Res 1986; 46:3842–3847.

90. Shaw JM, Shaw KV, Yanovich S, Iwanik M, Futch WS, Rosowsky A, Schook LB. Delivery of lipophilic drugs using lipoproteins. Ann NY Acad Sci 1987; 507:252–271.

91. Osborne JC Jr. Delipidation of plasma lipoproteins. In:Segrest JP, Albers JJ, eds. Methods of enzymology, vol. 128. New York:Academic Press, 1986:213–222.

92. Glass C, Pittman RC, Civen M, Steinberg D. Uptake of high-density lipoprotein-associated apoprotein A-I and cholesterol esters by 16 tissues of the rat *in vivo* and by adrenal cells and hepotocytes *in vivo*. J Biol Chem 1985; 260:744–750.
93. Pittman RC, Glass CK, Atkinson D, Small DM. Synthetic high density lipoprotein particles: application to studies of the apoprotein specificity for selective uptake of cholesterol esters. J Biol Chem 1987; 262:2435–2442.
94. Iwanik MJ, Shaw KV, Ledwith BJ, Yanovich S, Shaw JM. Preparation and interaction of a low-density lipoprotein:daunomycin complex with P388 leukemic cells. Cancer Res 1984; 44:1206–1215.
95. Yanovich S, Preston L, Shaw JM. Characteristics of uptake and cytotoxicity of a low-density lipoprotein–daunomycin complex in P388 leukemic cells. Cancer Res 1984; 44:3377–3382.
96. Rudling MJ, Collins VP, Peterson CO. Delivery of aclacinomycin A to human glioma cells *in vitro* by the low-density lipoprotein pathway. Cancer Res 1983; 43:4600–4605.
97. Candide C, Morliere P, Maziere JC, Goldstein S, Santus R, Dubertret L, Reyftmann JP, Polonovski J. In vitro interaction of the photoactive anticancer porphyrin derivative photofrin II with low density lipoprotein, and its delivery to cultured human fibroblasts. FEBS Lett 1986; 207:133–138.
98. Fielding CJ. Validation of a procedure for exogenous isotopic labeling of lipoprotein triglyceride with radioactive triolein. Biochim Biophys Acta 1979; 573:255–265.
99. Tall AR. Plasma lipid transfer proteins. J Lipid Res 1986; 27:361–367.
100. Craig IF, Via DP, Sherrill BC, Sklar LA, Mantulin WW, Gotto AM Jr, Smith LC. Incorporation of defined cholesteryl esters into lipoproteins using cholesteryl ester-rich microemulsions. J Biol Chem 1982; 257:330–335.
101. Blomhoff R, Drevon CA, Eskild W, Helgerud P, Norum KR, Berg T. Clearance of acetyl low density lipoprotein by rat liver endothelial cells: implications for hepatic cholesterol metabolism. J Biol Chem 1984; 259:8898–8903.
102. Badr M, Kodali DR, Redgrave TG. Effects of lipid composition on the association of plasma proteins with lipid emulsions. J Colloid Interf Sci 1986; 113:414–420.
103. Oswald B, Quarfordt S. Effect of apo E on triglyceride emulsion interaction with hepatocyte and hepatoma G2 cells. J Lipid Res 1987; 28:798–809.
104. Atkinson D, Small DM. Recombinant lipoproteins: implications for structure and assembly of native lipoproteins, Ann Rev Biophys Chem 1986; 15:403–456.
105. Parks JS, Martin JA, Johnson FL, Rudel LL. Fusion of low density lipoproteins with cholesterol ester–phospholipid microemulsions: prevention of particle fusion by apolipoprotein A-I. J Biol Chem 1985; 260:3155–3163.
106. Halbert GW, Stuart JFB, Florence AT. A low density lipoprotein–methotrexate covalent complex and its activity against L1210 cells in vitro. Cancer Chemother Pharmacol 1985; 15:223–227.
107. Edelstein C, Scanu AM. High-performance liquid chromatography of apolipoproteins. In:Segrest JR, Albers JJ, eds. Methods of enzymology, vol. 128. New York:Academic Press, 1986:339–354.
108. Cordle SR, Clegg RA, Yeaman SJ. Purification and characterization of bovine lipoproteins: resolution of high density and low density lipoproteins using heparin-Sepharose chromatography. J Lipid Res 1985; 26:721–725.

109. Lewis LA, Opplt JJ, eds. CRC Handbook of electrophoresis, vol. 1. Lipoproteins: basic principles and concepts. Boca Raton, FL:CRC Press, 1980.
110. Deckelbaum RJ, Shipley GG, Small DM. Structure and interactions of lipids in human plasma low density lipoproteins. J Biol Chem 1977; 252:744–754.
111. Lundberg B. Thermal properties of systems containing cholesteryl esters and triglycerides. Acta Chem Scand 1976; 30:150–156.
112. Hamilton JA, Morrisett JD. Nuclear magnetic resonance studies of lipoproteins. In:Segrest JR, Albers JJ, eds. Methods in enzymology; vol. 128. New York:Academic Press, 1986:472–515.
113. Krieger M, Peterson J, Goldstein JL, Brown MS. Mobility of apolar lipids of reconstituted low density lipoprotein as monitored by electron spin resonance spectroscopy. J Biol Chem 1980; 255:3330–3333.
114. Via DP, Smith LC. Fluorescent labeling of lipoproteins. In:Albers JJ, Segrest JP, eds. Methods of enzymology, vol. 129, New York:Academic Press, 1986:848–857.
115. Sklar LA, Craig IP, Pownall JH. Induced circular dichroism of incorporated fluorescent cholesteryl esters and polar lipids as a probe of human serum low density lipoprotein structure and melting. J Biol Chem 1981; 256:4286–4292.
116. Jonas A. Microviscosity of lipid domains in human serum lipoproteins. Biochim Biophys Acta 1977; 486:10–22.
117. Chen GC, Kane JP. Secondary structure in very low density and intermediate density lipoproteins of human serum. J Lipid Res 1979; 20:481–488.
118. Schonfeld G, Krul ES. Immunologic approaches to lipoprotein structure. J Lipid Res 1986; 27:583–601.
119. Basu SK, Goldstein JL, Anderson RGW, Brown MS. Degradation of cationized low density lipoprotein and regulation of cholesterol metabolism in homozygous familial hypercholesterolemia fibroblasts. Proc Natl Acad Sci USA 1976; 73:3178–3182.
120. Smith LC, Benson DM, Gotto AM Jr, Bryan J. Digital imaging fluorescence microscopy. In:Segrest JP, Albers JJ, eds. Methods in enzymology, vol. 129. New York:Academic Press, 1986:857–873.

4

Lipoproteins and Microemulsions as Carriers of Therapeutic and Chemical Agents

Alexander T. Florence
University of London, London, England

Gavin W. Halbert
University of Strathclyde, Glasgow, Scotland

I. INTRODUCTION

Earlier chapters discussed the desirability of achieving specificity in drug action, which has been recognized since the time of Ehrlich [1]. Drug targeting by means other than chemical has been attempted by using the characteristics of a variety of colloidal carriers rather than the drug to direct the active therapeutic agent to its site of action. Apart from those carrier systems that have, for example, attached to their surface monoclonal antibodies which target to tumor antigens, most carrier systems that have been investigated have been passive rather than active systems, relying on physical diversion in the body, such as uptake by the reticuloendothelial system or entrapment in the capillary beds of organs like the lung [2].

Although in most cases the intended site of action of a drug is anatomically separate from the site involved in toxicity, complete abolition of side effects can be achieved only if the carrier is totally target specific and if the drug does not leach out from the carrier during its passage to the target site. Few if any existing carriers fulfill these specifications. Our own work with albumin and other protein microspheres has demonstrated some of the problems not only in relation to the sometimes limited capacity of the carrier matrix for individual therapeutic agents but also in the ability of much of the load to leach out fairly rapidly in vivo [3,4].

The advantage of a naturally occurring, and hence fully biocompatible system with natural targeting abilities have already been delineated in this book. The advantages of the lipoproteins and low density lipoproteins (LDLs) in particular have been made clear. The purpose of this chapter is to examine the potential of lipoprotein particles (in particular LDL) as carriers for therapeutic agents. We examine not so much the techniques of incorporation of drugs with these particles, a topic covered in the Chapter 3, but the characteristics of these natural carriers and the drugs themselves which determine the uptake or solubilization, the capacity of the lipoprotein particles for drugs, and drug release mechanisms. In examining the literature it is obvious that the development of rules is difficult, since few systematic studies of lipoprotein uptake of homologous series of xenobiotics have been published. In a related field, the micellar solubilization of water-insoluble drugs, much more extensive studies have been carried out, yet a priori estimation of the extent of uptake is still a difficult task [5]; thus only general guidelines can be drawn here.

When using synthetic or semisynthetic colloidal drug carriers, modifications can be made to the system to optimize carrying capacity, biodegradability, and perhaps targeting potential. It is unlikely that such approaches can be made with lipoprotein particles, since the equilibrium state of the particle will be perturbed. In addition, we have to examine the extent to which drug incorporation alters the natural state of the particles. It is, of course, this obvious limitation of these natural carrier systems whose equilibrium structure can be perturbed by additives that has led to the search for synthetic alternatives. Because of the similarity of the LDL particle to an emulsion (of cholesteryl esters) stabilized by a mixed cholesterol–phospholipid monolayer, microemulsions of roughly equivalent particle dimensions have been explored as synthetic LDL analogs. This approach is discussed at the end of this chapter.

Clues as to the types of drug and other active moieties that can be incorporated into lipoproteins can be obtained from the natural functions of the lipoprotein particles and also from the interactions observed between administered drugs and other xenobiotics absorbed into the bloodstream and various lipoprotein particles. The primary function of serum lipoproteins is the transport of cholesterol, triacyl glycerols, and phospholipids [6]. The arrangement of the lipid and protein in LDL is similar to that in a microemulsion in which the polar lipids (phospholipids and cholesterol) solubilize or rather emulsify and stabilize the liquid core of water-soluble nonpolar lipids [7]. In the native particle the apoprotein is so arranged that approximately half of the molecule is exposed to the aqueous environment and the remainder resides in the polar lipid coat. Any disruption to this "coat," particularly through interaction with amphipathic molecules, might well alter the conformation and binding capacity of the apoprotein. The effects that additives such as drug molecules have on the function of the particle is to some extent dependent on changes in the relationship between the apoprotein and the particle, but they may be vital to the success of targeting.

In the first part of this chapter we examine some of the reports of the interaction of lipoproteins with compounds that have been absorbed into the blood. These reports have tended to be selective, concentrating, for example, on potential carcinogens or toxic molecules and how uptake into the lipoprotein might affect their distribution and metabolism.

II. BINDING OF BIOLOGICALLY ACTIVE MOLECULES TO LIPOPROTEINS IN VIVO

Drugs and other foreign molecules, such as vitamins, hormones, toxins, and pesticides, following absorption may interact with serum lipoproteins. There are at least four potential sites for interaction, excluding covalent interactions (Fig. 1), which have analogs in the sites of uptake into or onto micellar systems(s):

1. Adsorption onto the surface either to the protein moeity or to phospholipid headgroups.
2. Uptake into the phospholipid bilayer.
3. Uptake into the interface between the lipid core and the phospholipid bilayer.
4. Uptake into the lipid core.

Adsorption may be mediated by hydrophobic or electrostatic interactions. The principal hydrophobic interaction will be with the apoprotein, whereas electro-

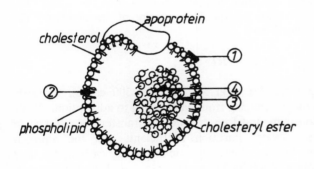

Figure 1 Sites of incorporation of drugs and other active (and inactive) moieties onto or into low density lipoprotein particles: (1) surface attachment; (2) penetration into the phospholipid–cholesterol monolayer; (3) solubilization at the cholesteryl ester–monolayer interface; or (4) solubilization in the lipid interior of the particle.

static interactions may occur between the phospholipid charged groups and oppositely charged solutes [2–4] as the result of partitioning between the largely aqueous external phase and the particle, the ultimate destination (i.e., depth of penetration) of the solute being determined by its polarity. Amphipathic solutes are likely to be sited in the phospholipid monolayer or in the lipid interior, provided they are miscible with the cholesteryl ester triglyceride core. Nonpolar molecules will partition into the lipid core, while lipophilic, polar molecules will straddle the core and phospholipid region.

In the following section we survey studies of interactions of absorbed agents with blood lipoproteins, exploring the nature of such interactions where possible.

A. Insecticide–lipoprotein Interactions

As lipoproteins have been implicated in the transport of chlorinated hydrocarbon insecticides in human serum [8], Maliwal and Guthrie [9] have investigated the interactions between a range of insecticides and high density lipoprotein (HDL) and LDL in vitro. DDT [1,1,-trichloro-2,2-bis(p-chlorophenyl)ethane]

DDT

and dieldrin were two of a group which displayed a high affinity for the lipoproteins, with association constants $(K) >> 10^5 M^{-1}$. The affinities of these two compounds and lindane, diazinon, and parathion for both HDL and LDL were similar. Measured values of the free energy of transfer into the lipoprotein particle, calculated from

$$\Delta G^{\circ}_{trans} = RT \cdot \ln K$$

revealed values ranging from 7.36 kcal for DDT to 300 cal for aldicarb, suggesting modes of interaction ranging from partitioning to adsorption. Table 1 lists the interaction constants and free energies of transfer for a range of insecticides and quotes oil/water partition coefficients and the aqueous solubilities of chlorinated hydrocarbons, organophosphates, carbamates, and nicotine. A good correlation is obtained between ΔG°_{trans} and the logarithm of the aqueous solubility of chlorinated hydrocarbons and carbamates, as can be seen in Fig. 2, even though the solubilities quoted are taken from a variety of sources. The data for the organophosphates deviate somewhat from the curve; this may be due to the provenance of the solubility measurements or it

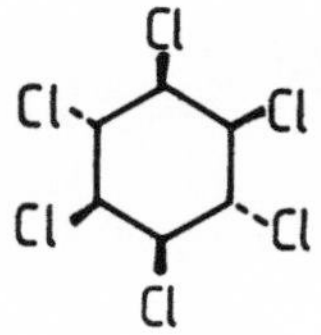

lindane

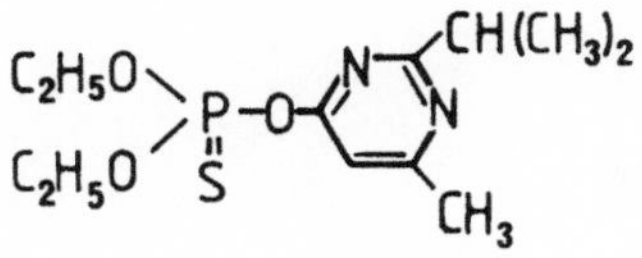

diazinon

Table 1 Distribution Constants (K) of Insecticides for LDL

Insecticide	Water solubility[a] (μg/ml)	K (M^{-1})	$\Delta G°_{trans}$ (kcal)	log $K°_w$
Chlorinated hydrocarbons				
DDT	0.0026	2.8 x 10^5	7.36	5.75
		3.05 x 10^5		
Dieldrin	0.07	1.35 x 10^5	6.94	4.41
Lindane	7	4.70 x 10^3	4.97	3.40
		5.90 x 10^3		
Organophosphates				
Parathion	24	3.9 x 10^3	4.86	3.40
		5.0 x 10^3		
Diazinon	40	2.2 x 10^3	4.52	—
		1.5 x 10^3		
Carbamates				
Carbaryl	40	6.4 x 10^1	2.45	2.36
		6.6 x 10^1		
Carbafuran	400	0.95 x 10^1	1.29	—
		1.10 x 10^1		
Aldicarb	6000	0.17 x 10^1	0.311	—
		0.25 x 10^1		
Others				
Nicotine	Miscible	0.21 x 10^1	0.44	—
		0.14 x 10^1		

[a] Solubility data were taken from various publications and are not to be considered absolute.
Source: Ref. 9.

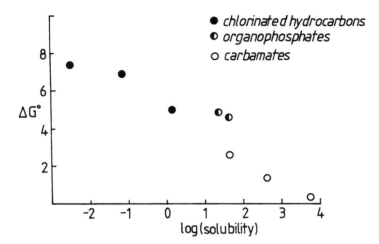

Figure 2 The free energy of transfer $\Delta G^\circ{}_{trans}$ (kcal) of a range of insecticides between an aqueous (0.01 M Tris-HC1) buffer at pH 7.0 and low density lipoprotein particles, as a function of the aqueous solubility of the solutes, drawn from the data of Maliwal and Guthrie [9]. In spite of the solubility data having been culled from several sources a clear correlation is seen.

may be a real effect due to structural differences. The trend is a clear one of decreasing free energy of transfer with increasing aqueous solubility.

The increase in $\Delta G^\circ{}_{trans}$ (Fig. 3) with increase in parachor (essentially molecular volume) [10] of the chlorinated hydrocarbons may simply reflect the increasing hydrophobicity of the series.

In discussing their results on insecticide uptake by LDL, Skalsky and Guthrie [8] wrote that, as a result of the findings of Brown and Goldstein [11] on the receptor-mediated control of cholesterol metabolism involving LDL, "xenobiotics attached to lipoprotein could obtain direct entry to a cell, effectively by-passing the cell membrane." They found no specific binding sites in their study of solute–lipoprotein interactions, which they claimed to be largely hydrophobic in nature, but they did not rule out the possibility of such sites.

The interaction of 2,4,5,2',4',5'-hexachlorobiphenyl (6-CB) with very low density lipoprotein (VLDL) was studied by Gallenberg and Voldicnik [12].

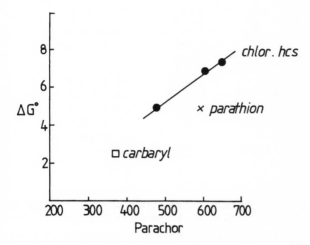

Figure 3 The free energy of transfer ΔG°_{trans} (kcal) for chlorinated hydrocarbons, parathion, and carbaryl as a function of the parachor, drawn from the free energy data of Maliwal and Guthrie [9] and parachor values from Dao, Lavy, and Dragun [10].

The 6-CB–VLDL complex was injected into mice and the distribution of the biphenyl was compared following an injection in a micellar surfactant (Emulphor) vehicle. It was found that 6-CB also interacts with LDL and HDL, but to lesser extent. Such differences in uptake by the different classes of lipoprotein might also provide clues as to factors determining uptake. It is obvious that this "natural" uptake of xenobiotic molecules is determined by partitioning or adsorption. The more drastic treatments possible in vitro, in which core esters are removed and replaced by lipophilic molecules, pose additional difficulties in interpretation of uptake mechanisms.

B. Other Agents

Binding sites were postulated in Nilsen's study [13] of serum quinidine interaction with chylomicrons, VLDL, LDL, HDL, and albumin (Table 2). Binding was affected by buffer composition, as one might expect in in vitro experiments, principally because pH will affect the ionization and thus the effective partitioning of the drug. There appeared to be one binding site on LDL which exhibited a cooperativity at lower concentrations of quinidine. The binding ratios (bound/free drug) range from 0.041 for chylomicrons to 0.955 for LDL.

The partitioning of a small series of lipophilic anticancer drugs was also investigated by Weinkam et al. [14], principally with a view to obtaining data on the effect of lipoprotein particles on the decomposition rate of chloroethyl nitrosoureas in serum. The protection offered to labile molecules by their

Table 2 Binding of Quinidine to Lipoprotein Particles and to Albumin

Lipoprotein/protein	Concentration (g/l)	Binding ratio (bound/free)	Number of binding sites per molecule
Chylomicrons	3.0	0.041	
VLDL	5.2	1.177	399.3 ± 27.9
LDL	9.9	0.955	128.5 ± 31.5
HDL	5.9	0.410	18.5 ± 3.6
Albumin	22.3	0.764	0.011 (lst site)
			1.29 (2nd site)

Source: Ref. 13.

solubilization in the interior of particles such as LDL is clear; it is akin to the protection offered by micellar solubilization. Weinkam et al. discuss the effect of size and steric factors in the uptake process, recognizing the ordered nature of the cholesteryl ester interior of LDL and the potential, particularly in HDL particles, whose lipid core is small, for the capacity of the core for large lipophilic molecules to be limited. In support of this concept, it has been found that large HDL particles carry more ß-carotene per unit weight than small particles [15]. Doubling the particle size increased the concentration of ß-carotene carried by a factor of 4. Both VLDL and LDL particles have been found to accommodate only four to six molecules of ß-carotene, which is a low number compared to other solutes.

ß-carotene

It is apparent, then, that factors other than lipophilicity influence the interaction of drugs with lipoproteins. For example, despite the highly lipophilic character of the long chain polyisoprenoid alcohol dolichol, it reacts very little with VLDL or LDL [16] but principally with HDL. The organization of the cholesteryl esters in the lipoprotein core is, to some extent, dictated by particle size. Whether the core is structured, random, or partially structured will affect the uptake of larger molecules, but other factors might well intrude.

The reasonable correlation obtained between lipoprotein/serum partition data and octanol/water partition coefficients (Table 3) for a limited range of drugs (BNCU, CCNU, and MeCCNU [14]) is echoed in the data for benzo(a)-pyrene and two derivatives [17]. In this case lipophilicity was measured by Rf values on a silica gel plate. Table 4 shows the molar uptake of benzo(a)-pyrene, its 3-hydroxy analog, and benzo(a)pyrene-7,8-dihydrodiol by VLDL, LDL, and HDL. The striking data are the values of molar uptake when normalized for lipoprotein lipid *volume* (in μmol/cc, Table 4), which suggest

Benzo(a)pyrene

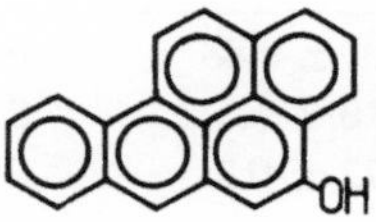

3 Hydroxybenzo(a)pyrene

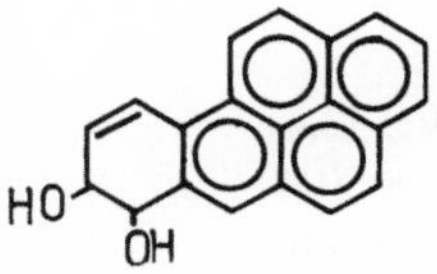

Benzo(a)pyrene 7,8 dihydrodiol

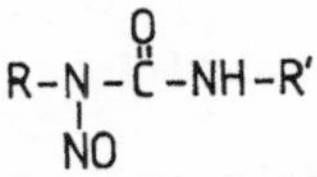

$$R-N-\overset{\overset{O}{\|}}{C}-NH-R'$$
$$\underset{NO}{|}$$

nitrosoureas

that solubility in the lipid phase is a primary factor determining uptake. The 3-hydroxybenzo(a)pyrene is solubilized to a greater extent in HDL than would have been anticipated and perhaps, like dolichol, this is a result of a specific interaction with HDL apolipoproteins or with the surface monolayer. Figure 4 shows a clear correlation between the number of molecules per LDL particle and the calculated lipophilicity (log$P$). The values of log$P$ were calculated (by the present authors) from Hansch coefficients, but the values for benzo(a)-pyrene agree with reported octanol/water partition coefficient data.

The suggestion was made that amphipathic drug molecules would enter the surface monolayer rather than the interior of lipoproteins as free fatty acids do. The binding of two such molecules, chlorpromazine and imipramine, to lipoproteins was studied in 1975 [18]. The surface activity of the phenothia-zines and other tricyclics [19] has led to a proliferation of binding studies of the drugs to biological molecules and macromolecules. Not surprisingly, Bickel

Table 3 Interaction of Chloroethylnitrosoureas with Serum Lipoproteins

Compound	$\underline{R}$	$\underline{R}'$	Octanol/water partition coefficient	Lipid/serum partition coefficient
MeCCNU	$ClCH_2CH_2-$	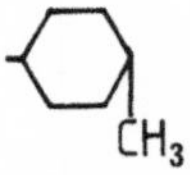 CH_3	2000	300
CCNU	$ClCH_2CH_2-$	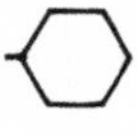	670	200
BCNU	$ClCH_2CH_2-$	$-CH_2CH_2Cl$	37	50

Source: Ref. 14.

Table 4 Uptake of Benzo(a)pyrene and Analogs by Lipoprotein

Compound	VLDL, 30–75 nm diameter	LDL, 17–25 nm diameter	HDL, 7–12 nm diameter	Rf value
Benzo(a)pyrene	2022 ± 370	232 ± 38	9.19 ± 0.88	0.695
3-Hydroxybenzo(a)pyrene	1878 ± 942	188 ± 54	16.6 ± 5.5	0.445
Benzo(a)pyrene-7,8-hydrodiol	225 ± 65	53 ± 18	1.72 ± 0.3	0.280
Molar uptake per lipoprotein lipid volume ($\mu mol/cc$)				
Benzo(a)pyrene	108 ± 19.7	114 ± 21.4	140 ± 13.4	
3-Hydroxybenzo(a)pyrene	100 ± 50.2	106 ± 15.2	254 ± 80.9	
Benzo(a)pyrene-7,8-hydrodiol	12 ± 3	29.9 ± 10.2	26.3 ± 4.6	

Source: Ref. 17.

[18] found that the affinity and capacity of lipoproteins for these agents was at least as high as that of albumin, and that albumin binding was equally inhibited by the presence of HDL, LDL, and VLDL and chylomicrons. Salicyclic acid, in contrast, did not bind to either erythrocytes or lipoproteins. Nortriptyline binding to lipoproteins and imipramine binding has also been reported by, respectively, Tillement et al. [20] and Danon and Chen [21], who also found that reserpine bound, albeit with large variability, to lipoproteins [22]. The binding correlated significantly with the triglyceride concentration, which is the major component of VLDL, and with the total lipid present in plasma.

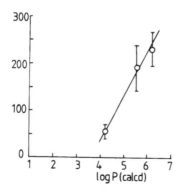

Figure 4 The number of molecules of benzo(a)pyrene and analogs (structures shown on p. 149) taken up by LDL as a function of calculated lipophilicity (plotted as log$P$).

Sgoutas et al. [23] found that 21–28% of cyclosporin A (CsA), a nonpolar cyclic undecapeptide of molecular weight 1203 and a low aqueous solubility (0.04 mg/ml; log$P$ octanol/water = 2.3) was associated to the extent of 21–28% with LDL and 33–43% with HDL in transplant patients. Chloroform/buffer and hexane/buffer partition coefficients of CsA are similar to those for progesterone.

These results can be compared with those of Lemaire and Tillement [24], who found that at therapeutic concentrations CsA was essentially taken up by erythrocytes with 21% bound to lipoprotein. The drug is normally available clinically as a solubilized preparation in a micellar solution of a nonionic surfactant at a concentration of 50 mg/ml in 32.9% alcohol. With its molecular weight of 1203 it is one of the larger molecules shown to associate with LDL.

Another lipophilic and poorly soluble molecule that associates with LDL is tetrahydrocannabinol (Δ^9THC) [25,26], which is bound to lipoproteins to the extent of 80–95%.

While the partition coefficients of many of the compounds discussed thus far may be as high as 1×10^6 [17], Rudman et al. [27] have suggested that lipophilic solutes with chloroform/buffer partition coefficients greater than only 11 dissolve in the lipid components of lipoproteins. They cite as examples diphenylhydantoin, bishydroxycoumarin, estradiol, testosterone, pentobarbital, and aldosterone.

III. LIPOPROTEINS AS CARRIERS OF EXTRINSIC PHARMACOLOGICAL AGENTS

Since the suggestions of the use of lipoproteins as carriers for drug molecules numerous attempts have been made to form lipoprotein–drug complexes. The approaches are dealt with in turn, including discussions of the surface attachment of radiolabels and surface chemical modifications, which give clues as to potential modes of drug attachment and incorporation.

A. Covalent Surface Attachment

The surface of the lipoproteins can most easily be regarded as a mosaic of phospholipid, free cholesterol, and apolipoprotein arranged in a monolayer around the hydrophobic internal core [28]. However, treatment of LDL with trypsin or pronase results in the loss of up to 30% of the total protein content [29], suggesting that part of the apoprotein is exposed to the external aqueous environment, a feature which allows molecules to be covalently bound to the exterior of the particle. The most common materials carried externally are radiolabels, which may be attached by a variety of techniques (see Chapter 8). Iodine has been bonded by the iodine monochloride method [30,31]. Other techniques have been developed [32] and a variety of radiolabels, including

[¹⁴C]sucrose [33], Tc⁹⁹ᵐ [34], and [¹⁴C]acetyl-LDL [35], are discussed in Chapter 8. This approach has not yet been used to allow the delivery of therapeutic doses of a radioactive substance to the target site, but it has the potential to be used this way.

Low molecular weight reactive molecules have also been reacted with the exterior of lipoproteins. These have mainly been used experimentally to block specific amino acid residues on the apoprotein and have ranged from acetyl groups [36,37], methyl groups [38], and gold [39] through to reaction with cyclohexanedione [40].

Variations on these themes have been used to attach drugs to the external surface of lipoproteins for site-specific delivery, but thus far work has been performed only on LDL. In the authors' laboratory [41] methotrexate (MTX) has been covalently bound to LDL apoprotein using a water-soluble crosslinking agent to give a complex that retained a degree of in vitro cytotoxic activity. The LDL complex showed a reduced activity compared to free MTX, a result found with other drug–protein conjugates [42] that employ direct attachment techniques. The use of an intermediate coupling agent also produced crosslinked LDL particles, but the effectiveness of the approach was reduced by the inherent tendency of the LDL to aggregate [43], a phenomenon reducing the quantity of drug attached to about 8 molecules MTX per LDL particle. Daunorubicin has also been attached to the surface using an activated spacer group that prevents the crosslinking of the particles [44], a technique which allows higher loading of drug to achieve a level of around 80 molecules per LDL particle. Conjugates of LDL produced by this method containing approximately 50 molecules of drug exhibited the same fate in vivo as native LDL when injected intravenously into mice [45], demonstrating that the complex is a suitable targeting system. Due to the ease of hydrolysis of the spacer molecule after cellular uptake of the complex [45], higher activity should be displayed.

The surface attachment of molecules has certain disadvantages. In the case of LDL, the LDL particles generally show a greater tendency to aggregate, possibly because of the loss of surface charge. The most serious problem is the potential interference with the receptor specificity of the system if drugs are bound to the apoprotein. It has long been recognized that covalent binding of small molecules to apoproteins alters or even destroys receptor binding [46], although this is dependent on the type and place of attachment of the ligand. The effect of daunorubicin linkage is dependent on the quantity of drug attached [44,47], but this limits the carrying capacity of the particle.

One of the proposed methods to circumvent this problem is to attach to the surface of the receptor protein molecule a nonspecific "piggyback" carrier containing the drug. This has been attempted by linking, through a specific coupling technique which maintains receptor specificity, small unilamellar liposomes to the LDL surface [47]. The drug may be incorporated into the liposomes and then attached to the LDL, allowing larger quantities of drug to be carried.   In this case the activity of the liposome-entrapped drug is increased by the receptor specificity of the LDL.   These liposome–LDL complexes have not been tested in vivo and it remains to be seen if the liposomes will be taken up by the reticuloendothelial system, thus limiting the utility of this approach.

B. Electrostatic and Noncovalent Adsorption to the Surface of Lipoprotein Particles

Using model probes such as 1-anilino-8-naphthalene sulfate (ANS) Ghosh and co workers [48] found only one type of binding site on the surface of LDL,

ANS

about 200 ANS molecules associated with each LDL particle.   At ANS concentrations about 10 mM the LDL structure was disrupted, delipidation taking place at around 0.1 M.   It is calculated that there are around 550 phospholipid groups on each LDL particle, so fewer than half appear to be associated with an ANS molecule.   Others [49] have suggested there are 362 potential binding sites for surface-attached molecules.   Electrostatic adsorption to amino acid residues on the apoprotein molecule or to phospholipids will be pH dependent.

Particle electrophoresis can distinguish sites of attachment.   Lipopolysaccharides (LPSs) form complexes with HDL after intravenous injection into rats, with resultant changes in the characteristic electrophoretic mobility of the particles suggesting surface adsorption.   The complexation takes place rapidly and the complex persists while the LPS circulates in the blood [50].   Electrophoretic studies of clioquinol (5-chloro-7-iodo-8-quinolinol) binding to lipoprotein [51] indicated that nonspecific solubilization rather than binding of this agent occurred, which is consistent with its structure.   When the amount of clioquinol exceeds the binding capacity of albumin, it is clear that VLDL, LDL, and HDL serve as auxiliary carriers for the drug: the lipoprotein binding *affinities* were less but the binding *capacity* was greater.

Clioquinol

C. Partitioning of Drugs into the Surface Monolayer

The external phospholipid monolayer is capable of accepting suitable amphipathic molecules that have the ability to reside at interfaces [5]. However, this approach is limited by the lack of suitable drug molecules. Many therapeutic molecules are amphipathic, possessing defined regions of hydrophobicity and hydrophilicity—phenothiazines, many local anesthetics, antihistamines, and other molecules [5]. The hydrophile–lipophile balance of many of these ionic species, on which surface activity depends, will be dependent on ionization. The unionized forms of the phenothiazine derivatives, for example, will be very hydrophobic and will partition into the lipid core. Partially ionized they will be likely to intercalate with the surface monolayer. Deliberate attempts can be made to achieve such effects. Cytotoxic analogs of phospholipids (ara-CDP-L-1,2-diacylglycerols) have been synthesized [52] and tested for their ability to be incorporated into HDL and LDL particles [53]. It was found that a complex of the phospholipid nucleoside formed rapidly with both human HDL and LDL; in the case of HDL, the complex formed was dependent on the interaction of the phospholipids in the particle and the nucleoside. The extent of interaction was related to the gel–liquid crystalline phase transition of the phospholipid nucleoside. No data are currently available on the biological activity of these complexes.

1, $R = - (CH_2)_{14}CH_3$
2, $- (CH_2)_{16}CH_3$
3 $- (CH_2)_7CH=CH-(CH_2)_7CH_3$

ara CDP-L-1,2-diacylglycerols [Ref. 52]

Related compounds prepared by Shaw and his co workers [54] include hexadecyl ara-CMP.

Hexadecyl ara CMP *[Ref. 54]*

The incorporation of some naturally occurring amphipathic agents has also been tested with studies performed on the interaction of ß-sitosteryl-*D*-glucopyranoside with a range of lipoproteins [55]. This material is an abundant

ß-Sitosteryl - ß-D-glucopyranoside

plant sterylglucoside with a variety of pharmacological effects including activity against P388 leukemic cells. Oral administration of the compound results in the appearance of around 2% of the administered dose in lymph chylomicrons. If the drug is incubated in vitro with plasma, 94% of the compound is associated with chylomicrons, VLDL, or LDL and produces similar pharmacological effects but HDL-associated drug has a reduced effect, even though there is a considerable interchange of the drug between the lipoprotein fractions after injection (see Chapter 9).

Molecules that are taken up into the monolayer also have the possibility of altering the distribution of the particle by (a) altering the surface charge or (b) adding new receptors to the surface. This technique has been used with the synthesis of a tris(galactosyloxymethyl)aminomethane-terminated cholesterol (Tris-Gal-Chol) [56], which interacts with galactose receptors. This molecule, which is soluble in water, associates in solution (in vitro) with liposomes and with LDL to form a complex, which on intravenous injection is rapidly taken up by the liver at a rate dependent on the quantity of Tris-Gal-Chol incorp-

GalOCH₂
│ H O   HO   O
GalOCH₂-C-N-C-CH₂N-C-(CH₂)₂C-O
│
GalOCH₂

Tris-Gal-Chol

orated [57] (also see Chapter 7). In the liver the particles are taken up mainly by the Kupffer cells, but there is also a slightly increased uptake by the parenchymal cells. This effect is due not to the apoprotein B receptor but to a galactose-mediated interaction of the tris-galactosyl cholesterol situated in the monolayer of the particle. Similar treatment of HDL with this compound leads to an increased uptake by the parenchymal cells, which again is dependent on the quantity of material incorporated into the particle [58]. These results and others ([59] and see Chapter 7) demonstrate that it is possible to divert lipoproteins from their original catabolic pathway and target them to other specific cellular sites. This is of obvious utility in the field of drug targeting and can be achieved by loading appropriately tailored molecules into the lipoprotein particle.

Residence of the drug in the LDL or other lipoprotein particle monolayer may not be without drawbacks since in the lipoprotein pathway it is known that exchange of surface constituents (apoproteins and phospholipid) takes place [60]. Material may move from the original lipoprotein in which it was incorporated to another. Additionally, the apoproteins in the surface layer may be affected by changes in the monolayer composition, which would undoubtedly alter the physicochemical properties of this layer [61] and might diminish any targeting specificity that the lipoprotein possesses.

D. Solubilization in the Internal Core

The major function of the lipoproteins is the transport of water-insoluble lipids in the blood; in addition, they seem to act as carriers for the fat-soluble vitamins, LDL apparently carrying around 70% of the ß-carotene present in blood [62]; it would therefore seem most logical to place drugs and other materials in the central core so that the external function of the particle remains the same. A variety of methods exist for incorporating agents into the central core, as discussed in Chapter 3; the applicability of these methods is dependent on the lipoprotein used in the incorporation technique. These are discussed here in an attempt to clarify some of the factors that lead to success.

E. Partitioning into Core and Surface Monolayer

The lipoproteins in solution represent a discrete colloidal lipid phase possibly in the form of a microemulsion. Water-insoluble, lipid-soluble materials should readily partition into the core of the lipoprotein particle, as discussed earlier in this chapter. Indeed, it appears that a range of lipid-soluble materials has this ability to partition into lipoprotein particles; simple partitioning does not seem to have been utilized to any great extent to transfer drugs into lipoproteins except for LDL, perhaps due to the greater interest in LDL as a possible targeting system for cancer chemotherapy for instance. More detailed studies therefore have been carried out on the incorporation of drugs into LDL.

Benzo(a)pyrene has been shown to associate in vitro with LDL simply by incubating LDL with glass beads coated with the benzo(a)pyrene in aqueous buffer [64]. However, when this complex was tested in vitro against LDL receptor positive and negative cell lines no difference in uptake could be determined between the cell lines, although the LDL did affect the extent of uptake. Other studies have shown that both LDL and HDL can reduce the mutagenicity of benzo(a)pyrene in vitro over a short period, but on prolonged exposure LDL increased mutagenesis [65], a feature the investigators ascribe to increased uptake of the material due to its association with LDL.

Daunomycin has been bound to LDL in a similar manner through the incubation of LDL with glass beads coated with the drug. The drug was found

Daunomycin

to be present both in the internal core of the particle and on the surface [66]. Around 8 nmol of daunomycin was associated for every milligram of apolipoprotein; the complex was stable at low temperatures. P388 leukemia cells take up the complex rapidly and the daunomycin becomes associated with the plasma membrane, microsomal-lysosomal-mitochondrial, and chromatin fractions, a distribution that was the same as that of labeled LDL. Further studies showed that the uptake of the complex was similar in daunomycin-sensitive and resistant cells and that uptake was through an LDL-mediated pathway [67]. The anthracyclines such as daunomycin have a degree of surface activity. Studies of the selectivity of anthracyclines for negatively charged phospholipid membranes [68] were consistent with a mode of binding involving both hydrophobic and electrostatic interactions, in agreement with the findings of Henry et al. [69]. The electrostatic interactions involve the amino group of the sugar moiety and the ionized phosphate groups of the phospholipid, hydrophobic interactions with the tetracyclic ring and the hydrocarbon interior of the bilayer. One might expect therefore some surface binding, which might account for the observed biological activity of complexes. However, the lipoprotein–drug complex also showed activity against the daunomycin-resistant cells with 45% inhibition of growth after 10 min exposure at 0.4 μg/ml daunomycin. Free drug under identical conditions has no effect [67].

Similar results have been achieved using aclacinomycin A, which forms a comparable LDL complex containing around 40% by weight of drug [70]. Cellular accumulation was dependent on drug uptake via the LDL pathway as the complex was not taken up by LDL receptor-deficient cells or when an excess of native LDL was present. The complex was also capable of inhibiting the growth of the glioma cells used in the study in a concentration-dependent manner but no information was presented on the activity relative to that of the native drug. The photoactive agent photofrin II has also been delivered to fibroblast cells using LDL and HDL [71]. The lipoproteins were more efficient than albumin complexes; the LDL process was saturable but HDL and albumin uptake by cells was a proces linear with solute concentration, indicating that uptake via the LDL pathway was receptor mediated and useful for the delivery of the photofrin II to LDL receptor positive cells.

F. Replacement of Central Core

One of the ways to achieve central location of drugs within the core of lipoproteins without great change in the size or structure of the particle is to replace the central core of the particle with new core constituents including the material to be carried. For example, reconstitution studies have shown that it is possible using the correct conditions to recombine the HDL apoproteins with lipids to form a particle similar to the native material [72,73]. Variation of the core lipids of the particle should allow the formation of HDL particles containing the desired entrapped material, the limitations being the miscibility of the core material and drug. Studies utilizing this type of technique for drug incorporation have, however, not yet been performed.

IV. LIPOPHILICITY, RECONSTITUTION, AND UPTAKE

Lipophilicity, as indicated earlier, is not the only determinant of uptake or reconstitution into LDL and other lipoprotein carriers. With amphipathic agents such as nonionic surfactants there is a more complex pattern of interaction, an action that can be measured by an increase in LDL diameter

[74]. Surfactant molecules would be expected to intercalate with the phospholipid–cholesterol monolayer and to adsorb onto hydrophobic sites on the apoprotein. At higher concentrations solubilization of the LDL occurs. Changes to the conformation of the apoprotein lead to an increased apparent hydrodynamic diameter if the protein unfolds, although some expansion would be expected if sufficient molecules are incorporated into the LDL particle.

We recently attempted to use nonionic surfactant–induced expansion of LDL to assist uptake of the oil-soluble dye Sudan III and thymol blue into LDL particles [75], but Fig. 5 shows that the effect is not simply due to the

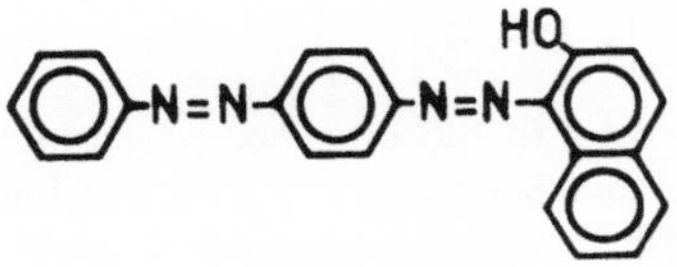

Sudan III

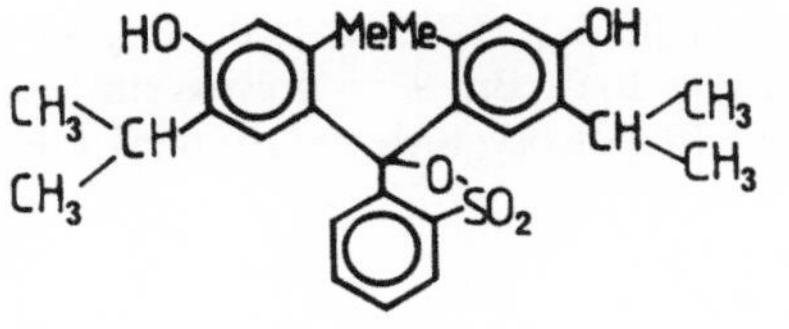

Thymol Blue

lipophilic nature of the surfactant but is related to its hydrophile–lipophile balance (HLB). Similar results are perhaps exemplified in the data of MacCoss et al. [53], who found that the uptake of three ara-CDP-L-1,2,-diacylglycerols was dependent on their acyl chain length and characteristics (see Table 5). The differences in uptake might reflect differences in the degree of aggregation of the parent compounds that form bilayer structures in solution, or they might be a reflection of the requirement elaborated by Firestone et al. [76] that the acyl chain of a solute for incorporation optimally should possess a cis double bond.

Physical difficulties such as the aggregation of the LDL particles on addition of adriamycin have prevented successful uptake of otherwise lipophilic but amphipathic compounds [66]. Such aggregation is anticipated if adsorption of solubilizate results in a reduction of surface charge on the lipoprotein colloid; some drugs have the potential to increase the surface charge by hydrophobic interaction with the LDL surface. These would maintain stable systems.

Results of Ashes et al. [15] on ß-carotene solubilization in HDL suggest that the particle size-related uptake is due not only to the fact that the interior of the very small particles is unable to accommodate as many molecules of solubilizate but also to changes in the physical state of the core lipids, which

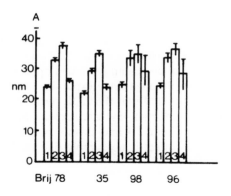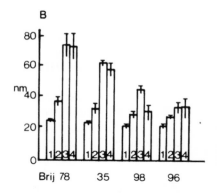

Figure 5 Increases in the diameter of LDL particles in the presence of a range of surfactants of the Brij series and thymol blue (A) and Sudan III (B). Thymol blue is located on the outside of the LDL, while Sudan III, being more liposoluble, partitions into the interior of the lipoprotein particle. The surfactant/LDL molar ratio is 140:1, while the thymol blue/LDL ratio is 6:1 and the Sudan III/LDL ratio is 7:1. 1 = LDL alone as control; 2 = LDL + surfactant; 3 = LDL + surfactant + dye; 4 = LDL + dye. Brij 78 = polyoxyethylene-(20)-stearyl ether [$C_{18}E_{20}$]; HLB = 15.0. Brij 35 = polyoxyethylene-(23)-dodecyl ether [$C_{12}E_{23}$]; HLB = 16.9. Brij 98 = polyoxyethylene-(20)-oleyl ether [$C_{18}E_{20}$]; HLB = 15.3. Brij 96 = polyoxyethylene-(10)-oleyl ether [$C_{18}E_{10}$]; HLB = 12.4.

affect uptake. If the particle diameter is greater than 14 nm, core cholesteryl esters can undergo an order–disorder transition above certain temperatures [60]. Fairly large molecules such as dolichol [78] bind to HDL even though the molecular length is 7 nm.

The concept of anchor moieties on molecules for reconstitution was proposed by Firestone and colleagues [76]: "The successful reconstitution of LDL is facilitated when the hydrophobic compound to be reconstituted contains certain groups that render it compatible with LDL's phospholipid coat and lipophilic core." Among these groups are oleyl, retinoyl, and cholesteryl moieties. However, the possession of these groups does not guarantee success: N-oleylimidazole, despite its oleyl group, could not be reconstituted into LDL, nor could the very lipophilic but perfluorinated C_8F_{17}-cholesteryl oleate derivative. Fluorinated hydrocarbons and aliphatic hydrocarbons are in fact

$-CH_2OCONHC_{12}H_{25}$

$-CH_2OCONHCH_2CH_2(CF_2)_7CF_3$

oleyl-O—   Perfluorinated cholesteryl oleate

incompatible. However, the trifluormethylated anthracycline AD-32 enjoys a much greater uptake than daunomycin. Not surprisingly, the cholesteryl

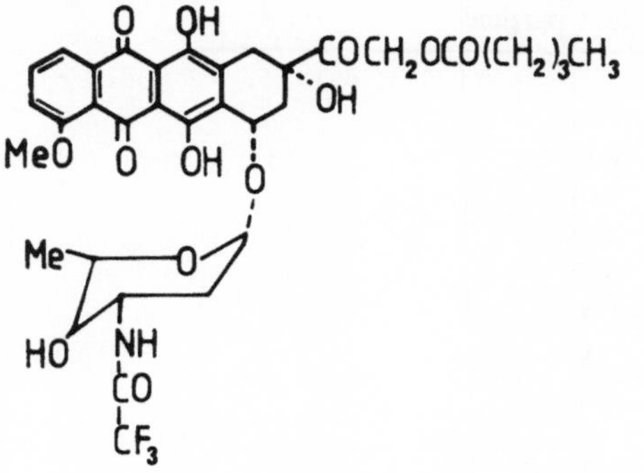

AD-32

nucleus forms the basis of a large number of probes and model solutes incorporated into LDL and HDL particles and of course it has been used as the anchor group on several therapeutic agents (Table 5).

Cholesterol nitroxide is so similar to cholesterol in its physical characteristics

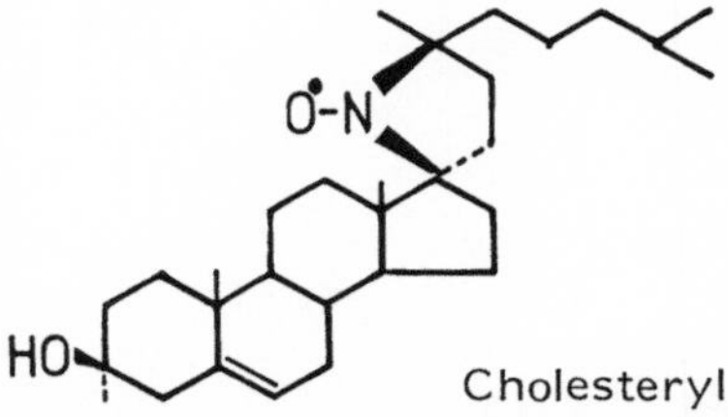

Cholesteryl nitroxide

that it can substitute for it in the LDL particle. PCMA cholesterol oleate replaces entirely the cholesteryl ester interior of LDL, while the 25-hydroxycholesteryl oleate is somewhat less successful, achieving a loading of about 380 molecules/LDL particle [77]. The possession of the hydroxyl group will to

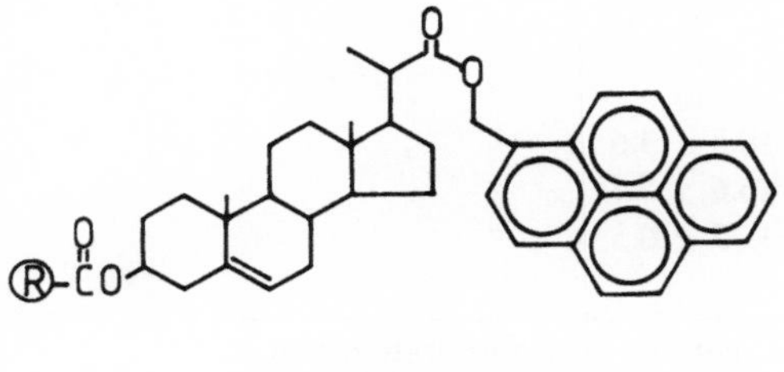

PCMA oleate

Ⓡ = $CH_3(CH_2)_7CH=CH(CH_2)_7$

Table 5 Molar Uptake of Lipophilic Compounds into LDL and HDL[a]

Drug	Moles drug/ lipoprotein particle
LDL	
MCA oleate	750
Triolein	680
Dioleyl fluorescein	440
Aclacinomycin	400
25 HO cholesteryl oleate	380
Benzo(a)pyrene	232 ± 38
ANS	200
N-retinoyl-L-leucyl DOX-14-linoleate	200
3-Hydroxybenzo(a)pyrene	188 ± 54[b]
prednimustine	163 ± 65[b]
estramustine	143 ± 84
MTP-PE	135
Photofrin II	130
β-Sitosteryl-D-gluco-pyranoside	126
Quinidine	128 ± 31.5
AD-32 (N-trifluoroacetyl) adriamycinol-14-valerate	129
Hexadecyl ara-CMP	90
Hexadecyl methotrexate	55
Benzo(a)pyrene-7,8-di hydrodiol	53 ± 18
Protoporphyrin	50
Daunomycin	8
Acetyl-LDL	
Muramyl tripeptide (MTP)-phosphatidyl ethanolamine	80
HDL	
Ara-CDP-L-1,2-dipalmitin	163
Ara-CDP-L-1,2-diolein	130
Ara-CDP-L-1,2-distearin	40
Quinidine	18.5 ± 3.6
3-hydroxybenzo(a)pyrene	16.6 ± 5.5
Benzo(a)pyrene-7,8-di-hydrodiol	1.72 ± 0.3

[a] From papers cited in text; different methods of incorporation may disallow direct comparison of data.
[b] Data from Ref. 104a,b.

some extent dictate the orientation of the solubilized molecules with the HO toward the exterior of the particle. The orientation in the bilayer or core or in some intermediate location is of obvious importance in relation to the stability of the incorporated drug, its potential effect on surface properties, and the targeting potential of the lipoprotein, as with Tris-Gal-Chol, which being amphipathic also orients with the hydrophilic head groups toward the exterior of the particle. The Δ^5-3β hydroxy cholesterol nitroxide synthesized by Keana et al. [78] resembles cholesterol in overall shape and polarity and probably substitutes for it in the bilayer. Spectral analysis is consistent with the interpretation that the lateral diffusion of the compound brings it in contact with apoprotein as there are both cholesterol–lipid and cholesterol–protein contacts. Craig and colleagues' [79] nitrobenzoxadiazole fluorophore (NBD-C) is a deri-

NBD-C

vative of cholesterol linoleate. Sudan cholesterol oleate [80] has also been reconstituted into LDL, and radioiodinated cholesterol has been incorporated into rat HDL [81].

Deckelbaum, Shipley, and Small [82] suggested that below the transition temperature for LDL lipids (45°C) the cholesterol esters are radially arranged in concentric layers approximately 14 nm in diameter (Fig. 6). Approximately 15% of the cholesterol and all of the triglycerides will be in the cholesterol ester core—a conclusion based on studies with the isolated lipids. Above the transition temperature they lose this smectic arrangement but still retain a degree of restriction. This means that partitioning of solutes into the cholesteryl ester core may well be restrained by this structural arrangement. Obviously the solubility of the additive in the cholesterol esters is important. It is likely that on reconstitution of LDL with other lipids the nature of the particle is changed. The surface characteristics may, as Deckelbaum suggests [83], be determined by the smectic nature of the core. As with micellar solubilization, studies in the component lipids do not always allow prediction of solubilization. Structured systems can differentiate between solutes of different molecular shape, whereas an isotropic liquid cannot. This may account for the fact that measured biomembrane/buffer partition coefficients for alcohols and phenols [84] in the red cell and nerve and muscle membranes are about one-fifth of the corresponding octanol/water partition coefficients.

According to Deckelbaum et al. [82], triglyceride abolishes the LDL phase transition when present above about the level of 20% in the core; the transi-

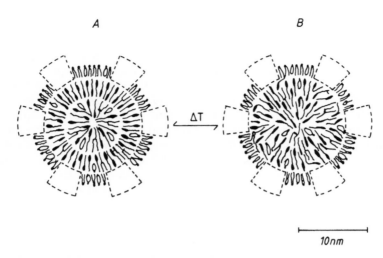

Figure 6 Schematic representation of the distribution of the lipids in LDL (A) below the transition in intact LDL, at 10°C and (B) above the transition, at 45°C. phospholipid molecules: ⟨ ; cholesterol: ⟨⟩ ; cholesterol ester: —•● ; triglyceride. Below the transition temperature the cholesterol esters are seen in A to be arranged in two concentric layers with a periodicity of approximately 36° in a smectic-like state. Above the transition this arrangement is lost, as shown in B, although the organization of the cholesterol esters is not completely random. While the cholesterol esters form the major part of the core, about 15% of the free cholesterol is found in the core along with all of the triglyceride load. (Redrawn from Ref. 82.)

tion temperature varies from 34° to 46° for LDL from subjects fed saturated or unsaturated fat diets [82,84]. Such differences may be the reason for experimental variability in uptake of lipophilic species. An additional factor determining the order in the cores of lipoproteins appears to be particle diameter [85]. In spite of the high content of triglycerides in IDL (20%) and β-VLDL, there appears to be some ordering of the components between 7° and 30°C.

Oeswein and Chun [86] discussed models for lipoproteins in which there is a sharply defined boundary between the hydrophobic core and the amphipathic layer surrounding it. Their calculations indicate that the percentages of total apolar lipid in the hydrophobic core are 28, 64, and 94% of LDL, HDL$_2$, and HDL$_3$, respectively, suggesting differences in the packing of the cholesteryl ester and triglyceride in these three classes. A schematic view of the HDL$_3$ particle in Fig. 7 shows the possible orientation of the various lipid species and the resultant surface irregularity [87], not unimportant in determining adsorption sites. However, studies of the reconstitution of LDL with esters of long chain fatty acids led Krieger et al. to state that "neither the differences

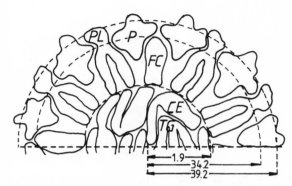

Figure 7 Structural model of HDL. PL, phospholipids; FC, free cholesterol; CE, cholesterol esters; TG, triglyceride. The surface irregularity due to the outward projection of amino acid side chains and polar head groups of the phospholipids is shown. (Redrawn Ref. 87.)

in solubility of saturated or unsaturated cholesteryl esters nor the differences in phase behavior are likely to account for the all-or-none differences in the incorporation of the saturated and unsaturated fatty acids into LDL. Rather the data are more compatible with the notion that the physical characteristics of the LDL particle itself favour the incorporation of unsaturated rather than saturated fatty acid esters, thus implying that differences in the three dimensional structures of these two classes of fatty acids are important. Consistent with this hypothesis is the observation that the cholesteryl ester of the *trans* isomer of the 18 carbon mono-unsaturated fatty acid (elaidate) was incorporated less well than the *cis* isomer (oleate), suggesting that the kink in the fatty acyl chain induced by the *cis*-configuration about the double bond plays some role in allowing the molecule to be incorporated into LDL" [88].

ß-carotene and dolichol are two molecules, already discussed, which possess polyunsaturated chains. The structure of ß-carotene was given earlier; dolichol possesses 17–22 isoprene units which are mainly cis, but it is bound principally to HDL [17], even though it has a molecular size equivalent to the diameter of the HDL core.

Apart from the cholesterol derivatives that associate with LDL and other lipoproteins, and compounds with long chain cis alkyl substituents, many of the molecules that have been found to partition into LDL are relatively small hydrophobic molecules, often remarkably similar in dimensions; these include DDT, 6-CB, 1,6-diphenyl-1,3,5-hexatriene, *N*-phenyl-1-naphthylamine, and the

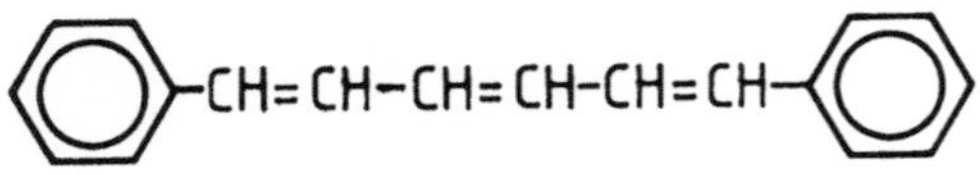

1,6-Diphenyl-1,3,5-hexatriene

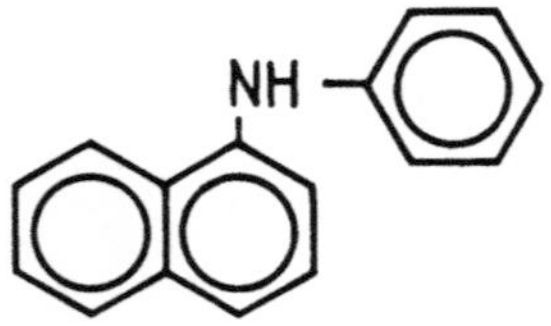

N-Phenyl-1-naphthylamine

drug probucol [89], which is a hypocholesterolemic and antiatherosclerotic agent, with a structure similar to that of butylated hydroxytoluene. It has been suggested that it associates with plasma LDL, preventing oxidative modification

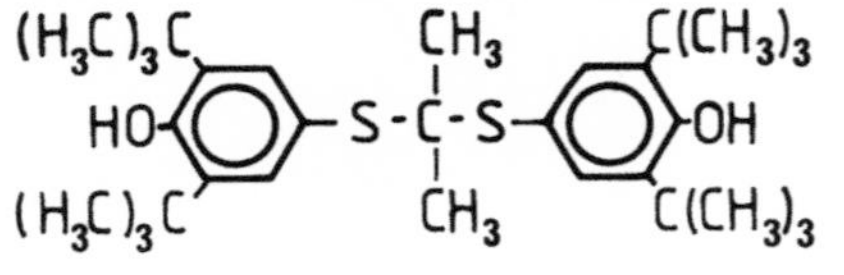

probucol

of the lipoprotein, which is believed to trigger its uptake by vessel walls. To some extent the problems emerging are dictated simply by the compounds that have been studied, as there have been few systematic investigations. The other clear class whose members are structurally diverse is the amphipathic drugs; they interact strongly with LDL as with other colloidal dispersions often in a nonspecific manner. The drugs in this class range from nonionic surfactants to the ara-CDP-L-1,2-diacylglycerols and hexadecyl ara-CMP discussed earlier but also include anthracyclines such as daunomycin and AD-32, which exhibit degrees of surface activity.

Whereas there have been few studies of homologous series, there have been more investigations of the interaction of individual solutes with VLDL, HDL, and LDL. Shu and Nichols's data on benzo(*a*)pyrene and its derivatives illustrate this well. Very low density lipoprotein takes up 9–12 times as much benzo(*a*)pyrene as LDL [90]; 6-CB also interacts with VLDL more than with LDL or HDL, but quinidine, cyclosporin A, and 3-hydroxybenzo(*a*)pyrene interacted to a greater extent with HDL than LDL. Shu and Nichols normalized the data for benzo(*a*)pyrene uptake into the lipoproteins per unit lipid volume and found, as discussed earlier, remarkably similar incorporation in VLDL, LDL, and HDL. The 3HO analog, however, exhibited a preference for the HDL particle, suggesting some specific interaction was occurring. It is likely that cyclosporin A behaves similarly, possibly adsorbing to the exterior of the particle. It has been suggested [86] that more of the apolar lipid extends into the outer shell of HDL than in LDL because of differences in the packing of cholesteryl esters and triglycerides in the two types, which might account for differences, but principally the proportion of phospholipid is higher in HDL. Yoo, Norman, and Busbee [91] found that benzo(a)pyrene uptake correlates best with the triglyceride levels in lipoproteins, using a series of VLDL, LDL, and HDL fractions from four subjects. While increasing triglyceride levels correlate with increasing cholesteryl ester levels in VLDL and LDL, the reverse is true with HDL, cholesteryl ester levels falling with increasing triglycerides [91]. Similar trends are found when triglyceride levels are plotted against total lipid levels (mg/dl). The trends in composition in HDL compared to VLDL and LDL may explain the fact that the 3HO derivative of benzo-(a)pyrene interacts preferentially with HDL: levels of phospholipid are also higher and the particle itself is therefore more polar.

V. MICROEMULSIONS AS SYNTHETIC ANALOGS OF LOW DENSITY LIPOPROTEIN PARTICLES

The obvious drawbacks of LDL particles as sources of clinically useful drug delivery systems have led to the search for synthetic analogs such as microemulsions comprising phospholipids, cholesteryl esters, and/or triglycerides and drug.

Microemulsions have droplet sizes in the range of 200 nm down to about 10 nm. They exist in a state between micellar dispersion and macroscopic emulsions [92], under a narrow set of physical conditions (T,P, composition). They are thermodynamically stable systems, although transfer of components to the continuous phase may destabilize the system. Ginsburg, Small, and Atkinson [93], having described the natural LDL particle as a microemulsion of cholesteryl esters stabilized by phospholipid and protein, prepared protein-free microemulsions of cholesteryl oleate or cholesteryl nervonate with egg yolk dimyristoyl- or dipalmitoyl-phosphatidylcholine by ultrasonication, producing particles with a Stokes' radius of about 10 nm. Protein-free microemulsion analogs of chylomicrons have also been produced [94] by ultrasonication of mixtures of triolein, cholesteryl oleate, phosphatidylcholine, and cholesterol dispersions, and of VLDL [95] and HDL [96,97] by similar techniques.

Somewhat smaller systems were prepared by Ginsburg et al. [93] and by the present authors [98], these having diameters of 22 and 43 nm, respectively. Table 6 gives details of the incorporation of a methotrexate diester into a microemulsion stabilized by egg yolk phosphatidylcholine. Etoposide (VP 16) failed to be incorporated into a stable microemulsion, whereas the diester was incorporated to a maximum of 61% of the added material, causing little or no change in the radius of the particle.

Table 6 Incorporation of Lipid-Soluble Cytotoxic Drugs into Egg Yolk
Phosphatidylcholine Cholesteryl Microemulsions

Drug	Quantity added	Percent in S2[a]	Radius (nm)
Etoposide	2	0	27.3
Methotrexate diester	2	12	20.2
Methrotexate diester	4	60	18.4
Methrotexate diester	10	38	30.1

[a] Methotrexate-benzyl-cholesteryl diester.
Source: Ref. 94.

Via et al. [99] produced cholesterol ester-rich microemulsions with particle diameters of 45 nm also containing cholesterol, phosphatidylcholine, and trioleyl-glycerol in the relative molar ratios found in LDL. Omission of the triolein or the use of a diunsaturated phosphatidylcholine destroyed the homogeneity of the microemulsions. The product was sufficiently "recognizable" in vivo that it could be used to donate cholesteryl esters to plasma lipoproteins and for investigating apoprotein–lipid interactions [100]. Artificial triacylglycerol emulsions can substitute for VLDL [101] in the lipid exchange process.

Incorporation of drug into the core of microemulsion systems will be dependent on the formation of a solution of drug in the core materials. The failure of etoposide uptake, while methotrexate-benzyl-cholesteryl diester is solubilized, will be related to miscibility of the drug substance with the cholesteryl ester core.

Too few solutes have been incorporated into synthetic lipoprotein dispersions to allow any sensible discussion of factors affecting uptake. It would seem to be, a priori, a question of dissolution of the drug or solute to a sufficient degree in the principal unstructured lipids of the artificial system, although some studies, like those of molecular motion and conformation of cholesteryl esters (in reconstituted HDL [102]), indicate that the physical state of the cores of both native and reconstituted particles may be similar [103]. If that is so, the same factors will determine solubilization and successful microemulsification.

REFERENCES

1.   Albert A. Selective toxicity, 5th ed. London:Chapman and Hall, 1973.
2.   Davis SS, Illum L. Colloidal delivery systems: opportunities and challenges. In Tomlinson E, Davis SS, eds. In Site-specific drug Delivery. Chichester: Wiley 1986:93–110.
3.   Willmott N, Cummings J, Stuart JFB, Florence AT. Adriamycin loaded albumin microspheres. Preparation, in vivo distribution and release in the rat. Biopharm Drug Dispos 1985; 6:91–104.

4. Chen Y, Willmott N, Florence AT. Comparison of albumin and casein microspheres as a carrier for doxorubicin. J Pharm Pharmacol 1987; 39:978–985.

5. Attwood D, Florence AT. Surfactant systems: Their chemistry, pharmacy and biology. London:Chapman and Hall, 1983.

6. Mahley RW, Innerarity TL. Lipoprotein receptors and cholesterol homeostasis. Biochim Biophys Acta 1983; 737:197–222.

7. Scanu AM. Structural studies on serum lipoproteins. Biochim Biophys Acta 1972; 265:471–508.

8. Skalsky H, Guthrie FE. Binding of insecticides to human serum proteins. Toxicol Appl Pharmacol 1978; 43:177–188.

9. Maliwal B, Guthrie FE. Interaction of insecticides with human plasma lipoproteins. chem Biol Interactions 1981; 35:177–188.

10. Dao TH, Lavy TL, Dragun J. Rationale of the solvent selection for soil extraction of pesticide residues. Residue Rev 1983; 87:91–104.

11. Brown MS, Goldstein JL. Receptor mediated control of cholesterol metabolism. Science 1976; 191:150–154.

12. Gallenberg LA, Voldicnik MJ. Potential mechanisms for redistribution of polychlorinated-biphenyls during pregnancy and lactation. Xenobiotica 1987; 17:299–310.

13. Nilsen OG. Serum albumin and lipoproteins as quinidine binding molecules in normal human sera. Biochem Pharmacol 1976; 25:1007–1012.

14. Weinkam RJ, Finn A, Levin VA, Kane JP. Lipophilic drugs and lipoproteins partitioning effects on chloroethylnitrosourea reaction rates in serum. J Pharm Exp Ther 1980; 214:318–323.

15. Ashes JR, Burley RW, Sidhu GS, Sleigh RW. Effect of particle size and lipid composition of bovine blood high density lipoprotein on its function as a carrier of ß-carotene. Biochim Biophys Acta 1984; 665:376–384.

16. Keenan RW, Kruczek ME, Fischer JB. The binding of (3-H)dolichol by plasma high density lipoproteins. Biochim Biophys Acta 1976; 486:1-9.

17. Shu HP, Nichols AV. Uptake of lipophilic carcinogens by plasma lipoproteins structure activity studies. Biochim Biophys Acta 1981; 665:376–384.

18. Bickel MH. Binding of chlorpromazine and imipramine to red cells, albumin, lipoproteins and other blood components. J Pharm Pharmacol 1975; 27:733–738.

19. Florence AT. Surface chemical and micellar properties of drugs in solution. Adv Colloid Interface Sci 1968; 2:115–149.

20. Tillement JP, Zini R, Boissier JR. Binding of chlorpromazine, imipramine and nortryptiline to plasma proteins. J Pharmacol 1974; 5(S2):99.

21. Danon A, Chen Z. Binding of imipramine to plasma proteins – effect of hyperlipoproteinemia. Clin Pharmacol Ther 1979; 25:316–321.

22. Chen Z, Danon A. Binding of reserpine to plasma albumin and lipoproteins. Biochem Pharmacol 1979; 28:267–271.

23. Sgoutas D, Macmahon W, Love A, Jerkunica I. Interaction of cyclosporin A with human lipoproteins. J Pharm Pharmacol 1986; 38:583–588.

24. Lemaire M, Tillement JP. Role of lipoproteins and erythrocytes in the *in vitro* binding and distribution of cyclosporin-A in the blood. J Pharm Pharmacol 1982; 34:715–718.

25. Wahlquist M, Nilsson IM, Sandberg F, Agurel S, Grandstand B. Binding of delta-9-tetrahydrocannabinol to human plasma proteins. Biochem Pharmacol 1970; 19:2579–2584.

26. Klausner HA, Wilcox HG, Dingell JV. The use of zonal ultracentrifuging action in the investigation of the binding of delta-9-tetrahydrocannabinol by plasma lipoprotein. Drug Metab Dispos 1975; 3:314–319.

27. Rudman D, Hollins B, Dixler TJ, Mosteller RC. Transport of drugs, hormones and fatty acids in lipaemic serum. J Pharmacol Exp Ther 1972; 180:797–810.

28. Assmann G, Brewer HB. A molecular model of high density lipoproteins. Proc Natl Acad Sci USA 1974; 71:1534–1538.

29. Margolis S, Langdon RG. Studies on human serum β-lipoproteins. III. Enzymatic modification. J Biol Chem 1966; 241:485–493.

30. MacFarlane AS. Efficient trace-labeling of proteins with iodine. Nature 1958; 182:53.

31. Bilheimer DW, Eisenberg S, Levy RI. The metabolism of VLDL proteins. Preliminary *in vitro* and *in vivo* observations. Biochim Biophys Acta 260:212–221.

32. Shepherd J, Bedford DK, Morgan HG. Radioiodination of human low density lipoprotein: comparison of four methods. Clin Chim Acta 1976:97–109.

33. Pittman RC, Carew TE, Attie AD, Witztum JL, Wantabe Y, Steinberg D. Receptor dependent and receptor independent degradation of low density lipoprotein in normal rabbits and in receptor deficient mutant rabbits. J Biol Chem 1982; 257:7994–8000.

34. Vallabhajosla S, Goldsmith SJ, Ginsberg HN, Badimon JJ, Paidi N-A, Fuster V. Tc-99m Labeled lipoproteins: new agents for studying lipoprotein metabolism and non-invasive imaging of atherosclerotic lesions. Nuklearmedizin (suppl) 1986; 22:606–608.

35. Nagelkerke JF, Barto KP, van Berkel ThJC. In vivo and in vitro uptake and degradation of acetylated low density lipoprotein by rat liver endothelial, Kupffer and parenchymal cells. J Biol Chem 1983; 258:12221–12227.

36. Alsat E, Mondon F, Rebourcet R, Bertheilier M, Erlich D, Cedard L, Goldstein S. Identification of specific binding sites for acetylated low density lipoprotein in microvillus membranes from human placenta. Mol Cell Endocrinol 1985; 41:229–235.

37. Dresel HA, Freidrich E, Via DP, Schettler G, Sinn H. Characterization of binding sites for acetylated low density lipoprotein in the rat liver in vivo and in vitro. EMBO J 4:1157–1162.

38. Carew TE, Pittman RC, Steinberg D. Tissue sites of degradation of native and reductively methylated [C-14]-labeled sucrose low density lipoprotein in rats: contribution of receptor-dependent and receptor-independent pathways. J Biol Chem 1982; 257:8001–8008.

39. Paavola LG, Strauss JF, Boyd CO, Nestler JE. Uptake of gold labelled and [H-3]-cholesterol lineolate labelled human low density lipoprotein by cultured rat granuloma cells: cellular mechanisms involved in lipoprotein metabolism and their importance to steroidogenesis. J Cell Biol 1985; 100:1235–1237.

40. Hynds SA, Welsh J, Stewart JM, Jack A, Soukop M, McArdle CS, Calman KC, Packard CJ, Shepherd J. Low density lipoprotein metabolism in mice with soft tissue tumors. Biochim Biophys Acta 1984; 795:589–595.

41. Halbert GW, Stuart JFB, Florence AT. A low density lipoprotein methotrexate covalent complex and its activity against L1210 cells in vitro. Cancer Chemother Pharmacol 1985; 15:223–227.

42. Halbert GW, Florence AT, Stuart JFB. Characterization of in vitro drug release and biological activity of methotrexate–bovine serum albumin conjugates. J Pharm Pharmacol 1987; 39:871–876.

43. Maudlin J, Fisher WR. pH and ionic strength dependent aggregation of serum low density lipoproteins. Biochemistry 1970; 9:2015–2020.

44. Masquelier M, Vitols SG, Peterson CO. LDL as a carrier of antitumoral drugs: in vivo fate of human LDL complexes in mice. Cancer Res 1986; 46:3842–3847.

45. Trouet A, Masqualier M, Baurain R, Deprez-de Campeneere D. A covalent linkage between daunorubicin and proteins that is stable in serum and reversible by lysosomal hydrolases, as required for a lysosomotropic drug–carrier conjugate: in vitro and in vivo studies. Proc Natl Acad Sci USA 1972; 79:626–629.

46. Mahley RW, Innerarity TL, Pitas RE, Weisgraber KH, Brown JH, Gross E. Inhibition of lipoprotein binding to cell surface receptors of fibroblasts following selective modification of arginyl residues in arginine rich and β–apoproteins. J Biol Chem 1977; 252:7279–7287.

47. Vidal M, Sainte-Marie J, Philippot JR, Bienvenue A. LDL-mediated targeting of liposomes to leukaemic lymphocytes in vitro. EMBO J 1985; 4:2461–2467.

48. Ghosh S, Basu MK, Schweppe JS. Interaction of 1-anilino-8-naphthalene sulphonate with human serum low density lipoprotein. Biochim Biophys Acta 1974; 337:395–403.

49. Muesing RA, Nishida T. Disruption of low and high density human plasma lipoproteins and phospholipid dispersions by 1-anilino naphthalene-8-sulphonate. Biochemistry 1971; 10:2952–2962.

50. Freudenberg MA, Bog-Hansen TC, Back U, Galanos C. Interaction of lipopolysaccharides with plasma high density lipoprotein in rats. Infect Immu 1980; 28:373–380.

51. Hobara N, Taketa K. Electrophoretic studies of clioquinol binding to human serum proteins. Biochem Pharmacol 1976; 25:1601–1606.

52. MacCoss M, Edwards JJ, Seed TM, Spragg P. Phospholipid nucleoside conjugates: the aggregation characteristics and morphological aspects of selected 1-β-D-arabinofuranosylcytosine-5'-diphosphate-L-1,2-diacylglycerols. Biochim Biophys Acta 1982; 719:544–555.

53. MacCos M, Edwards JJ, Lagocki P, Rahman Y-E. Phospholipid nucleoside conjugates. 5. The interaction of selected 1-β-D-arabinofuranosylcytosine-5' diphosphate-L-1,2-diacylglycerols with serum lipoproteins. Biochem Biophys Res Commun 1983; 116:368–374.

54. Shaw JM, Shaw KV, Yanovich S, Iwanik M, Futch WS, Rosowsky A, Schook LB. Delivery of lipophilic drugs using lipoproteins. Ann NY Acad Sci 1987; 507:252–271.

55. Seki J, Okita A, Watanabe M, Nakagawa T, Honda K, Tatewaki N, Sugiyama M. Plasma lipoproteins as drug carriers: pharmacological activity and disposition of the complex of β-sitostearyl-β-D-glucopyranoside with plasma lipoproteins. J Pharm Sci 1985; 74:1259–1264.

56.   Kempen HJM, Hoes C, van Boom JH, Spanjer HH, De Lange J, Langendoen A, van Berkel ThJC.   A water soluble cholesteryl containing tris-galactoside: synthesis, properties and use in directing lipid containing particles to the liver. J Med Chem 1984; 27:1306–1312.

57.   van Berkel ThJC, Kruijt JK, Spanjer HH, Nagelkerke JF, Harkes L, Kempen JHM.   The effect of a water soluble tris-galactoside-terminated cholesteryl derivative on the fate of low density lipoprotein and liposomes.   J Biol Chem 1985; 260:2694–2699.

58.   van Berkel ThJC, Kar Kruijt J, Kempen HJM.   Specific targeting of high density lipoprotein to liver hepatocytes by incorporation of a tris galactosyl terminated cholesteryl derivative.   J Biol Chem 1985; 260:12203–12207.

59.   Kempen HJM, Kuipers F, van Berkel ThJC, Vonk RJ.   Effect of infusion of tris-galactosyl-cholesterol on plasma cholesterol, clearance of lipoprotein cholesteryl esters and biliary secretion in the rat.   J Lipid Res 1987; 28:659–666.

60.   Tall AR, Small DM.   Body cholesterol removal: role of plasma high density lipoproteins. Adv Lipid Res 1980; 17:1–51.

61.   Schroeder F, Goh EH, Heimberg M.   Regulation of the surface physical properties of the very low density lipoprotein. J Biol Chem 1979; 254:2456–2463.

62.   Krinsky NI, Cornwell DG, Oncley JL.   The transport of vitamin A and carotenoids in human plasma. Arch Biochem Biophys 1985; 73:233–246.

63.   GC Chen, Kane JP, Hamilton RL.   Thermal behavior of cores of human serum triglyceride rich lipoproteins: a study of induced circular dichroism of β-carotene. Biochemistry 1984; 23:1119–1124.

64.   Ramsen JF, Shireman RB.   Effect of low density lipoprotein on the incorporation of benzo-a-pyrene by cultured cells.   Cancer Res 1981; 41:3179–3185.

65.   Yoo JSH, Norman JO, Joe CO, Busbee DL. High density lipoprotein decrease both DNA binding and mutagenicity of R-7, T-8-dihydroxy-T-9, 10-epoxy-7,8,9,10-tetrahydrobenzo-a-pyrene in V79 Chinese hamster cells. Mutat Res 1986; 159:83–89.

66.   Iwanik MJ, Shaw KV, Ledwith BJ, Yanovich S, Shaw JM. Preparation and interaction of a low density lipoprotein daunomycin complex with P388 leukemic cells. Cancer Res 1984; 44:1206–1215.

67.   Yanovich S, Preston L, Shaw JM. Characteristics of uptake and cytotoxicity of a low density lipoprotein daunomycin complex in P388 leukemic cells. Cancer Res 1984; 44:3377–3382.

68.   Burke TG, Sartorelli AC, Tritton TR. Selectivity of the anthracyclines for negatively charged model membranes: role of the amino group. Cancer Chemother Pharmacol 1988; 31:274–280.

69.   Henry N, Fantine EO, Bolard J, Garnier-Suillerot A. Interaction of adriamycin with negatively charged model membranes: evidence of two types of binding sites. Biochemistry 1985; 24:7085–7092.

70.   Rudling MJ, Collins VP, Peterson CO.   Incorporation and binding of anthracycline derivatives to low density lipoprotein: *in vitro* and *in vivo* studies on drug LDL conjugates. Cancer Res 1983; 43:4600–4605.

71.   Candide C, Morliere P, Maziere JC, Goldstein S, Santus R, Dubertret L, Reyftmann JP, Polonovski J. *In vitro* interaction of the photoactive anticancer

porphyrin derivative photofrin II with low density lipoprotein and its delivery to cultured human fibroblasts. FEBS Lett 1986; 207:133–138.

72.   Ritter MC, Scanu AM.   Role of apolipoprotein A-1 in the structure of human serum high density lipoproteins: reconstitution studies.   J Biol Chem 1977; 252:1208–1216.

73.   Ritter MC, Scanu AM.   Apolipoprotein A-II in the structure of human serum high density lipoprotein: approach by reassembly of techniques.   J Biol Chem 1979; 254:2517–2525.

74.   Tucker IG, Florence AT.   Interactions of ionic and non-ionic surfactants with plasma low density lipoprotein.   J Pharm Pharmacol 1983; 35:705–711.

75.   Eley JG, Halbert GW, Florence AT.   The incorporation of dyes into low density lipoprotein using non-ionic surfactants.   J Pharm Pharmacol 1987; 39 (suppl):1P.

76.   Firestone RA, Pisano JM, Falck JR, McPhaul MM, Krieger M.   Selective delivery of cytotoxic compounds to cells by the low density lipoprotein pathway. J Med Chem 1984; 27:1037–1043.

77.   Krieger M, Goldstein JL, Brown MS.   Receptor mediated uptake of low density lipoprotein with 25-hydroxycholesterol oleate suppresses coenzyme A reductase and inhibits growth of human fibroblasts.   Proc Natl Acad Sci USA 1978; 75:5052–5056.

78.   Keana JFW, Tamura T, McMillen DA, Jost PC.   Synthesis and characterization of a novel cholesterol nitroxide spin label.   Application to the molecular organization of human low density lipoprotein.   J Am Chem Soc 1981; 103:4904–4912.

79.   Craig IF, Via DP, Mantulin WW, Pownall HJ, Gotto AM, Smith LC.   Low density lipoprotein reconstituted with steroids containing the nitrobenzoxadiazole fluorophore.   J Lipid Res 1981; 22:687–696.

80.   Mosley ST, Goldstein JL, Brown MS, Falck JR, Anderson RGW.   Targeted killing of cultured cells by receptor dependent photosensitization.   Proc Natl Acad Sci USA 1981; 78:5717–5721.

81.   Counsell RE, Korn N, Pohland RC, Schwendner SW, Seevers RH.   Fate of intravenously administered high-density lipoprotein labelled with radio-iodinated cholesteryl oleate in normal and hypolipidaemic rats.   Biochim Biophys Acta 1983; 750:–503.

82.   Deckelbaum RJ, Shipley GG, Small DM.   Structure and interaction of lipids in human plasma low density lipoprotein.   J Biol Chem 1977; 252:744–754.

83.   Roll S, Seeman P.   The membrane concentrations of neutral and positive anaesthetics (alcohols, chlorpromazine, morphine) fit the Meyer–Overton rule of anaesthesia; negative narcotics do not.   Biochim Biophys Acta 1972; 255:207–219.

84.   Deckelbaum RJ, Tall AR, Small DM.   Interaction of cholesterol ester and triglyceride in human plasma very low density lipoprotein.   J Lipid Res 1977; 18:164–168.

85.   Chen GC, Kane JP, Hamilton RL.   Thermal behaviour of cores of human serum triglyceride rich lipoproteins: a study of induced circular dichroism of ß-carotene.   Biochemistry 1984; 23:1119–1124.

86.   Oeswein JW, Chun PW, Density and composition models for lipoproteins. Biophys Chem 1981; 14:233–245.

87.   Edelstein C, Kezdy FJ, Scanu AM, Shen BW.   Apolipoproteins and the

structural organization of plasma lipoproteins: human plasma HDL-3. J Lipid Res 1979; 20:143–152.

88. Krieger M, McPhaul MJ, Goldstein JL, Brown MS. Replacement of neutral lipids of low density lipoprotein with esters of long chain unsaturated fatty acids. J Biol Chem 1979; 254:3845–3853.

89. Suckling KE, Groot PHE. Cholesterol: a question of balance. Chem Brit 1988; 24:436.

90. Shu HP, Nicholes. Benzo-a-pyrene uptake by human plasma lipoproteins *in vitro*. Cancer Res 1979; 39:1224–1230.

91. Yoo J-SH, Norman JP, Busbee DL. Benzo-a-pyrene uptake by serum lipids: correlation with triglyceride concentration. Proc Soc Exp Biol Med 1984; 177:434–440.

92. Prince LM. Microemulsions: theory and practice. New York:Academic Press, 1977.

93. Ginsburg GS, Small DM, Atkinson D. Microemulsions of phospholipids and cholesteryl esters: protein free analogues of low density lipoprotein. J Biol Chem 1982; 251:8216–8227.

94. Redgrave TG, Maranhao RC. Metabolism of protein free lipid emulsion models of chylomicrons in rats. Biochim Biophys Acta 1985; 835:104–112.

95. Hamilton JA, Small DM, Parks JS. H-1-NMR studies of lymph chylomicra and very low density lipoproteins from non-human primates. J Biol Chem 1983; 258:1172–1179.

96. Pittman RC, Glass CK, Atkinson D, Small DM. Synthetic high density lipoprotein particles: application to studies of the apoprotein specificity for selective uptake of cholesteryl esters. J Biol Chem 1987; 262:2435–2442.

97. Pittman RC, Knecht TP, Rosenbaum MS, Taylor CA. Nonendocytic mechanisms for the selective uptake of high density lipoprotein associated cholesterol esters. J Biol Chem 1987; 262:2443–2450.

98. Halbert GW, Stuart JFB, Florence AT. The incorporation of lipid soluble antineoplastic agents into microemulsions: protein free analogues of low density lipoprotein. Int J Pharm 1984; 21:219–232.

99. Via DP, Craig IF, Jacobs GW, van Winkle WB, Charlton SC, Gotto AM, Smith LC. Cholesteryl ester rich microemulsions: stable protein free analogs of low density lipoprotein. J Lipid Res 1982; 23:570–576.

100. Parks JS, Mastin JA, Johnson FL, Rudel LL. Fusion of low density lipoproteins with cholesterol ester phospholipid microemulsions: prevention of particle fusion by apoprotein A-1. J Biol Chem 1985; 260:3155–3163.

101. Granot E, Deckelbaum FJ, Eisenberg S, Oschvy Y, Bengtsson-Olivecrona G. Core modification of human low density lipoprotein by artificial triacylglycerol emulsions. Biochim Biophys Acta 1985; 833:308–315.

102. Parmar YI, Gorrissen H, Wassall SR, Cushley RJ. Molecular motion and conformation of cholesteryl esters in reconstituted high density lipoprotein by deuterium magnetic resonance. J Biol Chem 1983; 258:2000–2004.

103. Lundberg B, Suominen L. Preparation of biologically active analogues of serum low density lipoprotein. J Lipid Res 1984; 25:550–558.

104. Eley JG, Halbert GW, Florence AT. The incorporation of estramustine into low density lipoprotein and its activity in tissue culture, Int J Pharm 1990; 63:121–127; b. Incorporation of prednimustine into LDL: activity against P388 cells in tissue culture, Int J Pharm. 1990; 65:219–224.

5

Lipoproteins, Malignancy, and Anticancer Agents

Curt Peterson, Michèle Masquelier, Mats Rudling,*
Kristina Söderberg, and Sigurd Vitols
Karolinska Hospital, Stockholm, Sweden

I. INTRODUCTION

Cholesterol has both structural and metabolic functions in animal cells. It is considered to regulate the membrane fluidity, thereby maintaining a microenvironment appropriate for the operation of membrane-linked enzymes and transport proteins. Cholesterol is also a precursor in the synthesis of steroid hormones in the adrenals and the gonads and of bile acids in the liver. Body requirements of cholesterol can be fulfilled by de novo synthesis or by uptake from the diet.

In humans, about two-thirds of plasma cholesterol can be recovered in the low density lipoprotein (LDL) fraction, mostly in esterified form [1]. Plasma LDL has a half-life of 2–3 days [2]. Low density lipoprotein is derived from very low density lipoprotein (VLDL), which is secreted from the liver. The major function of LDL is to transport cholesterol to peripheral tissues, where it is taken up by cells via both receptor-mediated and nonspecific endocytosis.

Normally, LDL receptors seem to be downregulated since it is possible to increase the receptor activity of freshly isolated white blood cells by incubating them in a medium containing lipoprotein-deficient serum [3]. The cells respond by an upregulation of the number of LDL receptors and an increased

Present Affiliation:
* Huddinge University Hospital, Huddinge, Sweden

activity of 3-hydroxy-3-methylglutaryl coenzyme A reductase (HMG-CoA reductase), the rate-limiting enzyme in the de novo synthesis of cholesterol. On reexposure of the cells to LDL, the number of LDL receptors declines according to first order kinetics and cholesterol synthesis is suppressed through inhibition of HMG-CoA reductase. As cellular requirements for LDL are saturated, excess intracellular cholesterol is stored in reesterified form.

Attempts have been made to estimate the contributions of receptor-mediated and nonspecific cellular uptake of LDL to the overall LDL catabolism. Shepherd and co-workers prepared chemically modified LDL, which is not recognized by LDL receptors [4]. After injections of mixtures of ^{125}I-labeled native and ^{131}I-labeled modified LDL, it was possible to calculate separately the receptor-mediated and nonspecific uptake rates by comparisons of the disappearance rates of native and modified LDL from plasma. It was found that about two-thirds of cellular LDL uptake in the body occurs via the receptor pathway. (For detailed discussion, see Chapter 1.)

II. CHOLESTEROL AND CANCER

Several epidemiologic studies have been performed during the last decades in order to study the role of plasma cholesterol in the development of cardiovascular disease. Concomitant analysis of cancer incidence or cancer mortality in some of these studies has given unexpected results. In addition to the correlation between a high serum cholesterol and a high risk for cardiovascular disease, a correlation between a low serum cholesterol and an increased cancer frequency has been found in many studies (see Ref. 5). This finding was recently explored in a large cohort (about 360,000) of men aged 35–57 years who were screened for possible randomization to the Multiple Risk Factor Intervention Trial in the United States [6]. Mortality follow-up revealed a significant excess of cancer in the lowest decile of serum cholesterol during the early years of follow-up. This was most pronounced for cancer in the stomach, lymphatic tissue, hematopoietic tissue, colon, and lung.

If a low level of plasma cholesterol were a cause of increased cancer risk, then efforts to lower cholesterol levels for the purpose of reducing the incidence of cardiovascular disease might be questioned. However, the results of other types of studies on the role of environmental factors for cancer development do not support the hypothesis that hypocholesterolemia per se is a risk factor for cancer. On the contrary, comparisons of international mortality statistics with per capita consumption of fat have shown that cancer at several sites (breast, prostate, colon, rectum, and endometrium) is associated with a high fat intake [7]. The role of fat in carcinogenesis has also been extensively studied in animals. A high fat intake enhances the development of several cancer types [8]. The results of some of the epidemiologic studies might be questioned since they have been performed in selected materials at elevated risk for cardiovascular disease (middle-aged men living in industrialized areas). This might suggest that the association between hypocholesterolemia and increased cancer mortality is due to competing risks since men with high plasma cholesterol levels are more likely to die from cardiovascular disease than are those with low cholesterol. Consequently, the latter group will have a higher death rate from other causes.

During the years 1971–1975, the National Health and Nutrition Examination Survey was carried out on a probability sample of the civilian noninstitutional population of the United States [9]. A total of 5125 men (yielding 459 cancers) and 7363 women (398 cancers) was initially examined and followed for a median of 10 years. An examination of age-adjusted cancer incidence rates by serum cholesterol level showed an inverse association between cholesterol and all cancer, lung, colorectal, pancreatic, bladder cancer, and leukemia.

During 1968–1972 more than 39,000 men and women aged 15–99 participated in multiphasic screening examinations in Finland [10]. Total populations, or random samples of populations, were invited to the study in rural, semiurban, and industrial communities in different parts of the country. During a median follow-up of 10 years, 1381 cancer cases were diagnosed. Serum cholesterol level was inversely associated with cancer incidence among nonsmokers. The strongest negative association was found during the first years of follow-up, especially for rapidly developing cancers. The associations were found not to be confounded by serum vitamins A or E, serum selenium, or several other factors.

In 1961, the Swedish National Board of Health and Welfare initiated a health survey in certain districts of middle Sweden. The aim was to study the possibility to identify preclinical disease through analyses of blood samples and some physical variables. Serum cholesterol was determined in about 92,000 individuals between 25 and 75 years of age. The cholesterol concentrations of these individuals 25 years ago were recently matched with data from the Swedish Cancer Register and the Swedish Cause of Death Register [11]. During the follow-up period (about 20 years), 27,000 of the 92,000 persons died and 9000 cases of cancer were reported in the cohort. Of these about 5000 had died at the end of the follow-up period. Total cancer incidence and total cancer mortality were inversely correlated to serum cholesterol. This was most pronounced during the first 2 years of follow-up. With increased length of follow-up, the inverse correlation approached normal risk level. Different sites of cancer were associated with serum cholesterol in different ways.

The apparently conflicting results on the relation between cholesterol and cancer have caused an intensive debate. What is the cause–effect relationship for the observed correlations? It has been suggested that individuals maintaining a low blood cholesterol have an increased biliary sterol excretion and that bacterial metabolism of these sterols in the gut could result in an increased production of carcinogenic sterols [12]. In many studies, the association between low plasma cholesterol and increased cancer risk disappears with increased length of observation time, suggesting that a low plasma cholesterol level is not a risk factor for cancer but could be the result of an existing cancer undiagnosed at the time of blood sampling [13]. This would mean that biochemical effects of an ongoing cancer development can be observed years before the diagnosis has been settled. This opens up a new challenge for early cancer screening perhaps by sequential serum cholesterol analyses.

The opinion that hypocholesterolemia may be a consequence of cancer is also supported by the results of a number of studies on cancer patients. Thus Chao et al. made serial measurements of serum cholesterol levels in patients with different kinds of malignancies [14]. In many patients, the cholesterol level declined with first order kinetics during the course of the disease and these patients usually died when their cholesterol values were lower than half the initial values. Malignancies with short cholesterol half-lives included acute myelogenous leukemia and cancer of the stomach, kidney, and colon.

In another study, plasma total, LDL, and high density lipoprotein (HDL) cholesterol concentrations were determined in 32 patients with acute myelogenous leukemia [15]. As seen in Fig. 1, cholesterol levels were markedly reduced in the leukemia patients at diagnosis compared with sex-matched healthy controls. Remission was associated with a significant increase in total and LDL cholesterol (Fig. 2). On relapse, LDL cholesterol fell markedly in two patients.

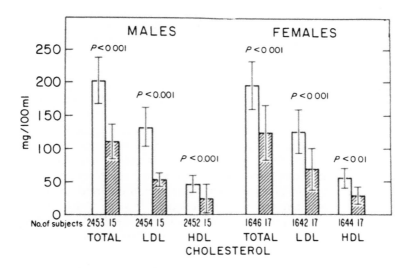

Figure 1 Total and lipoprotein cholesterol concentrations (mean ± SD) of leukemia patients (hatched bars) and healthy subjects (open bars). (From Ref. 15.)

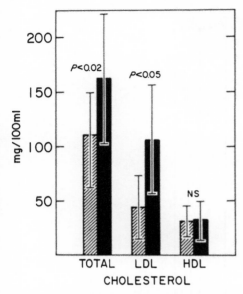

Figure 2 Total and lipoprotein concentration (mean ± SD) of 6 leukemia patients studied at baseline (hatched bars) and during remission (solid bars). (From Ref. 15.)

Serum lipids and lipoproteins were measured in 103 consecutive cancer patients and 100 age-matched noncancer controls [16]. Cancer patients as a group demonstrated significantly lower total cholesterol and LDL cholesterol values compared with the controls. Breast cancer proved to be an exception associated with increased serum total cholesterol and LDL cholesterol. The reason for this difference is not known. Plasma cholesterol levels have been reported in 103 consecutive patients with prostatic carcinoma [17]. The mean level was significantly lower in 30 patients with metastases as compared to those without detectable metastases. Furthermore, the fractional catabolic rate of autologous ^{125}I-LDL was higher in three patients with metastases than in five patients without metastases.

III. LDL RECEPTORS IN MALIGNANT CELLS

In 1978, Ho et al. reported that leukemic cells isolated from peripheral blood of a few patients with acute myelogenous leukemia had 3–100 times higher LDL receptor activity than normal white blood cells [18]. This finding has been extended by a series of studies in our laboratory. Vitols et al. characterized the LDL receptor activity in leukemic cells isolated from peripheral blood and bone marrow aspiration biopsies from about 50 patients with acute

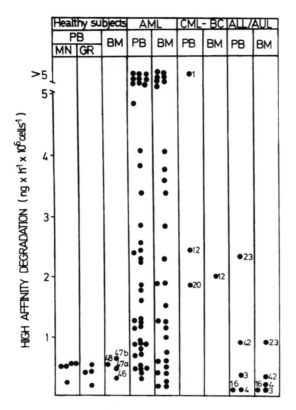

Figure 3 LDL receptor activity determined as high affinity degradation of ^{125}I-LDL by various types of malignant and normal white blood cells. PB, cells isolated from peripheral blood; BM, cells isolated from bone marrow aspiration biopsies; MN, mononuclear cells; GR, granulocytes; AML, acute myeloblastic leukemia; CML-BC, chronic myelogenous leukemia in blast crisis; ALL/AUL, acute lymphoblastic or undifferentiated leukemia. (From Ref. 19.)

myeloblastic leukemia, acute lymphoblastic leukemia, acute undifferentiated leukemia, and chronic myelogenous leukemia in blast crisis [19]. For comparison, the LDL receptor activity was also studied in mononuclear and granulocytic cells from healthy subjects as well as in hematopoietic precursor cells from donors at bone marrow transplantations. As seen in Fig. 3, leukemic cells from most patients with acute myeloblastic leukemia had elevated LDL receptor activity as compared to nucleated blood and bone marrow cells from healthy subjects. The LDL receptor activity varied between patients over a wide range. Cells with a monocytic or myelomonocytic differentiation (M4 and M5 subclasses according to the classification of the French-American

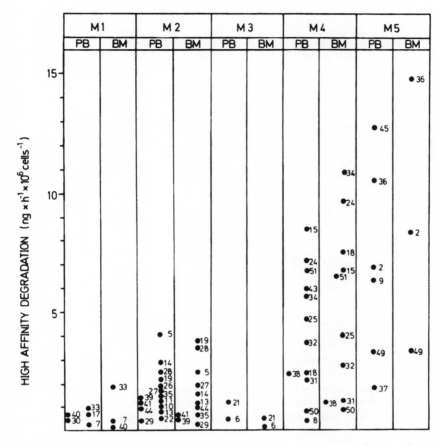

Figure 4 LDL receptor activity of leukemic cells isolated from patients with acute myeloblastic leukemia in relation to the Fab subclasses of the disease. (From Ref. 19.)

British acute leukemia cooperative study group) had the highest values, up to 50 times that of normal cells (Fig. 4) All three patients with chronic myelogenous leukemia in blast crisis had cells with high LDL receptor activity, whereas cells from patients with acute lymphoblastic leukemia had values similar to normal cells (Fig. 3). The reasons for these differences are not understood. One possibility could be that leukemic cells from patients with acute lymphoblastic leukemia satisfy their cholesterol need by an enhanced de novo synthesis.

In 59 patients with acute leukemia, the plasma cholesterol concentration at diagnosis showed an inverse correlation with the LDL receptor activity of leukemic cells isolated from peripheral blood [20]. The lowest plasma cholesterol concentrations were found in patients with very high cell counts and

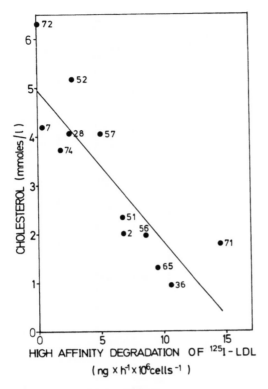

Figure 5 Plasma cholesterol concentrations in patients with acute leukemia with an initial white blood cell count exceeding 90 x 10⁹ per liter in relation to the LDL receptor activity of cells isolated from peripheral blood. (From Ref. 20.)

high LDL receptor activity of the leukemic cells (Fig. 5). During chemotherapy, plasma cholesterol levels increased concomitantly with the fall in cell count in patients with leukemic cells expressing high LDL receptor activity (Fig. 6). However, this was not seen in patients with low cell count or low leukemic cell LDL receptor activity.

A correlation was obtained between an elevated LDL receptor activity and the occurrence of an extra chromosome 8 in the leukemic cell clone [21]. Evidence was presented that a gene on the long arm of chromosome 8 is of importance for the regulation of the LDL receptor activity. This is of interest since the oncogenes c-myc and c-mos are located in the same region of chromosome 8 [22]. One possibility might be that the malignant transformation leads to a concomitant activation of a gene regulating the LDL receptor activity.

It has been reported that leukemic cells from blood and bone marrow specimens, in addition to an elevated LDL receptor activity, also have an

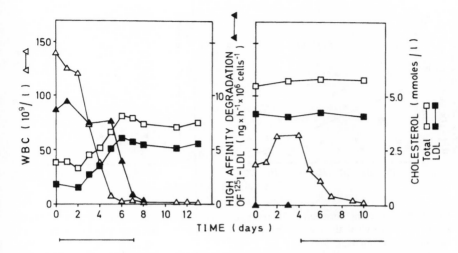

Figure 6 Time course for the white blood cell count, total and LDL cholesterol, and LDL receptor activity of leukemic cells isolated from peripheral blood during treatment of leukemia patients. Bars represent the time periods chemotherapy was given. (From Ref. 20.)

elevated activity of HMG-CoA reductase and thereby an increased de novo synthesis of cholesterol [23]. In spite of this, leukemic cells have a reduced cholesterol content compared to normal white blood cells [24]. What is the fate of the cholesterol supplied to the leukemic cells? One possibility could be that cholesterol is released from the cells as a result of a rapid membrane turnover and that HDL acts as a cholesterol acceptor [25]. High density lipoprotein is believed to promote the removal of cholesterol from the cells and mediate its transport centripetally to the liver from where it can be excreted in the bile. However, since the plasma cholesterol concentration also is low in these patients, this process must be very rapid. To our knowledge, this putative flux of cholesterol has not been studied in leukemia patients.

It is possible that hypocholesterolemia in patients with solid tumors may also be caused by a high LDL uptake by the tumor cells. Indeed, evidence for high LDL uptake in various tumor cells has been found both in vitro and in vivo. Gal et al. reported that a number of human gynecological cancer cell lines grown in culture exhibited a higher uptake of LDL than did the corresponding normal tissues [26,27].

In order to quantitate LDL receptors in tissues, we developed a binding assay for LDL receptors in crude homogenates based on the ability of heparin to specifically release receptor-bound LDL [28]. Using this assay, LDL receptors have been quantitated in normal and malignant human tissues ob-

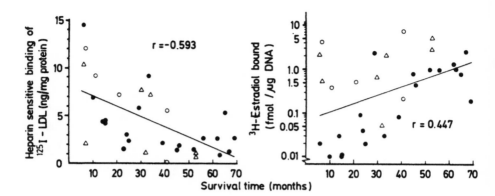

Figure 7 Relations between LDL receptor content (left), estrogen receptor content (right), and survival time in 32 breast cancer patients who died within six years of diagnosis. (From Ref. 30.)

tained at surgery [29]. Of the normal tissues studied, the adrenal gland and the ovary have the highest receptor binding of LDL per gram of tissue. Taking the weights of the organs into account, it could be calculated that the liver contributed to about 50% of the total LDL catabolism in the body. In certain cases, tumors from the stomach and the parotid gland expressed very high LDL binding. However, in other patients, tumors from the same organs expressed binding values similar to those of the corresponding normal tissues.

In patients who died from breast cancer, an inverse correlation was found between the number of LDL receptors in the tumor tissue and the survival time of the patients [30] (Fig. 7). In brain tumors, the LDL binding was often low, although the binding to tumor tissue was overall significantly higher than the biding to surrounding brain [31]. Some tumors had enhanced activity of HMG-CoA reductase. There was no correlation between the LDL receptor binding and the HMG-CoA reductase activity, indicating that the receptor and the enzyme are regulated independently of each other.

In the near future, improved techniques will be available for comparisons of the LDL receptor activities in tumor and normal tissues. The gene for the LDL receptor protein has been identified and cloned [32]. Thereby, it will be possible to determine whether the gene is amplified and/or overexpressed in tumors.

The uptake of LDL in animal tumors has been studied in a number of experimental systems. It must be remembered, however, that all these animal studies have been performed with human LDL. Even if human LDL is recognized by the animals' LDL receptors [33], the competition from endogenous LDL is definitely different from the human situation since there are pronounced species differences in the plasma lipoprotein pattern [34]. Therefore, the results must be interpreted with caution. Norata et al. [35] injected ^{125}I-LDL

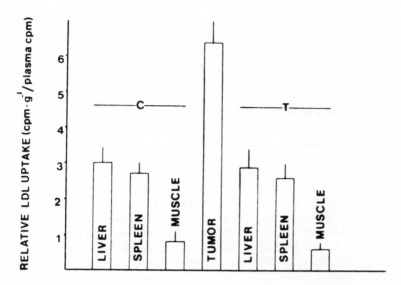

Figure 8 Relative uptake of ^{125}I-LDL in tissues from BALB/c mice bearing a fibrosarcoma tumor. The radiolabeled LDL was injected 8–10 days after the injection of 10^6 cells in the hind leg and the mice were killed 24 h later. Means ± SEM for 10 animals are shown (From Ref. 35.)

into mice with a transplantable fibrosarcoma tumor. Twenty-four hours after the injection, the radioactivity was much higher in the tumor than in liver, spleen, and muscle (Fig. 8). Hynds et al. found that the tumor was surpassed only by the adrenals and the liver with regard to LDL accumulation per gram of tissue in mice bearing a soft tissue tumor [36]. High tumor uptake of LDL has also been reported by Dresel and co-workers [37]. Using gamma camera imaging, they studied the distribution of radioactivity after injection of ^{131}I-LDL in solid tumor–bearing rats. With time up to 6 h postinjection, there was a pronounced accumulation of radioactivity in the tumor.

Since iodinated LDL is rapidly degraded in cells and tissues and the iodinated fragments are then extruded, the techniques used in the studies described can be expected to underestimate the real tissue uptake and degradation of LDL. Therefore, we are currently exploring the possibility of using ^{14}C-sucrose–LDL to study the in vivo tumor uptake of LDL in patients undergoing surgery. Animal experiments have shown that sucrose-LDL remains in tissues with a half-live of 2–3 days [38]. We recently completed a study in 10 patients with acute myeloblastic leukemia [39]. At diagnosis, leukemic cells were isolated from peripheral blood and the LDL receptor activity was determined by incubation with ^{125}I-LDL for 4–6 h as previously described [19]. Before chemotherapy, the patients also received an intravenous

injection of ^{14}C-sucrose–LDL. Blood samples were then collected at certain
time intervals and the radioactivity in the leukemic cell pellet was determined
as long as a sufficient cell number could be obtained. There was a linear
correlation between the in vivo accumulation of ^{14}C-sucrose in leukemic cells
and the high affinity degradation of ^{125}I-LDL by isolated leukemic cells.
Interestingly, it took 2–3 days for the cellular accumulation of ^{14}C-sucrose to
reach the peak concentration. One patient died about a week after the
injection of ^{14}C-sucrose–LDL. At autopsy, samples of various tissues could be
obtained for determination of radioactivity. As expected, the highest
concentration was found in the adrenal cortex. High concentrations were also
found in the liver and the bone marrow.

Since there is strong evidence for a preferential LDL accumulation in
tumor cells, LDL might be of interest for tumor imaging. A technique has
been described for labeling LDL with technetium-99m, a radionuclide ideal for
imaging [40]. Studies in rabbits and monkeys showed that the biodistribution
of ^{99m}Tc-LDL was similar to that of ^{131}I-tyramine cellobiose–LDL, a previously
validated trapped radioligand [41]. In contrast, ^{125}I-LDL was rapidly deiodinat-
ed in the liver with subsequent excretion of radioactivity into the intestine.
After injection of ^{99m}Tc-LDL in patients with myeloproliferative disease, the
tissue distribution indicated that the spleen and bone marrow are the major
sites of LDL catabolism in this patient group [42].

There is strong evidence that most tumor cells have an enhanced
cholesterol supply, either by uptake of preformed cholesterol in LDL or by de
novo synthesis. The reason for this high cholesterol demand is not clear.
Cholesterol is normally used for membrane synthesis. However, compared to
growing normal cells, e.g., in the bone marrow and in the intestinal mucosa,
human tumor cells do not exhibit a rapid growth rate. There is evidence that
tumor cells have a less rigid cell membrane than normal cells. This could be
associated with a rapid membrane turnover, and more cholesterol thus might
be required.

Shiroeda et al. recently reported that a human lung cancer cell line
produced factors which increased the LDL receptor activity of other cells [43].
By gel filtration, two fractions were isolated which stimulated the LDL receptor
activity, one with a molecular weight of 140,000 and the other with a molecular
weight exceeding 850,000. These fractions also stimulated the incorporation
of tritiated thymidine into DNA as well as the de novo synthesis of cholesterol.

IV. CURRENT PROBLEMS IN CANCER CHEMOTHERAPY

Many highly cytotoxic drugs have been introduced for cancer chemotherapy
[44]. About a dozen cancer forms (e.g., testicular cancer and certain forms of

leukemias and lymphomas) can now be considered to be curable by chemotherapy even at advanced stages. Unfortunately, these tumors account for only about 10% of all cancers. The most common types—breast cancer, lung cancer, gastric cancer and colorectal cancer—are less sensitive to chemotherapy. The main reason for the limited success in cancer chemotherapy is the insufficient selectivity of the antineoplastic drugs, causing toxic effects not only on tumor cells but also on normal cells, particularly those dividing in the bone marrow and the gastrointestinal mucosa, for example. Consequently, there is much interest in new treatment strategies to target cytotoxic drugs to malignant cells in order to enhance the therapeutic effects and/or reduce the toxic side effects.

At the beginning of the twentieth century Paul Ehrlich proposed that molecules with an affinity for certain tissues be used as "magic bullets," carrying the therapeutic agent to the desired cells. A number of carriers have been tested experimentally and some also clinically. Drugs have been attached to DNA, hormones, and monoclonal antibodies and encapsulated in liposomes [45,46]. However, clinical success thus far has been limited. Doxorubicin administered as a DNA complex seems to be superior to the free drug in the treatment of acute myeloblastic leukemia in adults [47]. However, the complex probably dissociates readily in plasma and therefore does not act as a true targeting device but rather as a slow release preparation of doxorubicin. This changes the pharmacokinetics of the drug, leading to a higher drug uptake in leukemic cells after administration of doxorubicin-DNA as compared to doxorubicin alone [48].

One potential advantage with liposomes as drug carriers is that they are easy to load with drugs, both lipid- and water-soluble. Early in vitro experiments with liposomes were encouraging but when drug-containing liposomes were administered to animals, the liposomes were recognized as "non-self" and hence rapidly removed from the circulation by cells of the reticuloendothelial system [49]. Nevertheless, this has been taken advantage of in the treatment of fungal infections in the liver of immunocompromised patients by liposome encapsulation of amphotericin B [50,51]. Furthermore, it is now possible to prepare liposomes with a prolonged circulation time [52]. However, a major problem with liposomes as drug carriers for cancer treatment is the lack of a homing device directing the liposomes to the tumor cells.

From the selectivity point of view, antibodies are very attractive as drug carrier candidates. Evidence has been presented that the administration of murine monoclonal antibodies to cancer patients can inhibit tumor growth [53]. It seems likely that the effectiveness of such a therapeutic approach could be enhanced by linking cytotoxic drugs or toxins to the monoclonal antibodies. However, there are many problems involved in such an approach. True tumor-specific antigens probably do not exist. Another problem is the existence of circulating antigens that could bind the antibodies before they reach the tumor site. Attempts to use radiolabeled antibodies for tumor imaging have not been able to visualize smaller tumor masses than is possible with other diagnostic means like CT scans. Furthermore, the coupling of drug molecules to antibodies may interfere with antigen recognition or with the activity of the drug.

Liposomes are attractive as drug carriers because they are easy to load with drugs with different physicochemical properties. Antibodies have another attractive property since they interact very specifically with the antigen. Ideally, one should try to combine the attractive features of these two carrier systems. At least some of the problems encountered with liposomes and antibodies as drug carriers could be avoided by using LDL as a "natural liposome," taken up in cells by receptor-mediated endocytosis, which is a specific process. Since several malignant cell types express more LDL receptors than normal cells, selectivity should be achieved, at least to a certain extent. Moreover, LDL is known to pass through vascular endothelial barriers [54]. Indeed, the physiological function of LDL is to serve as a carrier of cholesterol to cells in various parts of the body according to their need for cholesterol for, say, membrane synthesis. One prerequisite for the successful use of LDL as a drug carrier in cancer chemotherapy is that a sufficient number of cytotoxic drug molecules can be linked to or incorporated into each LDL particle so that toxic effects are exerted on the malignant cells after uptake of drug-containing LDL. The incorporation must also give a stable complex and the drug must be resistant to the lysosomal milieu. The conformation of apoprotein B should be maintained to allow efficient binding to LDL receptors. At least 85% of the lysine residues of LDL are required for a normal receptor interaction [55]. Another important criterion is that the structure of the particle must not be changed in such a way that it is rendered foreign and recognized by the cells of the reticuloendothelial system.

V. INCORPORATION OF EXOGENOUS SUBSTANCES INTO LDL

Since the physiological function of LDL is to transport water-insoluble cholesterol in the body, it can be expected that exogenous lipophilic substances ingested in the body could also be transported in the oily core of LDL particles.

Remsen and Shireman have presented evidence that benzo(a)pyrene, a lipophilic, potentially carcinogenic substance, is taken up to cells bound to LDL [56]. However, LDL-associated benzo(a)pyrene entered cells from patients with familial hypercholesterolemia to the same extent as cells with normal LDL receptor activity, indicating that cellular uptake did not occur by receptor-mediated endocytosis. The insecticide toxaphene is a lipophilic, ubiquitous pollutant of the environment. Incubation of ^{14}C-toxaphene with human plasma in vitro resulted in a relatively homogeneous distribution of the radioactivity among the lipoprotein fractions [57]. In rat plasma, the distribution was different, which shows that there are species differences in the transport of lipophilic substances. In the rat, toxaphene accumulated in the adrenal cortex and when added to rat adrenocortical cells in vitro, toxaphene inhibited ACTH-stimulated corticosterone synthesis. However, there are strong indications that the cellular uptake of toxaphene did not occur via receptor-mediated endocytosis of LDL. Thus there was no difference in the tissue distribution of radioactivity in rats after injection of ^{14}C-toxaphene associated with native and acetylated LDL. Acetylation of LDL abolishes its ability to recognize LDL receptors [55].

In 1977 Shireman et al. reported that a soluble complex is formed between apoprotein B and albumin after removal of the lipid components of LDL [58]. The complex retained the ability to bind to LDL receptors of human fibroblasts. This finding suggested that it might be possible to remove the cholesteryl ester core of LDL and replace it with exogenous lipophilic substances without altering the ability of the LDL particle to bind to the LDL receptor. As a first test of this idea, Krieger et al. described a method by which more than 99% of the cholesteryl esters in the core of the LDL particle could be removed by heptane extraction and replaced with exogenous cholesteryl linoleate [59]. The reconstituted LDL particle was taken up and degraded by cells and regulated HMG-CoA reductase activity like native LDL. It was also taken up at a reduced rate in mutant fibroblasts from a patient with homozygous familial hypercholesterolemia. In the next step, Krieger et al. used the same technique to incorporate fluorescent probes into LDL [60]. It was thereby possible to visualize the receptor-mediated endocytosis of LDL in cultured fibroblasts. When fibroblasts from healthy subjects were incubated with the fluorescent LDL preparations, fluorescent granules appeared in the cells. In contrast, no fluorescent granules could be seen in fibroblasts from patients with the homozygous form of familial hypercholesterolemia.

VI. CYTOTOXIC DRUG DELIVERY VIA LDL RECEPTORS IN VITRO

A cytotoxic effect was obtained on exposure of fibroblasts to LDL in which the cholesteryl esters had been replaced by 25-hydroxycholesteryl oleate, a toxic cholesterol analog ester [61]. The reconstituted LDL suppressed the activity of HMG-CoA reductase in fibroblasts, and in the absence of a source of cholesterol, the cells developed an abnormal morphology, their growth was inhibited, and the cells died. Cells from a patient with the homozygous form of familial hypercholesterolemia were far less sensitive. By incorporation of pyrene, a photosensitizing agent, into LDL it was possible to obtain a receptor-mediated killing of cells in culture on subsequent exposure to ultraviolet light [62]. Normal fibroblasts but not fibroblasts from a patient with familial hypercholesterolemia in homozygous form were killed.

Firestone and co-workers incorporated a number of compounds into LDL and studied the selective cytotoxicity of the complexes formed [63]. Most of the compounds studied either were impossible to incorporate by the Krieger method or were not toxic after successful incorporation. However, two compounds were found that reconstituted well and exerted a selective cytotoxicity via the LDL receptor pathway at reasonably low concentrations. One was nitrogen mustard carbamate lined to the cholesterol nucleus esterified with oleic acid. At a concentration of 10 μg/ml, this compound incorporated into LDL arrested growth of most of the normal cells but killed only half of them. The toxic effects could not be enhanced by increasing the dose, probably due to saturation of the LDL pathway. When an additional methylene group was introduced at the nitrogen mustard, a compound was obtained that reconstituted well, was completely selective for cells expressing LDL receptors, and killed all test cells at about 5 μg/ml and above.

Other techniques have also been tried for preparing cytotoxic drug-LDL complexes. An aclacinomycin A–LDL complex containing 150–450 drug molecules per LDL particle could be prepared simply by incubating LDL with a large excess of the drug dissolved in buffer at 40°C [64]. On incubation with a human glioma cell line, cell growth was inhibited. It could be demonstrated that drug uptake was dependent on the expression of LDL receptors. However, cellular drug uptake also occurred by receptor-independent pathways, probably after leakage from the particles.

Iwanik and co-workers could incorporate about 7 molecules of daunorubicin per LDL particle [65]. The incorporation method consisted of using a dry film or powdered drug and stirring with LDL at 37–40°C for 2 h. Judging from data on fluorescence quenching by DNA, about 60% of the drug was located in the core of LDL. Native LDL but not HDL competed with drug uptake in P388 cells [66]. The complex exerted cytotoxicity also on a daunorubicin-resistant P388 subline, which is of great interest in regard to overcoming drug resistance by modification of the drug delivery pathway. With this incorporation method, it was possible to increase the incorporation by using a more lipophilic drug. Thus about 100 molecules of AD-32 (N-trifluoroacetyladriamy-cin-14-valerate), a very lipophilic derivative of the clinically widely used drug doxorubicin [67], could be incorporated per LDL particle [68]. The LDL complexes with daunorubicin and AD-32 showed precipitation of greater than 70% drug when treated with antibodies to apoprotein B-100. Hynds [69] and Kerr and Kaye [70] used the same method for incorporation of daunorubicin into LDL and tested the cytotoxicity on lung tumor cell spheroids. They found

LDL-incorporated daunorubicin was more effective than free daunorubicin both when measured by spheroid growth delay and in a clonogenic assay after disaggregation of the spheroids.

Hynds also used another approach in the attempts to load LDL with cytotoxic drugs [69]. In a first step, chlorambucil was esterified with cholesterol, thereby making the drug more lipophilic. Then the drug ester was loaded into HDL by exchange from micelles. Finally, the chlorambucil ester was incorporated into LDL by exchange from HDL, utilizing the cholesteryl ester transfer protein from human plasma. However, the complex obtained had virtually no cytotoxicity due to drug degradation during the incorporation procedure. Therefore, the method was simplified so that exchange was performed directly from a drug–lipid microemulsion. Up to 30% of the chlorambucil used was recovered in LDL by this method. When the inhibition of protein synthesis was tested on human glioma cells in culture, the complex was more effective than chlorambucil itself.

A similar technique was used to incorporate 9-methoxy-ellipticin into LDL [71]. The drug was incorporated into dimirystoyl phosphatidylcholine, cholesteryl oleate stabilized microemulsion, and the latter fused with human LDL. The drug–LDL complex was more effective than the free drug to kill L1210 and P388 leukemic cells in vitro. This is surprising, since it is difficult to imagine that receptor-mediated endocytosis could be a more rapid drug transport mechanism than diffusion. However, several lines of evidence were presented indicating that the cytotoxic activity of the drug–LDL complex depended on the LDL receptor.

Lundberg used an alternative method to incorporate a lipophilic cytotoxic mustard carbamate into LDL [72]. In this method, LDL was first delipidated with a detergent or a sterolester hydrolase. The solubilized apoprotein B was then added to a sonicated microemulsion consisting of the compound to be incorporated and egg yolk phosphatidylcholine. Then LDL–drug complexes were isolated by ultracentrifugation (see Chapter 3). The complex killed 100% of LDL receptor possessing cells in culture. Heparin, which is known to inhibit uptake of LDL by the receptor mechanism, abolished the cytotoxicity of the complex. This incorporation method gave a recovery of about 60% compared to less than 10% with the Krieger method and the simple incubation of LDL with drug.

Using the reconstitution method of Krieger et al. [60], we could incorporate AD-32 into LDL [73]. The complex contained approximately 500 drug molecules per LDL particle. When isolated white blood cells were incubated with the AD-32–LDL complex, cellular drug accumulation widely exceeded that which could be calculated from the uptake and degradation of the LDL part of the complex. This indicates that in addition to the receptor-mediated drug uptake, there was also nonspecific drug delivery to the cells. Indeed, AD-32 does not fulfill the structural requirements pointed out by Krieger et al. [60] for substances to be incorporated into LDL. We therefore synthesized a linoleyl-retinoyl derivative of doxorubicin and could incorporate 100–200 drug molecules per LDL particle [74]. Studies of normal and mutant (receptor-deficient) human fibroblasts revealed that this drug, incorporated into LDL, was selectively taken up by cells via the LDL receptor pathway. Thus the presence of excess native LDL inhibited the uptake and degradation of the LDL part of the complex and to the same extent also the drug uptake. In contrast, methylated LDL, which does not bind to LDL receptors [55], did not inhibit cellular uptake and degradation of LDL, nor did it reduce cellular drug accumulation. It could be calculated that about 80% of the total drug uptake occurred via receptor-mediated uptake of drug–LDL complex.

Complexes have also been prepared between LDL and highly polar compounds. Methotrexate was covalently linked to the apoprotein B part of LDL [75]. About 10 drug molecules were attached to each LDL particle, resulting in an increase in the radius and polydispersity of the particles. The cytotoxic effect of LDL-linked methotrexate on L1210 murine leukemia cells in vitro was 30 times less than that of the free drug. There is no information on the expression of LDL receptors on the cells used.

VII. IN VIVO FATE OF DRUG-LDL COMPLEXES

Hynds tested drug–LDL complexes in vivo [69] using the subrenal capsule assay first described by Bogden et al. [76]. In this method, fragments of human solid tumors are implanted under the renal capsule of normal immunocompetent mice. The animals are then treated and the effect on tumor size is determined 6 days after tumor implantation. Treatment of the mice with daunorubicin–LDL caused a smaller size increase than treatment with free daunorubicin, but histopathological examination showed that the effect was due mainly to greater inhibition of lymphocyte infiltration and not to delay of tumor cell growth.

By using the LDL reconstitution procedure of Krieger et al. [60], cytotoxic compounds can be delivered to cells in vitro by the LDL receptor pathway. However, after intravenous injections in mice of complexes prepared according to this method, we found that the modified LDL was removed from the blood much more rapidly than was native LDL [77]. The modified lipoprotein accumulated in the liver and the spleen. This was also the case with cholesteryl oleate incorporated into LDL with the Krieger method. The method involves LDL being subjected to freezing, freeze-drying, and extraction with heptane in the presence of potato starch to protect the protein. Just freezing and thawing of LDL greatly affected the in vivo fate of LDL [77].

By using sucrose instead of starch as protective agent, we found that it was possible to incorporate AD-32 into LDL without altering the in vivo fate of the lipoprotein [77]. Using the quasielastic light scattering technique, the hydrodynamic radius of LDL–AD-32 prepared under protection of sucrose was

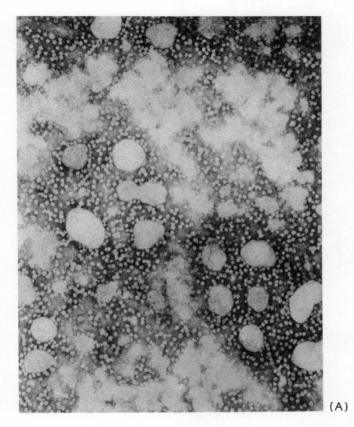

(A)

Figure 9 Electron micrographs of native LDL (A) and of LDL-WB 4291 prepared with the Krieger technique (B) and with the technique described in Ref. 77 (C). The LDL preparations were applied to carbon–formvar membranes and negatively stained with 2% phosphotungsten solution. They were examined on a Philips EM301 instrument; magnification 40,000.

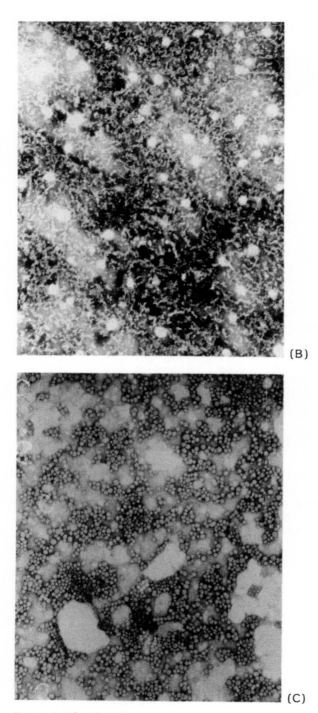

(B)

(C)

Figure 9 (Continued)

150 Å as compared to 130 Å for native LDL and 300 Å for LDL–AD-32 prepared by the Krieger technique. It has also been possible to incorporate WB-4291, a lipophilic nitrogen mustard derivative [78], into LDL without interfering with the in vivo fate of LDL. Figure 9 presents electron micrographs of native LDL (A) and LDL–WB 4291 complex prepared by the Krieger method (B) and the new method (C). The latter complex has potent antitumor effects in leukemic mice.

Using our technique, Eley et al. succeeded in incorporating about 400 molecules of estramustine, a nornitrogen mustard derivative of estradiol, per LDL particle [79]. The complex diameter was only 3% larger than the diameter of native LDL. In contrast, they could incorporate only about 13 molecules of prednimustine, prednisolone linked to chlorambucil, per LDL particle. The diameter of the prednimustine-containing particle increased by 700% compared to native LDL. It was speculated that the reason why prednimustine but not estramustine perturbs the structure of the particle could be that prednimustine has two free secondary hydroxyl groups, whereas estramustine has only one. Estramustine incorporated into LDL retained cytotoxic activity.

VIII. FUTURE OF LDL AS A DRUG CARRIER IN CANCER CHEMOTHERAPY

Use of LDL as a drug carrier in cancer chemotherapy may help to solve two current problems. The armamentarium of cytotoxic drugs could be extended considerably by using LDL as a solvent for lipophilic drugs which cannot be used today because of lack of a suitable vehicle. The second and even more important step would be to increase the drug concentration in malignant cells expressing a higher number of LDL receptors than most normal cells. However, it will not be possible to obtain an absolute selectivity since normal cells also have LDL receptors. The normal organs most likely to suffer from LDL-incorporated chemotherapy are the adrenals and the liver. The adrenal cortex has high LDL receptor activity because of the high rate of steroid hormone synthesis. It will probably be possible to downregulate the LDL uptake in the adrenals by pretreatment with corticosteroids. In fact, the results of animal experiments by Shepherd and co-workers support this possibility [36]. The liver also has high LDL receptor activity but since most drugs are metabolized to less toxic compounds in the liver, this problem might be easier to handle. Furthermore, it is possible to downregulate the LDL receptors in the liver by infusion of bile acids [36].

ACKNOWLEDGMENTS

The studies performed in our laboratory were supported by the Swedish Cancer Society and the Karolinska Institute. We thank Dr. Marianne Anderson, University of Lund, Sweden, who carried out the quasielastic light scattering measurements, and Dr. Peter Collins, Department of Tumor Pathology, Karolinska Hospital, who prepared the electron micrographs of the LDL complexes.

REFERENCES

1. Goldstein JL, Brown MS. Atherosclerosis: the low-density lipoprotein receptor hypothesis. Metabolism 1977; 26:1257–1275.
2. Brown MS, Faust JR, Goldstein JL. Role of the density lipoprotein pathway in regulating the content of free and esterified cholesterol in human fibroblasts. J Clin Invest 1975; 55:783–793.
3. Goldstein JL, Brown MS. Progress in understanding the LDL receptor and HMG-CoA reductase, two membrane proteins that regulate the plasma cholesterol. J Lipid Res 1984; 25:1450–1461.
4. Shepherd J, Bicker S, Lorimer AR, Packard CJ. Receptor-mediated low density lipoprotein catabolism in man. J Lipid Res 1979; 20:999–1006.
5. Feinleib M. Review of the epidemiological evidence for a possible relationship between hypocholesterolemia and cancer. Cancer Res 1983; 43:2503s–2507s.
6. Sherwin RW, Wentworth DN, Cutler JA, Hulley SB, Kuller LH, Stamler J. Serum cholesterol levels and cancer mortality in 361662 men screened for the multiple risk factor intervention trial. JAMA 1987; 357:943–948.
7. McMichael AJ, Jensen OM, Parkin DM, Zaridze DG. Dietary and endogenous cholesterol and human cancer. Epidem Rev 1984; 6:192–216.
8. Carroll KK, Braden LM, Bell JA, Kalamegham R. Fat and cancer. Cancer 1986; 58:1818–1825.
9. Schatzkin A, Hoover RN, Taylor PR, Ziegler RG, Carter CL, Albanes D, Larson DB, Licitra LM. Site-specific analysis of total serum cholesterol and incident cancer in the national health and nutrition examination survey. I. Epidemiologic follow-up study. Cancer Res 1988; 48:452–458.
10. Knekt P, Reunanen A, Aromaa A, Heliövaara M, Hakulinen T, Hakama M. Serum cholesterol and risk of cancer in a cohort of 39000 men and women. J Clin Epidemiol 1988; 41:519–530.
11. Törnberg S. Cancer risks in relation to serum levels of cholesterol and beta-lipoprotein. An epidemiologic study. Thesis. Karolinska Institutet, Stockholm. 1987.
12. Reddy BS. Dietary fat and its relationship to large bowel cancer. Cancer Res 1981; 41:3700–3705.
13. Rose G, Shipley MJ. Plasma lipids and mortality: a source of error. Lancet 1980; i:523–526.
14. Chao FC, Efron B, Wolf P. The possible prognostic usefulness of assessing serum proteins and cholesterol in malignancy. Cancer 1975; 35:1223–1229.

15. Budd D, Ginsberg H. Hypocholesterolemia and acute myelogenous leukemia. Association between disease activity and plasma low-density lipoprotein cholesterol concentrations. Cancer 1986; 58:1361–1365.

16. Alexopoulos CG, Blatsios B, Averginos A. Serum lipids and lipoprotein disorders in cancer patients. Cancer 1987; 60:3065–3070.

17. Eriksson M, Angelin B, Berglund L, Eriksson S, Henriksson P, Rudling M, Stege R. Low serum cholesterol in metastasizing prostate carcinoma: due to receptor-mediated catabolism of LDL? Proc XIIVth Annual Meeting Swedish Medical Association, 1987:182. (in Swedish)

18. Ho YK, Smith GS, Brown MS, Goldstein JL. Low density lipoprotein (LDL) receptor activity in human myelogenous leukemia cells. Blood 1978; 52:1099–1104.

19. Vitols S, Gahrton G, Öst Å, Peterson CO. Elevated low density lipoprotein receptor activity in leukemic cells with monocytic differentiation. Blood 1984; 63:1186–1193.

20. Vitols S, Gahrton G, Björkholm M, Peterson C. Hypocholesterolemia in malignancy due to elevated LDL receptor activity in tumor cells; evidence from studies in leukemic patients. Lancet 1985; 23:1150–1154.

21. Lindquist R, Vitols S, Gahrton G, Öst Å, Peterson C. Low density lipoprotein receptor activity in human leukemic cells: relation to chromosome aberrations. Acta Med Scand 1985; 217:553–558.

22. Neel BG, Jhanwar SC, Chaganti RSK, Hayward WC, Two human c-oncogenes are located on the long arm of chromosome 8. Proc Natl Acad Sci USA 1982; 79:7842–7846.

23. Betteridge DJ, Krone W, Ford JM, Galton DJ. Regulation of sterol synthesis in leukaemic blast cells: a defect resembling familial hypercholesterolemia. Eur J Clin Invest 1979; 9:439–441.

24. Klock JC, Pieprzyk JK. Cholesterol, phospholipids, and fatty acids of normal immature neutrophils: comparison with acute myeloblastic leukemia cells and normal neutrophils. J Lipid Res 1979; 20:908–911.

25. DeLamatre J, Wolfbauer G, Phillips MC, Rothblat GH. Role of apolipoproteins in cellular cholesterol efflux. Biochim Biophys Acta 1986; 875:419–428.

26. Gal D, MacDonald PS, Porter JC, Smith JW, Simpson ER. Effect of cell density and confluency on cholesterol metabolism in cancer cells in monolayer culture. Cancer Res 1981; 41:473–477.

27. Gal D, Ottashi M, MacDonald PC, Buschbaum HJ, Simpson ER. Low density lipoprotein as a potential vehicle for chemotherapeutic agents and radionucleotides in the management of gynecologic neoplasms. Am J Obstet Gynecol 1981; 139:877–885.

28. Rudling MJ, Peterson CO. A simple binding assay for the determination of low-density lipoprotein receptors in cell homogenates. Biochim Biophys Acta 1985; 833:359–365.

29. Rudling MJ, Reihner E, Einarsson K, Ewerth S, Angelin B. Low density lipoprotein receptor binding activity in human tissues: quantitative importance of hepatic receptors and evidence for regulation of their expression in vivo. Proc Natl Acad Sci USA. In press.

30. Rudling MJ, Stihle L, Peterson CO, Skoog L. Content of low density lipoprotein receptors in breast cancer tissue related to survival of patients. Brit Med J 1986; 292:580–582.

31. Rudling MJ, Angelin B, Peterson CO, Collins VP. Low density lipoprotein receptor activity in human intracranial tumors and its relation to the cholesterol requirement. Cancer Res 1990; 50:483–487.

32. Yamamoto T, Davis CG, Brown MS, Schneider WJ, Casey ML, Goldstein JL, Russel DW. The human LDL receptor: a cysteine-rich protein with multiple Ala sequences in its mRNA. Cell 1984; 39:27–38.

33. Ho WKK, Leung AM. Binding of homologous versus heterologous low density lipoproteins by different tissues of the rat. Comp Biochem Physiol; 72B:547-549.

34. Chapman MJ. animal lipoproteins: chemistry, structure, and comparative aspects. J Lipid Chem 1980; 21:789–852.

35. Norata G, Canti G, Ricci L, Nicolin A, Trezzi E, Catapano AL. In vivo assimilation of low density lipoprotein by a fibrosarcoma tumour line in mice. Cancer Lett 1984; 25:203–208.

36. Hynds SA, Welsh J, Stewart JM, Jack A, Soukop M, McArdle CS, Calman KC, Packard CP, Shepherd J. Low-density lipoprotein metabolism in mice with soft tissue tumors. Biochim Biophys Acta 1984; 795:589-595.

37. Dresel HA, Friedrich E, Via DP, Schettler G, Sinn H. Characterisation of binding sites for acetylated low density lipoprotein in the rat liver in vivo and in vitro. EMBO J 1985; 4:1157–1162.

38. Pittman RC, Attie AD, Carew TE, Steinberg D. Tissue sites of degradation of low density lipoprotein: application of method for determining the fate of plasma proteins. Proc Natl Acad Sci USA 1979; 76:5345–5349.

39. Vitols S, Angelin B, Ericsson S, Gahrton G, Juliusson G, Masquelier M, Paul C, Peterson C, Rudling M, Söderberg-Reid K, Tidefeldt U. Uptake of low density lipoproteins by human leukemic cells in vivo: relation to plasma lipoprotein levels and possible relevance for selective chemotherapy. Proc Natl Acad Sci USA 1990; 87:2598–2602.

40. Lees RS, Garabedian HD, Lees AM, et al. Technetium-99m low density lipoproteins: preparation and biodistribution. J Nucl Med 1985; 26:1056–1062.

41. Vallabhajosula S, Paidi M, Badimon JJ, Le N-A, Goldsmith SJ, Fuster V, Ginsberg H. Radiotracers for low density lipoprotein biodistribution studies in vivo: technetium-99m low density lipoprotein versus radioiodinated low density lipoprotein preparations. J Nucl Med 1988; 29:1237–1245.

42. Vallabhajosula S, Gilbert HS, Goldsmith SJ, Paidi M, Hanna MH, Ginsberg HN. Low-density lipoprotein (LDL) distribution shown by ^{99m}technetium-LDL imaging in patients with myeloproliferative diseases. Ann Intern Med 1989; 110:208-213.

43. Shiroeda O, Yamaguchi N, Kawai K. Stimulation of low density lipoprotein receptor activity by conditioned medium from a human cancer cell line. Cancer Res 1987; 47:4630–4633.

44. Krakoff IH. Cancer chemotherapeutic agents. Chem Abstr 1987; 37:93–105.

45. Poznansky MJ, Juliano RJ. Biological approaches to the controlled delivery of drugs: a critical review. Pharmcol Rev 1984; 36:277–336.

46. Shaw JM, Shaw KV, Schook LB. Drug delivery particles and monoclonal antibodies. In:Schook LB, ed. Monoclonal antibody production techniques and applications. New York:Marcel Dekker, 1987:285–310.

47. Paul C, Björkholm M, Christenson I, Engstedt L, et al. Induction and intensive

consolidation in acute nonlymphoblastic leukemia (ANL) with combinations containing doxorubicin-DNA. Prolonged remission duration and survival. Proc 14th Int Cancer Congress, Budapest, 1986, Lectures and Symposia, 1987; 10:101–108.

48. Peterson C, Paul C, Gahrton G. anthracycline-DNA complexes as slow-release preparations in the treatment of acute leukemia. In:Lewis DE, ed. Controlled release of pesticides and pharmaceuticals. New York:Plenum, 1981:49–65.

49. Poste G. Liposome targeting in vivo: problems and opportunities. Biol Cell 1983; 47:19–39.

50. Lopez-Berestein G, Fainstein V, Hopfer R, et al. Liposomal amphotericin B for the treatment of systemic fungal infections in patients with cancer, a preliminary study. J Infect Dis 1985; 151:704–710.

51. Sculier J-P, Coune A, Meunier F, Brassinne C, Laduron C, Hollaert C, Collette N, Heymans C, Klastersky J. Pilot study of amphotericin B entrapped in sonicated liposomes in cancer patients with fungal infections. Eur J Cancer Clin Oncol 1988; 24:527–538.

52. Geho W, Lau J. Masking of liposomes from RES recognition. US Pat 4,501,728 (1985).

53. Foon KA, Schroff RW, Bunn PA. Monoclonal antibody therapy for patients with leukemia and lymphoma. Dev Oncol 1985; 38:139–160.

54. Brown MS, Goldstein JL. Receptor-mediated endocytosis: insights from the lipoprotein receptor system. Proc Natl Acad Sci USA 1979; 76:3330–3337.

55. Weisgraber KH, Innerarity TL, Mahley RW. Role of lysine residues of plasma lipoproteins in high affinity binding to cell surface receptors on human fibroblasts. J Biol Chem 1978; 253:9053–9602.

56. Remsen J, Shireman R. Effect of low density lipoprotein on the incorporation of benzo(a)pyrene by cultured cells. Cancer Res 1981; 41:3179–3185.

57. Mohammed AB. Studies on the fate and biological effects of toxaphene (polychlorinated terpenes) in vitro and in vivo. Thesis. University of Uppsala. 1988.

58. Shireman R, Kilgore LL, Fisher WR. Solubilization of apolipoprotein B and its specific binding by the cellular receptor for low density lipoprotein. Proc Natl Acad Sci USA 1977; 74:5150–5154.

59. Krieger M, Brown MS, Faust JR, Goldstein JL. Replacement of endogenous cholesteryl esters of low density lipoprotein with exogenous cholesteryl linoleate. Reconstitution of a biologically active lipoprotein particle. J Biol Chem 1978; 253:4093–4101.

60. Krieger M, Smith LC, Anderson RGW, Goldstein JL, Kao YJ, Pownall HJ, Gotto AM Jr, Brown MS. Reconstituted low density lipoprotein: a vehicle for the delivery of hydrophobic fluorescent problems to cells. J Supramol Struct 1979; 10:467–478.

61. Krieger M, Goldstein JL, Brown MS. Receptor-mediated uptake of low density lipoprotein reconstituted with 25-hydroxycholesteryl oleate suppresses 3-hydroxy-3-methylglutaryl-coenzyme A reductase and inhibits growth of human fibroblasts. Proc Natl Acad Sci USA 1978; 75:5052–5056.

62. Mosley ST, Goldstein JL, Brown MS, Falck JR, Anderson RGW. Targeted killing of cultured cells by receptor-dependent photosensitization. Proc Natl Acad Sci USA 1981; 78:5717–5721.

63. Firestone RA, Pisano JM, Falck JR, McPhaul MM, Krieger M. Selective delivery of cytotoxic compounds to cells by the LDL pathway. J Med Chem 1984; 27:1037–1043.

64. Rudling MJ, Collins VP, Peterson C. Delivery of aclamycin A to human glioma cells in vitro by the low-density lipoprotein pathway. Cancer Res 1983; 43:4600–4605.

65. Iwanik M, Shaw KV, Ledwith B, Yanovich S, Shaw JM. Preparation and interaction of a low-density lipoprotein:daunomycin complex with P388 leukemic cells. Cancer Res 1984; 44:1206–1215.

66. Yanovich S, Preston L, Shaw JM. Characteristics of uptake and cytotoxicity of a low-density lipoprotein–daunomycin complex in P388 leukemic cells. Cancer Res 1984; 44:3377–3382.

67. Israel M, Modest EJ, Frei E. N-trifluoroacetyl-adriamycin-14-valerate: an analog with greater experimental antitumor activity and less toxicity than adriamycin. Cancer Res 1975; 35:1365–1368.

68. Shaw JM, Shaw KV, Yanovich S, Iwanik M, Futch WS, Rosowsky A, Schook LB. Delivery of lipophilic drugs using lipoproteins. Ann NY Acad Sci 1987; 507:252–271.

69. Hynds SA. Cancer chemotherapy: use of low density lipoproteins as targeting vehicles for treatment. Thesis. University of Glasgow. 1986.

70. Kerr DJ, Kaye SB. Aspects of cytotoxic drug penetration with particular reference to anthracyclines. Cancer Chemother Pharmacol 1987; 19:1–5.

71. Samadi-Baboli M, Favre G, Blancy E, Soula G. Preparation of low density lipoprotein–9-methoxy-ellipticin complex and its cytotoxic effect against L1210 and P388 leukemic cells in vitro. Eur J Cancer Clin Oncol 1989; 25:233–241.

72. Lundberg B. Preparation of drug–low density lipoprotein complexes for delivery of antitumoral drugs via the low density lipoprotein pathway. Cancer Res 1987; 47:4105–4108.

73. Vitols S, Gahrton G, Peterson C. Significance of the low-density lipoprotein (LDL) receptor pathway for the in vitro accumulation of AD-32 incorporated into LDL in normal and leukemic white blood cells. Cancer Treat Rep 1984; 68:515–520.

74. Vitols SG, Masquelier M, Peterson CO. Selective uptake of a toxic lipophilic anthracycline derivative by the low-density lipoprotein receptor pathway in cultured fibroblasts. J Med Chem 1985; 28:451–454.

75. Halbert GW, Stuart JF, Florence T. A low density lipoprotein–methotrexate covalent complex and its activity on L1210 cells in vitro. Cancer Chemother Pharmacol 1985; 15:223–227.

76. Bogden AE, Cobb WR, Lepage DJ, et al. Chemotherapy responsiveness of human tumour as a first transplant generation xenografts in the normal mouse: six-day subrenal capsule assay. Cancer 1981; 48:10–20.

77. Masquelier M, Vitols S, Peterson C. Low-density lipoprotein as a carrier of antitumoral drugs: in vivo fate of drug–human low-density lipoprotein complexes in mice. Cancer Res 1986; 46:3842–3847.

78. Berenbaum MC. Time-dependence and selectivity of immunosuppressive agents. Immunology 1979; 36:355–365.

79. Eley JG, Halbert GW, Florence AT. The incorporation of steroidal alkylating agents into low density lipoprotein. J Pharm Pharmacol 1989; 41:4p.

6

Acetylated Low Density Lipoprotein and the Delivery of Immunomodulators to Macrophages

Mark S. Rutherford,* W. Stewart Futch, Jr.,[†] and Lawrence B. Schook
University of Illinois, Urbana, Illinois

I. INTRODUCTION

The development of biological response modifiers and the identification of lymphokines that regulate macrophage (Mϕ) antimicrobial and antitumor activities have attracted great interest [1–3]. Although many of these factors are efficacious *in vitro*, therapeutically relevant doses are often toxic to normal cells or may cause side effects such as fever, nausea, fatigue, or alterations in white blood cell counts, as is observed when interferon gamma (IFN-γ) [4] or tumor necrosis factor alpha (TNF-α) [5] is injected freely into subjects. An additional consideration is that many of the modulators are subject to rapid bioinactivation or excretion. One such example is that of muramyl dipeptide (MDP; *N*-acetylmuramyl-*L*-alanyl-*D*-isoglutamine), a component of the mycobacterium cell wall. Although it is a powerful inducer of Mϕ antitumor [6] and antimicrobial [7] activities, MDP is excreted from the body within 60 min [8], thereby limiting its efficacy for therapy *in vivo*. However, liposomes containing MDP or its lipophilic derivative muramyl tripeptide phosphatidylethanolamine (MTP-PE, amide composed of *N*-acetylmuramyl-*L*-alanyl-*D*-

Present Affiliation:
* St. Jude Children's Research Hospital, Memphis, Tennessee
† Eastern Virginia Graduate School of Medicine, Norfolk, Virginia

Table 1 Cells Lines and Tissues Used in Acetyl-LDL Binding Studies

Cell type	Derivation/elicitation
T lymphocyte	Primary isolation from thymus
B lymphocyte	X63-AG8.653 myelomas
Endothelial cell	Human umbilical vein endothelial (HUVE) cell
HL-60	Myeloid, promyelocytic leukemia
BMDM	CSF-1–induced bone marrow–derived macrophage
Resident macrophage	Adherent cells obtained after saline lavage of peritoneum
Responsive macrophage	Adherent proteose peptone-elicited peritoneal exudate cells
Elicited macrophage	Adherent thioglycolate-elicited peritoneal exudate cells
Fully activated macrophage	Adherent *C. parvum* (*P. acne*)-elicited peritoneal exudate cells

isoglutamyl-*L*-alanine and dipalmitoyl phosphatidylethanolamine; see Table 1) showed effective activation of Mφs to destroy both tumor cells [9] and fatal herpes simplex virus type 2 (HSV-2)–infected cells without causing the lysis of uninfected cells [8].

Liposomes interact with a wide variety of cell types and are therefore nonspecific for Mφs. Furthermore, they are often unable to pass through the vascular endothelium, hence they bypass resident tissue or tumor-associated Mφ. Thus their use for Mφ-specific delivery of biological response modifiers is inefficient. Extensive investigations have revealed that Mφ possess mechanisms that allow the uptake and digestion of cholesterol-containing lipoproteins [10]. In particular, low density lipoprotein (LDL) that has been modified by chemical acetylation (acetyl-LDL) is taken up with extremely high efficiency by Mφs via receptor-mediated endocytosis [10]; thus acetyl-LDL is now being examined for use as a Mφ-specific drug delivery vehicle. The acetyl-LDL endocytotic pathway (also called scavenger pathway) occurs in a restricted number of cell types, primarily Mφs and certain endothelial cells (e.g., liver endothelial cells, Chapter 7, Ref. 9), and the uptake of acetyl-LDL by Mφs and blood monocytes has been well characterized [9,10]. Acetyl-LDLs pass through the vascular endothelium into tissues; therefore, they are suitable for site delivery to resident tissue and tumor-associated Mφs. Furthermore, it is now possible to sequester a number of lipophilic drugs into lipoproteins [9,11,12]. This chapter describes recent work in the development of this Mφ-specific acetyl-LDL drug delivery system, preparation of acetyl-LDL–biological response modifier complexes, and applications for induction of Mφ antitumor and antimicrobial activities.

II. MACROPHAGE IMMUNE FUNCTION

Macrophages are bone marrow–derived mononuclear cells with established roles in host protection against facultative and obligate bacterial pathogens [2,7], viruses [8,13], protozoan parasites [14], and neoplastic cells [2,11]. However, Mϕs are not constitutively competent to destroy pathogenic organisms or neoplastic cells. Studies have shown that induction of cytocidal activity proceeds by a series of defined developmental steps [2,15] initiated when IFN–γ binds to its specific cell surface receptor [2] on Mϕ. When stimulated by IFN-γ, the Mϕ undergoes several biochemical and functional changes and is now considered to be "primed". Primed Mϕs display increased size, metabolism, and phagocytosis, enhanced MHC class II antigen expression and reactive oxygen product formation, augmented antigen presentation, and the ability to bind, but not kill, tumor cells [2,15].

Upon interaction with a second "triggering" signal, primed Mϕs differentiate further to a fully activated state characterized by the ability to kill facultative intracellular parasites and tumor cells, reduced antigen presentation and proliferation, and the secretion of cytolytic proteases and TNF–α [2,15]. Activating signals include endotoxin, phorbol esters, granulocyte/Mϕ-colony stimulating factor (GM-CSF), and muramyl di- and tripeptides [1,2,6–9,14–17]

The bacteriocidal function of the Mϕ has been studied, and Fc receptor-mediated phagocytosis appears to be the major immune-mediated mechanism for attachment of microbes to Mϕs. This mechanism is especially important for encapsulated bacteria such as *Streptococcus pneumonae* and *Neisseria meningitides* [18]. Complement receptors also contribute to the uptake of microbes during primary challenge with an organism [2]. It has furthermore been reported that nonimmune receptors such as those recognizing glycoproteins with terminal mannose or fructose residues participate in phagocytosis of *Salmonella tryphimurium, Klebsiella aerogenes*, and *Corynebacterium parvum* [19]. Following uptake by the Mϕ, the mechanisms for killing microorganisms include the production of the oxygen metabolites superoxide anion (O_2) and hydrogen peroxide (H_2O_2), lowered phagolysosome pH, and the production of cationic peptides [2,7,20].

Macrophages activated for microbicidal functions are also cytotoxic for tumor cells but not for normal, untransformed cells. Macrophage-mediated tumor cell cytotoxicity is a multistep process that involves the specific binding of neoplastic targets for extended periods and the secretion of cytotoxic products such as TNF–α, cytolytic protease, H_2O_2, and arginase [2,21]. Only fully activated Mϕs secrete these factors and show significant levels of cytotoxicity against tumor cells. It has been shown that resistance to the cytotoxic action of the Mϕ is dependent on the tumor target as certain targets are resistant to the cytolytic mechanisms of activated Mϕs [22].

Much evidence indicates that cytolytic protease is important in Mϕ-mediated lysis of neoplastic cells. Only fully activated Mϕs secrete cytolytic protease,

and secretion of this factor correlates strongly with tumoricidal function of the
Mφ [23]. Low molecular weight protease inhibitors which block cytolytic
protease activity also inhibit tumoricidal activity [24]. Furthermore, the kinetics
of neoplastic cell destruction *in vitro* with cytolytic protease are similar to that
of Mφ-mediated tumor cell cytotoxicity [23]. Although H_2O_2 is unable to
mediate tumor cell lysis, it synergizes with cytolytic protease [2].

A second product of activated Mφs which mediates tumor cell lysis is TNF-α
(cachectin). It is produced exclusively by Mφs and vascular smooth muscle
cells [25,26], and its synthesis is greatly enhanced by treatment with IFN-γ and
endotoxin [27,28], Fc receptor crosslinking [29], and signals that increase
intracellular cGMP levels [21]. The mechanism of TNF-α-mediated tumor cell
cytotoxicity is unclear at present but appears to involve changes in target cell
gene expression [30], which may result from alterations of enzymes essential
for physiological function and cell viability.

III. RECEPTOR-MEDIATED ENDOCYTOSIS OF ACETYL-LDL BY MACROPHAGES

Macrophages take up lipoprotein-bound cholesterol either by phagocytosis of
whole cells or membrane fragments or via receptor-mediated endocytosis of
plasma lipoproteins. Whereas LDL receptors are present on a variety of cell
types, normal tissue Mφs express few receptors for native LDL [31] and take
up native LDL very slowly *in vitro* [10,32]. In contrast, LDL that had been
reacted with acetic anhydride in vitro to form acetyl-LDL is taken up rapidly
by a receptor-mediated endocytotic pathway in Mφs [10,11,31]. It is important
to note that freshly isolated blood monocytes express receptors for native LDL
[10,32,33]. However, the monocyte takes up far greater amounts of acetyl-
LDL than native LDL.

The acetyl-LDL receptor pathway exists on a restricted number of cell
types, primarily Mφs, monocytes, and certain endothelial cells [9], thereby
offering a distinct targeting advantage over the LDL receptor system (Fig. 1).
The acetyl-LDL receptor has been found on Mφs from every source and
species thus far examined, including resident and elicited peritoneal Mφs
[11,32], Kupffer cells [32], monocyte-derived Mφs [32,33], and established lines
of murine Mφ tumors [34]. The receptor is a 260 kilodalton, trypsin-sensitive
glycoprotein with no calcium requirement and a dissociation constant for
acetyl-LDL of $3 \times 10^{-8} M$ [9]. Low density lipoprotein will not bind the acetyl-
LDL receptor and does not interfere with acetyl-LDL uptake by Mφs [9–11].
The large net negative charge of acetyl-LDL determines its ability to bind to
the Mφ acetyl-LDL receptor. Because it is unlikely that acetyl-LDL exists
extracellularly in the body, the *in vivo* function of the pathway has been
suggested as that of a scavenger for LDLs that have been modified by
oxidation products of arachidonic acid or naturally occurring aldehydes [10].

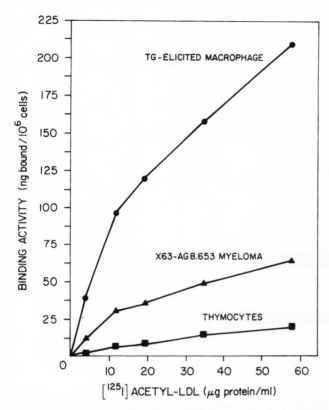

Figure 1 Binding activity of acetyl-LDL to cells of the immune system. [^{125}I]Acetyl-LDL was added to 10^6 cells at 4°C for 2 h. After washing, the amount of bound acetyl-LDL was determined. The thioglycolate (TG)-elicited macrophages were purified by adherence. A differential cell count demonstrated this cell preparation to be ≥90% macrophages, with approximately 10% small lymphocytes. The X63-AG8.653 myeloma cell line (obtained from ATCC) served on a source of B cells.

In effect, any modification of LDL that leads to a large increase in the number of negatively charged amino acid residues causes the LDL to lose its ability to bind the classic LDL receptor and instead bind the acetyl-LDL receptor[10].

Macrophages exist within the body as heterogeneous populations in various stages of functional activation [35,36]. Consideration of the effect of activational status on receptor expression and activity is therefore a prerequisite when choosing a receptor-mediated endocytotic system for delivery of immunomodulators to Mφs. Many receptors, such as those for mannose-conjugated proteins, the Fc region of immunoglobulins, and complement vary markedly with Mφ activational status. In addition, the receptors for LDL and ß-migrating very low density lipoprotein (ß-VLDL) are subject to feedback

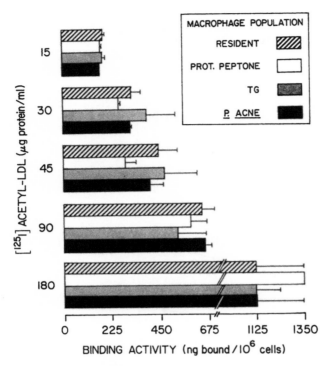

Figure 2 Acetyl-LDL binding activity to macrophages in various phenotypic stages of antitumor activity. [^{125}I]Acetyl-LDL as shown (15, 30, 45, 90, and 180 μg/ml protein) was added to macrophages isolated after elicitation with the indicated agents. After incubation at 4°C for 2h, free acetyl-LDL was washed away, and bound [^{125}I]acetyl-LDL was determined. No significant change in the binding profile was observed if the activity was expressed on a cellular protein basis (ng acetyl-LDL bound/g macrophage protein).

inhibition when cells accumulate large amounts of cholesterol [10]. In contrast, Mϕs of all activational states express amounts of acetyl-LDL receptor similar to those expressed by unstimulated resident Mϕs [10,11] (Fig. 2), and acetyl-LDL receptors remain constant in number even when Mϕs accumulate extremely high amounts of cholesterol [32]. These data suggest that the Mϕ will constitutively take up acetyl-LDL–immunomodulator complexes regardless of activational status. Moreover, the Mϕ will not become refractory to acetyl-LDL–immunomodulator complexes, thus allowing the specific delivery of significantly large doses of the drug in question.

[^{125}I]acetyl-LDL has been used to examine acetyl-LDL binding and uptake by Mϕs [11]. Results showed that both binding and uptake were threefold greater at 37°C than at 4°C. Interaction of [^{125}I]acetyl-LDL with the Mϕ was

Table 2 Induction of Tumoricidal Activity by Acetyl-LDL–MTP-PE

Effector-to-target cell ratio	Cytotoxicity (%)			
	No treatment	Acetyl-LDL–MTP-PE	MTP-PE	Acetyl-LDL
40:1	5 ± 2	20 ± 8	7.5 ± 7	3.5 ± 2
20:1	1.2 ± 2.5	12 ± 2.2	4.3 ± 1.2	6 ± 2
10:1	0.8 ± 2.5	14 ± 3	8.6 ± 4	4 ± 2
5:1	0.5 ± 1	3.5 ± 2	6 ± 4	4.5 ± 4
1:1	0.0 ± 0.6	3 ± 3	3 ± 4	0.5 ± 2

Thioglycolate-elicited Mϕs at different cell numbers were incubated with acetyl-LDL–MTP-PE (18.0 μg of protein per 3.6 μg of MTP-PE per well), free MTP-PE (3.6 μg per well), acetyl-LDL (18.0 μg protein per well), or media alone for 24 h. Macrophages were washed; then [³H]thymidine-labeled P815 mastocytoma cells (1 x 10⁴) were added. After 24 h incubation, culture supernatants were counted for radioactivity, and the percentage of specific release (percentage of cytotoxicity) was calculated. Data represent the mean ± SD of two separate experiments performed in triplicate. Significance in percentage of cytotoxicity of acetyl-LDL–MTP-PE-treated Mϕs at 10:1, 20:1, and 40:1 target cell ratios was $P < 0.05$ compared to controls receiving no treatment. The addition of 10 ng of lipopolysaccharide during Mϕ incubations had no effect on the cytotoxic activity (data not shown).

inhibited only by the addition of unlabeled acetyl-LDL and not by excess LDL or liposomes. When the Mϕ activator MTP-PE and DiI, a lipophilic carbocyanine fluorescent dye, were partitioned into the acetyl-LDL particle, neither the binding specificity nor the binding affinity of the particle was altered. Furthermore, the acetyl-LDL–MTP-PE complex efficiently activated thioglycolate-elicited peritoneal Mϕs for cytostatic and tumoricidal activity (Table 2). The addition of a second activator, endotoxin, during acetyl-LDL–MTP-PE activation of Mϕs showed no further enhancement for tumoricidal function, and free acetyl-LDL or free MTP-PE failed to activate the Mϕs.

To date, no other Mϕ immunomodulators have been examined by partition into and delivery via acetyl-LDL. Potentially any factor with sufficient lipophilicity and reasonably small molecular size (i.e., < 3 kilodaltons) could be used. Candidate immunomodulatory factors include the immunosuppressant corticosteroids and prostaglandin E₂, phorbol esters such as phorbol myristic acetate (PMA) [15,16,76], the calcium ionophore A23187 [15,16], cyclophosphamide [77], and endotoxin [2,15]. In particular, PMA and A23187 are known to dramatically increase Mϕ tumoricidal activity [16,76], and both compounds are thought to interact directly with cell membranes, thus indicating their lipophilicity. Endotoxin is a powerful activator of Mϕ bactericidal [2] and tumoricidal [21] functions and is also readily lipophilic [37]. Extreme care must be taken, however, to ensure proper dosages of these compounds. Phorbol myristic acetate is a carcinogen, and many endotoxins induce the synthesis of TNF-α [27,28], which could lead to cachexia following prolonged administration.

Protein biological response modifiers are now being produced in great quantities via molecular cloning techniques and hold promise for use in management of immunological responses. These factors, however, face two major obstacles not encountered in the partitioning of organic compounds into acetyl-LDL. First, these proteins are large molecules (> 3 kilodaltons) and would therefore be difficult to stably associate with the carrier, regardless of their lipophilicity. The partitioning process (Section IV) itself may destroy the proteins. Second, large molecules would project far out of the surface of the delivery molecules, granting easy access to proteolytic enzymes. Furthermore, the delivery efficiency of the acetyl-LDL–protein may be reduced by binding of the protein to its normal cellular receptor on other cell types.

One obvious strategy would be to use only the active fragment of the protein. The smaller peptide would likely be accommodated by the carrier, provided it remains sufficiently lipophilic. Toward this end, the three-dimensional structure of TNF-α was recently determined [79]. Also, the biological activity of the 144 amino acid GM-CSF molecule was recently found to be dependent on two distinct regions of the molecule, encompassing stretches of only 11 and 17 amino acid residues [78]. Moreover, glycosylation is not required for its function, further reducing the bulk of the protein. It may prove possible to link these peptide sequences with a lipophilic molecule and stably associate them with the acetyl-LDL carrier. Further advances in protein structural determinations and protein engineering should make it possible to identify and produce the biologically active fragments of many protein immunomodulators, thus permitting their use in the acetyl-LDL delivery system.

However, several factors will exclude the use of many immunodulators. Upon binding to receptors, all lipoproteins are rapidly internalized and delivered, intact, to lysosomes [10,32]. Here the protein and cholesteryl esters are rapidly hydrolyzed, with the liberated cholesterol being freed from the lysosome. When [^{125}I]acetyl-LDL is used as a substrate, 50% of the initial cell-bound radioactivity is degraded within 30 min, and [^{125}-I]monoiodotyrosine is excreted from the cell [32]. Therefore, any potential immunomodulator must retain its activity despite the rigors of the lysosome. This is especially true for peptides which may not normally undergo receptor-mediated endocytosis and catabolism but instead function via the generation of second messenger molecules following surface receptor binding. Interferon-γ, the major product of T cells that alters Mϕ function [37], induces Ca^{2+} fluxes and increases the potential of protein kinase C (PrK-C) within minutes of binding to its receptor [37,38]. Several biological effects of IFN-γ can be mimicked when cellular PrK-C is pharmacologically stimulated with phorbol esters [37]. Although IFN-γ is catabolized by Mϕs [38], an additional Prk-C–independent signal transduction mechanism appears to exist [39]. Both of these mechanisms are likely to be required for full $M\phi$ response to IFN-γ. Delivery of IFN-γ (or active fragments of IFN-γ) via acetyl-LDL (or any lipoprotein carrier) would bypass the events mediated by IFN-γ receptor occupancy, thereby reducing the efficiency of acetyl-LDL–IFN-γ induction of $M\phi$ activity. The same applies to TNF-α, which binds to specific surface receptors [40], and endotoxin, which intercalates into the cell membrane lipid bilayer to initiate the phosphatidylinositol-4,5-bisphosphate hydrolysis cascade of second messenger signal transduction and PrK-C activation [37]. In this regard, endotoxin incorporated into liposomes shows diminished ability to induce both TNF-α secretion and tumoricidal function in murine $M\phi$s [41], thereby indicating intralysosomal degradation and inactivation of the endotoxin [80] and suggesting that interaction of the lipid A moiety of endotoxin with the $M\phi$ plasma membrane is required to transduce proper signals for cell activation.

Conversely, uptake of certain peptides via the acetyl-LDL receptor may enhance their bioactivity by overcoming the species specificity of their receptors [42]. For example, Kleinerman et al. [43] reported that introduction of encapsulated human $M\phi$-activating factor (MAF) into rat $M\phi$s via phagocytosis enhanced tumoricidal function. Free human MAF had no effect due to the species barrier at the level of the receptor. Cross-species activation of $M\phi$s by liposome-encapsulated IFN-γ has also been observed [74]. The ability to use a single protein immunomodulator across species barriers eliminates the need to clone all of the relevant genes for these proteins for all of the relevant animal species. Liposome–MAF was also more efficient than free MAF in activating $M\phi$s *in vitro* [44]. Although the nature of the intracellular target for MAF is unclear, the MAF receptor transduces no signals required for activation. This likely is not true for all lymphokines.

IV. PREPARATION OF ACETYL-LDL AND ACETYL-LDL–IMMUNOMODULATOR COMPLEXES

Low density lipoproteins are complex particles that contain several classes of neutral and charged lipids as well as the 500 kilodalton apoprotein B-100, a highly water-insoluble and unstable glycoprotein (Fig. 3). Acetylation of LDL removes positive charges from the ϵ-amino groups of lysine residues. This converts the particle into a strongly anionic lipoprotein, and acetyl-LDL loses its ability to bind the LDL receptor of non-$M\phi$ cells. The particle, however remains precipitable by antibodies to native LDL [10]. Although the identity of the negatively charged residues on acetyl-LDL which mediate binding to the receptor is not known, competition studies with [^{125}I]acetyl-LDL indicate that binding to the receptor requires a large number of negatively charged residues arranged in such a way as to produce a high charge density [10].

Exhaustive acetylation of isolated LDL is achieved by using acetic anhydride in the presence of sodium acetate at 0°C [45]. Acetyl-LDL is then purified by size exclusion chromatography, dialysis against argon-purged sodium phosphate/sodium chloride, and 0.45 μm filtration. Other chemical modifications of LDL that remove positive charges from lysine residues can also convert the lipoprotein into a species that now binds the acetyl-LDL receptor. Acetoacetylated LDL, maleylated LDL, succinylated LDL, and malondialdehyde-treated LDL [10,11,32] all act as ligands for the acetyl-LDL receptor. Again, extensive modification is required, and malondialdehyde treatment of LDL requires that at least 30 molecules of malondialdehyde be incorporated per LDL particle in order to generate sufficient net negativity to bind the acetyl-LDL receptor [10].

Several methods exist for preparing lipoprotein–drug complexes [9]. The lipophilic core of the particle can be extracted with organic solvents, and the extracted lipids are replaced by the addition of lipophilic drugs. This procedure, however, is time consuming and suffers from poor reproducibility. A second method, detergent dialysis, involves solubilization of the lipophilic drug in octylglucoside, a nonionic detergent and surfactant, addition of acetyl-LDL, and dialysis of the mixture at 25°C. This method is often not satisfac-

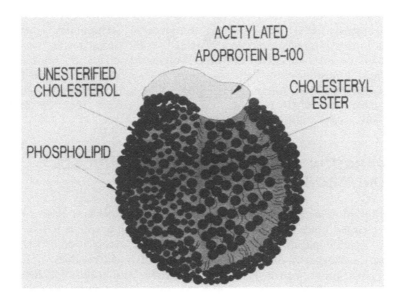

Figure 3 Low density lipoprotein and acetylated low density lipoprotein. A spherical water-soluble particle, it is 20–25 nm in diameter with molecular mass of 2.5 x 10^6 daltons. The oily core of the particles is composed mainly of cholesteryl ester (81 mol%) with lesser amounts of triglyceride (19 mol%), whereas the surface monolayer is composed of phospholipid (58 mol%) and unesterified cholesterol (42 mol%) and contains a major glycoprotein, apoprotein B-100 (550 kilodaltons).

tory because removal of surfactant from the particles is incomplete even after extensive dialysis and gel permeation chromatography. Aqueous addition of water-soluble drugs to lipoproteins is useful for drugs that partition in the phospholipid–cholesterol monolayer or bind to the surface of lipoproteins (method 3). The drug is added dropwise to lipoprotein during stirring at 37°C. Unbound drug is removed by gel permeation chromatography and filtration through 0.45 μm membranes. The immunomodulators that can be used in this method are limited, and partitioning efficiency is low [9].

The method of choice in our hands for preparing acetyl-LDL–drug complexes is a "dry-film" stirring technique [9]. Here the drug in solvent is dried as a film on a glass surface followed by the slow addition of lipoprotein in argon-saturated sodium phosphate/sodium chloride buffer. The mixture is gently stirred in the dark under argon and purified by gel permeation chromatography and 0.45 μm filtration. The recovery of acetyl-LDL protein for the entire procedure is typically 85–90%. Using this method, a range of 90–140 MTP-PE molecules could be partitioned in the acetyl-LDL particle [11], and partition of MTP-PE into acetyl-LDL did not alter the ability of the particle to bind the Mφs or the ability of MTP-PE to activate the Mφ [11].

As discussed in the previous section, peptide fragments rather than whole proteins will likely be used for partition into the acetyl-LDL carrier. This will require either that the peptides be covalently linked to lipophilic molecules or that the peptides themselves be made sufficiently lipophilic. Efforts to increase the lipophilicity of peptides have generally taken two approaches. The first method is that of peptide latentiation or conversion of water-soluble functional groups on the peptide to lipid-soluble derivatives [46]. Processes such as methylation of amide nitrogens or the substitution of aliphatic amino acids for acidic or basic residues serve to eliminate hydrogen bonding with water. Obviously, extensive modification of the peptide may alter its biological activity. This technique may, however, be promising when small peptides need to be delivered to Mφs, as is the case for synthetic peptide vaccines (see Section V). Alternatively, coupling a peptide to a lipophilic protein may generate a protein chimera with sufficient lipophilicity to be stably partitioned into lipoprotein particles. Unfortunately, no information on increasing lipophilicities is available at this time concerning any of the known Mφ immunomodulatory proteins.

V. APPLICATIONS FOR MACROPHAGE IMMUNOMODULA-TION USING THE ACETYL-LDL DELIVERY SYSTEM

A. Destruction of Tumors and Virus-Infected Cells

The enhancement of Mφ activity in the defense against tumors and infection has obvious clinical advantages. The systemic activation of Mφs has been shown to cause destruction of cancer metastases [47,48] and impart resistance

Table 3 Activities of Muramyl Peptides on Macrophages In Vitro and In Vivo

---

Morphological and Biochemical Changes
    Increased adherence and spreading
    Increased glucose metabolism
    Increased production of oxygen intermediates
    Increased PGE_2 synthesis
    Increased plasminogen activator and collagenase production
    Decreased DNA polymerase activity
    Induction of colony-stimulating activity
    Induction of IL-1

Functional changes
    Increased antitumor activity
    Increased host resistance to parasites
    Increased host resistance to viral challenges
    Increased phagocytic and bacteriocidal activity

Source: Adapted from Ref. 75.

to systemic viral diseases [8,13]. The drug MTP-PE has thus far been examined in greatest detail for use as an *in vivo* immunomodulator (Table 3), largely due to its proven Mϕ-activating properties and ready lipophilicity [11]. Furthermore, while acetyl-LDL has a targeting advantage for Mϕs, MTP-PE also offers the added benefit of exerting its major actions on Mϕs and monocytes [49]. Consequently, the combination of specificity on the part of both the drug and its carrier for a particular tissue or cell type provides a useful approach for the therapeutic management of disease states.

Although it has been shown that acetyl-LDL–MTP-PE activates Mϕs for tumoricidal function *in vitro* [11], it is as yet unclear how acetyl-LDL–MTP-PE will function *in vivo* in the reduction of tumor mass and prevention of metastasis. Preliminary data indicate that intraperitoneal administration of acetyl-LDL–MTP-PE generates tumoricidal Mϕs [11]. This supports the finding of Koff *et al.* [8], who demonstrated protection of mice against HSV-2 infection following a single injection of liposome–MTP-PE, and Edwards *et al.* [7], who observed decreased *Mycobacterium* numbers in tissues from mice treated with MDP. Fidler *et al.* [47] also observed systemic activation of Mϕs following oral administration of free MTP-PE. However, significant retention of the drug by numerous organs 24 h postadministration was observed, underscoring the need for Mϕ-specific delivery vehicles such as acetyl-LDL. This approach is nevertheless limited. There are not enough Mϕs in the system to cope with large tumor burdens [44], thus necessitating the use of combined therapies: surgery to reduce tumor load followed by systemic activation of resident Mϕs to scavenge any remaining neoplasms. Alternatively, irradiation followed by several weeks of *in vivo* Mϕ activation with MTP-PE has been shown to significantly increase the survival of mice injected with fibrosarcoma cells [50]. Chemotherapy would likely reduce the capabilities of Mφs, hence its use should be delayed or negated. Finally, systemic activation of Mφs will not affect survival when tumors are resistant to the effector phase of tumoricidal Mφs [51].

Presently, no information exists concerning the delivery of other Mφ-immunomodulatory compounds or peptides (see Section II) via the acetyl-LDL system. It should be possible, however, to improve upon the efficacy observed with other delivery systems due to increased site delivery of the acetyl-LDL system. Upon intravenous injection, liposomes are found largely in the fixed Mφs of the liver and spleen, but they suffer from an inability to cross capillaries and enter the extravascular space. Macrophage-activating factor encapsulated into liposomes has been shown *in vitro* to activate Mφs to selectively destroy HSV-2–infected cells [52]. Treatment with encapsulated MAF also activated Mφs to kill A375 melanoma cells [43], and IFN-γhas been used in a liposome form to treat murine hepatitis [53]. It is expected, then, that partitioning MAF or IFN-γ(or their active peptide fragments) into acetyl-LDL particles will similarly impart protection against viral diseases and tumors while minimizing any toxicities or side effects due to nonspecific targeting of the immunomodulator [4,5].

Reduction of side effects is a major concern for clinicians. The advent of recombinant DNA technology has brought biological response modifiers to the forefront in the treatment of cancer. Clinical trials with IFN-α_{2b} [54], IFN-γ[4], and TNF-α {5} have demonstrated that infusion of these Mφ activators results in toxicities and mild to severe side effects such as nausea, fatigue, anorexia, headache, and diarrhea, even at low doses. Furthermore, these proteins display a wide range of activities on a variety of cell types, adding to the generation of undesirable side effects [55,56]. Their partition, if possible, into the Mφ-specific acetyl-LDL delivery system would allow the administration of overall smaller dosages due to the increased retention of the biological response modifier.

B. Treatment of Pathogenic Organisms

One area receiving considerable attention is the use of lipophilic drugs and delivery systems in the treatment of infectious diseases. Many of the agents used therapeutically in parasitic infections are toxic, and many infections are refractory to conventional drug therapy. Individuals with intracellular parasitic diseases may be cured with single small doses of relevant drugs entrapped in liposomes [53]. Avoiding the requirement for repeated doses of the drug in the free form increases the feasibility of treating these diseases in the developing world. Consider *Leishmania*, a protozoan parasite endemic in many tropical areas. The organism enters and colonizes Mφs, residing within endocytotic vacuoles without being destroyed. Liposomes have been used to deliver antimonial drugs to the infected cells [44], usually resulting in 30- to 40-fold improvements in therapeutic indexes. At the same time, however, the liposome–drug complexes were more toxic than the free drug, probably due to liposomal-enhanced phagocytosis by non-Mϕ cells. Thus acetyl-LDL particles would be better suited to deliver drugs to Mϕs because fewer cell types take up acetyl-LDL.

Since IFN-γ has been observed *in vitro* to stimulate killing of intracellular *Leishmania* [57,58], this peptide is a logical choice for partition into acetyl-LDL and administration to infected individuals. Macrophage resistance to *Trypanosoma cruzi*, a protozoan parasite that replicates in the cytoplasm of Mϕs, has been reported to be enhanced by *in vitro* treatment of infected Mϕs with IFN-γ [59], TNF-α plus endotoxin [14], and GM-CSF [59], further suggesting a role for Mϕ-specific acetyl-LDL delivery of biological response modifiers in the treatment of these intracellular parasites. Additionally, organisms belonging to the *Mycobacterium avium* complex, the most common bacterial pathogens in patients with AIDS, invade and replicate within Mϕs but are killed when the Mϕ is treated with TNF-α [60]. Other diseases in which microorganisms reside intracellularly in the Mϕ and which are candidates for treatment with acetyl-LDL–immunomodulator complexes include *Mycobacterium tuberculosis* [53], *M. intracellulare* [7], and human immunodeficiency virus (HIV) [61]. Again, combined therapies in which both antimicrobial drugs and biological response modifiers are concomitantly delivered to Mϕs via the acetyl-LDL particle may prove most effective and safe for the treatment of certain immunodeficiency diseases [62] or immunosuppressed individuals such as the aged, those with AIDS, or transplant recipients who suffer from numerous systemic infections.

Aminoglycosides such as streptomycin have been incorporated into liposomes for delivery to Mϕ endosomes in the treatment of brucellosis, a widespread disease among cattle [44]. As in the case of leishmaniasis, these gram-negative bacteria reside within the endocytotic vacuoles of Mϕs. Because of this, the organism responds very poorly to common antibiotics. The enhanced therapeutic index observed with liposome–aminoglycoside complexes [44] should be improved upon by specifically targeting the drug to Mϕs via the acetyl-LDL receptor provided that the more lipophilic aminoglycosidic derivates can be used.

C. Vaccinations

An area in which particles such as acetyl-LDL hold great promise is potentiation of immune responses by acting as immunological adjuvants. The adjuvants used today, including complete and incomplete Freund's adjuvant, bacterial endotoxins, and mineral adsorbents, are unacceptable for human use because they frequently induce local and systemic toxicity or form granulomas [53]. It has been known for over a decade, however, that liposome encapsulation of normally weakly immunogenic proteins can induce far greater antibody responses than can free antigen used as an inoculum [63]. The recognition of antigen by helper T cells (T_H) is a key event in the induction of humoral immune responses. T_H cells recognize a processed form of the antigen presented in conjunction with class II MHC molecules on the surface of antigen-presenting cells. Macrophages participate in processing and presentation events [2,64] and are thus primary targets for antigen delivery to initiate an immune response leading to protection. Richards *et al.* [65] showed that a nonimmunogenic peptide from *Plasmodium falciparum*, the human malaria parasite, was made highly immunogenic by encapsulation into liposomes, and the antibodies produced reacted with the native protein and with living parasite. When lipid A was incorporated into the liposome–peptide complex to activate the Mϕs, antibody titers remained high for an extended time period. Obviously, long term protection against pathogens is the goal of any vaccine. Because primed Mϕs more actively and efficiently present antigen [2], and because endocytosed acetyl-LDL is delivered intracellularly to lysosomes [10,32], antigen presentation, elicitation of an immune response, and long term protection against a pathogen may be accomplished via administration of acetyl-LDL particles that contain IFN-γ as well as the peptide. The IFN-γ enhances the amount of class II MHC molecules and IL-1 produced by the Mϕs in response to secondary signals [66]. Interleuken-1 stimulates the clonal expansion of T cells; thus coadministration of the peptide and IFN-γ or other Mϕ-priming signals such as MTP-PE would augment the immune response.

Successful immunizations with liposomal antigens include diphtheria toxoid, cholera toxin, hepatitis B antigen, and rubella hemagglutinin [53]. Recent data also indicate a critical role of antigen-presenting cells in generating tumor rejection [81], suggesting a novel approach to induce tumor-specific immunity by delivering immunogenic tumor peptides to Mϕs via the acetyl-LDL system. Since many proteins from pathogens and neoplasms have been cloned, it is becoming increasingly likely that vaccinations will be accomplished using recombinant or synthetic peptides, thus eliminating the inherent dangers of live modified and heat-killed organisms, inactivated toxins, or tumor cells themselves. Peptide inoculums often show reduced immunogenicity and fail to elicit adequate protection against the native protein or live organism. For this reason it becomes desirable both to systemically prime the Mϕs for maximum presenting function and to ensure greatest delivery of the peptide to the Mϕs. It should be noted that the same limitations apply to synthetic peptide use in acetyl-LDL delivery vehicles as apply to partitioning protein biological response modifiers: the peptides must remain relatively small and must retain immunogenic activity following partition or the addition of lipophilic molecules or residues. These potential problems are eased somewhat by the fact that many of the peptides that have been studied are actually small and lipophilic in nature since the areas of the native protein which appear to be recognized by class II MHC molecules and the T_H receptors are amphoteric stretches of

10–15 amino acids [82]. Moreover, retention of natural peptide determinant conformation is not necessarily critical to generate an immune response since many determinants require extensive processing by the antigen-presenting cell prior to association with class II MHC molecules [64]. The acetyl-LDL particle may also provide a solution to the large amounts of peptide required for immunizations because the particle selectively binds Mϕs. However, while the acetyl-LDL particles themselves have not been shown to be immunogenic, protracted coadministration of Mϕ-priming peptides such as IFN-γ or MAF may lead to the generation of immune responses to these proteins as well as to the inoculum peptide. Also, the site of administration and the amount of protein administered must be investigated for each inoculum to ensure the best response.

D. Gene Transfer Delivery Vehicles

Recently, vehicles for the *in vitro* and *in vivo* delivery of genetic information have been examined. Liposomes were originally employed, and efficient gene transfers have been achieved [71]. Any gene transfer system is required to preferentially bind target cells as well as encapsulate significant quantities of the genes in question without compromising their biological integrity. Acetyl-LDL particles would not appear suitable as gene transfer vehicles due to the large size and hydrophilic nature of DNA. However, short stretches of oligonucleotides derivatized with hydrophobic anchors may prove useful in delivery vehicles such as acetyl-LDL. Endocytosed acetyl-LDL particles are delivered to lysosomes, however, necessitating interference with normal lysosomal function of the target cell to prevent nucleic acid hydrolysis. Lysosomotropic agents have been used to enhance liposome-mediated gene transfers [71] and could possibly be partitioned into the acetyl-LDL particles, but other procedures are likely to be more useful for *in vivo* delivery. Transient expression of genes has been observed when the DNA was complexed with asialoglycoproteins [72]. Presumably the DNA was protected from complete enzymatic digestion and escaped into the cytoplasm. Cationic lipids that fuse spontaneously with membranes may also enhance DNA delivery to the cytoplasm following acetyl-LDL particle degradation in the lysosome if they can be attached to the nucleic acid fragment. Because many tissue Mϕs are relatively long-lived, *in vivo* gene transfer for the purpose of correcting Mϕ genetic defects [62] may become therapeutically feasible if stable gene transfer efficiencies are great enough. Periodic additional transfers would maintain a population of Mϕs expressing the introduced gene. Conversely, the expression of specific genes by Mϕs could be transiently blocked by the delivery of antisense nucleic acids [73], thereby attenuating undesired responses such as inflammation. The negative regulation of a gene may prove highly feasible because only short oligomeric stretches of antisense nucleic acids are required to block specific mRNA translation [73].

E. Potential Problems

Perhaps the greatest potential problem with the use of acetyl-LDL particles as Mφ-specific delivery vehicles concerns the fate of the particle itself. As stated previously, endocytosed acetyl-LDL is delivered to lysosomes [10]. The cholesteryl esters are hydrolyzed, and the liberated cholesterol enters the cytoplasm. Here it is either reesterified or excreted from the cell when cholesterol acceptors such as HDL are present [10]. Loading Mφs with cholesteryl esters induces the synthesis and secretion of apolipoprotein E (apo E), a normal constituent of cholesterol-rich high density lipoproteins which functions in "reverse cholesterol transport," whereby excess cholesterol is transported to the liver [10,67]. Apo E is also known to have potent suppressive activity for lymphocytic stimulation [68]. Thus excessive uptake of acetyl-LDL particles and subsequent apo E secretion could negatively influence immune responses, especially humoral responses. Furthermore, lipid-laden Mφs, monocytes, and polymorphonuclear leukocytes (PMN) can produce free radicals such as hydrogen peroxide, superoxide anion, hydroxyl radical, and singlet molecular oxygen that are capable of oxidizing LDL to make it cytotoxic to fibroblasts [69,83,84]. Activated monocytes show enhanced free radical production; thus the conversion of LDL to its cytotoxic, oxidized form would also be enhanced when Mφ immunomodulators are delivered via acetyl-LDL particles. A growing body of evidence also suggests that following oxidative modification of LDL, the oxidized LDL rapidly enters Mφs via the acetyl-LDL or scavenger receptor pathway, giving rise to foam cells ([84]; Chapter 1). Furthermore, Tanabe et al. [70] showed that LDL oxidized by PMN inhibited NK cell activity. Natural killer cells have a cytolytic function against tumor- and virus-infected cells and therefore play an important role in host defense mechanisms. However, coaddition of general free radical scavengers prevented the formation of oxidized LDL [69], indicating that acetyl-LDL particles can be used as delivery vehicles if dosage schedules are properly managed and with the coadministration of free radical scavengers such as vitamin E and glutathione.

VI. CONCLUDING REMARKS

Acetyl-LDL particles appear to have potential for *in vitro* and *in vivo* Mφ-specific delivery of immunomodulatory peptides and drugs. A distinct targeting advantage can be gained with the acetyl-LDL receptor as it is not subject to feedback inhibition and resides on a restricted number of cell types, primarily Mφs, monocytes, and endothelial cells. Enhanced site delivery to Mφs would be expected to increase the efficacy of the compound in question while simultaneously lowering its toxicity or undesired side effects generated from interactions with non-Mφ cells. The acetyl-LDL particles themselves are nonimmunogenic, do not alter the partitioned drug's bioactivity, and are capable of associating significant quantities of compounds. The combination of specificity on the part of both the drug and its carrier provides a useful approach for the therapeutic management of disease states.

ACKNOWLEDGMENTS

We extend our appreciation to Dr. William Regelson, Medical College of Virginia, for his encouragement and support during the development of this work. The authors acknowledge the generous gift of MTP-PE from Drs. D. G. Braun, R. Scartazzini, and K. Scheibli of CIBA-Geigy Ltd., Basel, Switzerland. We also thank Helen Gawthorp for her assistance in the preparation of this chapter. This work was supported in part by a grant from the NIH (ES-03468) and the American Heart Association, Virginia Affiliate.

REFERENCES

1. Belosevic M, Davis CE, Meltzer MS, Nacy CA. Regulation of activated macrophage antimicrobial activities. J Immunol 1988; 141:890–896.
2. Adams DO, Hamilton TA. The cell biology of macrophage activation. Ann Rev Immunol 1984; 2:283–318.
3. Sampson-Johannes A, Carlino JA. Enhancement of human monocyte tumoricidal activity by recombinant M-CSF. J Immunol 1988; 141:3680–3686.
4. Schiller JH, Storer B, Bittner G, Willson JKV, Borden EC. Phase II trial of a combination of interferon-β_{ser} and interferon-γ in patients with advanced malignant melanoma. J Interferon Res 1988; 8:581–589.
5. Creaven PJ, Plager JE, Dupere S, Huben RP, Takita H, Mittelman A, Proefrock A. Phase I clinical trial of recombinant human tumor necrosis factor. Cancer Chemother Pharmacol 1987; 20:137–144.
6. Roche AC, Bailly P, Midoux P, Monsigny M. Selective macrophage activation by muramyldipeptide bound to monoclonal antibodies specific for mouse tumor cells. Cancer Immunol Immunother 1984; 18:155–159.
7. Edwards C, K, III, Hedegaard HB, Zlotnik A, Gangadharam PR, Johnston RB, Jr, Pabst MJ. Chronic infection due to *Mycobacterium intracellulare* in mice: association with macrophage release of prostaglandin E_2 and reversal by injection of indomethacin, muramyl dipeptide, or interferon-γ. J Immunol 1986; 136:1820–1827.
8. Koff WC, Showalter SD, Hampar B, Fidler IJ. Protection of mice against fatal herpes simplex type 2 infection by liposomes containing muramyl tripeptide. Science 1985; 228:495–496.
9. Shaw JM, Shaw KV, Yanovich S, Iwanik M, Futch WS, Rosowsky A, Schook LB. Delivery of lipophilic drugs using lipoproteins. Ann NY Acad Sci 1987; 507:252–271.
10. Brown MS, Goldstein JL. Lipoprotein metabolism in the macrophage. Ann Rev Biochem 1983; 52:223–261.

11. Shaw JM, Futch WS, Jr, Schook LB. Induction of macrophage antitumor activity by acetylated low density lipoprotein containing lipophilic muramyl tripeptide. Proc Natl Acad Sci USA 1988; 85:6112–6116.

12. Yanovich S, Prestion L, Shaw JM. Characteristics of uptake and cytotoxicity of a low density lipoprotein–daunomycin complex in P388 leukemic cells. Cancer Res 1984; 44:3377–3382.

13. Ishihara C, Hamada N, Yamamoto K, Iida J, Azuma I, Yamamura Y. Effect of muramyl dipeptide and its stearoyl derivatives on resistance to sendai virus infection in mice. Vaccine 1985; 3:370–374.

14. Wirth JJ, Kierszenbaum F. Recombinant tumor necrosis factor enhances macrophage destruction of *Trypanosoma cruzi* in the presence of bacterial endotoxin. J Immunol 1988; 141:286–288.

15. Hamilton TA, Adams DO. Molecular mechanisms of signal transduction in macrophages. Immunol Today 1987; 8:151–158.

16. Somers SD, Weiel JE, Hamilton TA, Adams DO. Phorbol esters and calcium ionophore can prime murine peritoneal macrophages for tumor cell destruction. J Immunol 1986; 136:4199–4205.

17. Hart PH, Whitty GA, Piccoli DS, Hamilton JA. Synergistic activation of human monocytes by granulocyte-macrophage colony stimulating factor and IFN-γ. J Immunol 1988; 141:1516–1521.

18. Mosser D, Edelson P. Mechanisms of microbial entry and endocytosis by mononuclear phagocytes. Contemp Top Immunobiol 1984; 13:71–96.

19. Weir D. Surface carbohydrates and lectins in cellular recognition. Immunol Today 1980; 1:45–51.

20. Freund M, Pick E. The mechanism of action of lymphokines. J Immunol 1986; 137:1312–1318.

21. Renz H, Gong JH, Schmidt A, Nain M, Gemsa D. Release of tumor necrosis factor-α from macrophages. J Immunol 1988; 141:2388–2393.

22. Philip R. Cytolysis of tumor necrosis factor (TNF)-resistant tumor targets. J Immunol 1988; 140:1345–1349.

23. Johnson WJ, Somers SD, Adams DO. Activation of macrophages for tumor cytotoxicity. Contemp Top Immunobiol 1983; 14:127–146.

24. Johnson WJ, Marino PA, Schreiber RD, Adams DO. Sequential activation of murine mononuclear phagocytes for tumor cytolysis: differential expression of markers by macrophages in several stages of development. J Immunol 1983; 131:1038–1043.

25. Pennica D, Shalaby MR, Palladino MA, Jr. Tumor necrosis factors alpha and beta. In:Gillis S, ed. Recombinant lymphokines and their receptors. New York:Marcel Dekker, 1987:301–317.

26. Warner SJC, Libby P. Human vascular smooth muscle cells. Target for and source of tumor necrosis factor. J Immunol 1989; 142:100–109.

27. Beutler B. Tkacenko V, Milsark I, Krochin N, Cerami A. Effect of interferon on cachectin expression by mononuclear phagocytes. J Exp Med 1986; 164:1794–1796.

28. Myers MJ, Pullen JK, Ghildyal N, Eustis-Turf E, Schook LB. Regulation of IL-1 and TNF-α gene expression during the differentiation of bone marrow-derived macrophage. J Immunol 1989; 142:153–160.

29. Debets JMH, van der Linden CJ, Dieteren IEM, Leeuwenberg JFM, Buurman WA. Fc-receptor crosslinking induces rapid secretion of tumor necrosis factor

(cachectin) by human peripheral blood monocytes. J Immunol 1988; 141:1197-1201.

30. Schutze S, Scheurich P, Schluter C, Ucer U, Pfizenmaier K, Kronke M. Tumor necrosis factor-induced changes in gene expression in U937 cells. J Immunol 1988; 140:3000–3005.

31. Goldstein JL, Ho YK, Brown MS, Innerarity TL, Mahley RW. Cholesteryl ester accumulation in macrophages resulting from receptor-mediated uptake and degradation of hypercholesterolemic canine ß-very low density lipoproteins. J Biol Chem 1980; 255:1839–1848.

32. Goldstein JL, Ho YK, Basu SK, Brown MS. Binding site on macrophages that mediates uptake and degradation of acetylated low density lipoprotein, producing massive cholesterol deposition. Proc Natl Acad Sci USA 1979; 76:333–337.

33. Traber MG, Kayden JH. Low density lipoprotein receptor activity in human monocyte-derived macrophages and its relation to atheromatous lesions. Proc Natl Acad Sci USA 1980; 77:5466–5470.

34. Traber MG, Defendi V, Kayden HJ. Receptor activities for low density lipoprotein and acetylated low density lipoprotein in a mouse macrophage cell line (IC21) and in human monocyte-derived macrophages. J Exp Med 1981; 154:1852–1867.

35. Sorg C. Heterogeneity of macrophages in response to lymphokines and other signals. Molec Immunol 1982; 19:1275–1278.

36. Akagawa KS, Kamoshita K, Tokunaga T. Effects of granulocyte-macrophage colony-stimulating factor and colony-stimulating factor-1 on the proliferation and differentiation of murine alveolar macrophages. J Immunol 1988; 141:3383–3390.

37. Adams DO, Hamilton TA. Molecular transductional mechanisms by which IFN-γ and other signals regulate macrophage development. Immunol Rev 1987; 97:5–27.

38. Celada A, Schreiber RD. Role of protein kinase C and intracellular calcium mobilization in the induction of macrophage tumoricidal activity by interferon-γ J Immunol 1986; 137:2373–2379.

39. Radzioch D, Varesio L. Protein kinase C inhibitors block the activation of macrophages by IFN-ß but not by IFN-γ. J Immunol 1988; 140:1259–1263.

40. Bakouche O, Ichinose Y, Heicappel R, Fidler IJ, Lachman LB. Plasma membrane-associated tumor necrosis factor. J Immunol 1988; 140:1142–1147.

41. Dijkstra J, Larrick JW, Ryan JL, Szoka FC. Incorporation of LPS in liposomes diminishes its ability to induce tumoricidal activity and tumor necrosis factor secretion in murine macrophages. J Leuk Biol 1988; 43:436–444.

42. Fransen L, Ruysschaert MR, van der Heyden J, Fiers W. Recombinant tumor necrosis factor: species specificity for a variety of human and murine transformed cell lines. Cell Immunol 1986; 100:260–267.

43. Kleinerman ES, Fogler WE, Fidler IJ. Intracellular activation of human and rodent macrophages by human lymphokines encapsulated in liposomes. J Leuk Biol 1985; 37:571–584.

44. Poznansky MJ, Juliano RL. Biological approaches to the controlled delivery of drugs: a critical review. Pharmacolog Rev 1984; 36:277–336.

45. Basu SK, Goldstein JL, Anderson RG, Brown MS. Degradation of cationized low density lipoprotein and regulation of cholesterol metabolism in homozygous familiar hypercholesterolemia fibroblasts. Proc Natl Acad Sci USA 1976; 73:3178-3182.

46. Pardridge WM. Receptor-mediated peptide transport through the blood-brain barrier. Endocrine Rev 1986; 7:314–330.
47. Fidler IJ, Fogler WE, Brownbill AF, Schumann G. Systemic activation of tumoricidal properties in mouse macrophages and inhibition of melanoma metastases by the oral administration of MTP-PE, a lipophilic muramyl dipeptide. J Immunol 1987; 138:4509–4514.
48. Fidler IJ, Sone S, Fogler WE, Barnes ZL. Eradication of spontaneous metastases and activation of alveolar macrophages by intravenous injection of liposomes containing muramyl dipeptide. Proc Natl Acad Sci USA 1981; 78:1680–1684.
49. Oldham RK. Biologicals and biological response modifiers: new approaches to cancer treatment. Cancer Invest 1985; 3:53–70.
50. Dukor P, Schumann G. Modulation of non-specific resistance by MTP-PE. In:Majde JA, ed. Immunopharmacology of infectious diseases: vaccine adjuvants and modulators of non-specific resistance. New York:Alan R. Liss, 1987:255–265.
51. Bakouche O, Lachman LB, Knowles RD, Kleinerman ES. Cytotoxic liposomes: membrane interleukin 1 presented in multilameller vesicles. Lymphokine Res 1988; 7:445–456.
52. Koff, WC, Showalter SD, Chakrabarty MK, Hampar B, Ceccorulli LM, Kleinerman ES. Human monocyte-mediated cytotoxicity against herpes simplex virus-infected cells: activation of cytotoxic monocytes by free and liposome-encapsulated lymphokines. J Leuk Biol 1985; 37:461–472.
53. Gregoriadis G. Liposomes for drugs and vaccines. Trends Biotechnol 1985; 3:235–241.
54. Smith D, Wagstaff J, Thatcher N, Scarffe H. A phase I study of rDNA alpha$_{2b}$-interferon as a 6 week continuous intravenous infusion. Cancer Chemother Pharmacol 1987; 20:327–331.
55. Grunfeld C, Verdier JA, Neese R, Moser AH, Feingold KR. Mechanisms by which tumor necrosis factor stimulates hepatic fatty acid synthesis *in vivo*. J Lipid Res 1988; 29:1327–1335.
56. Sugarman BJ, Aggarwal BB, Hass PE, Figari IS, Palladino MA Jr, Sheppard HM. Recombinant human tumor necrosis factor-α: effects on proliferation of normal and transformed cells *in vitro*. Science 1985; 230:943–945.
57. Murray HW, Rubin BF, Rothermel CD. Killing of intracellular *Leishmania donovani* by lymphokine-stimulated human mononuclear phagocytes. J Clin Invest 1983; 72:1506–1512.
58. Meltzer MS, Crawford RM, Gilbreath MJ, Finbloom DS, Davis CE, Fortier AH, Schreiber RD, Nacy CA. Lymphokine regulation of non-specific macrophage cytotoxicity against neoplastic and microbial targets. In:Majde JA, ed. Immunopharmacology of infectious diseases: vaccine adjuvants and modulators of non-specific resistance. New York:Alan R. Liss, 1987:27–39.
59. Reed SG, Nathan CF, Pihl DL, Rodricks P, Shanebeck K, Conlon PJ, Grabstein KH. Recombinant granulocyte/macrophage colony-stimulating factor activates macrophages to inhibit *Trypanosoma cruzi* and release hydrogen peroxide. J Exp Med 1987; 166:1734–1746.
60. Bermduez LEM, Young LS. Tumor necrosis factor, alone or in combination with IL-2, but not IFN-γ, is associated with macrophage killing of *Mycobacterium avium* complex. J Immunol 1988; 140:3006–3013.

61. Folks TM, Kessler SW, Orenstein JM, Justement JS, Jaffe ES, Fauci AS. Infection and replication of HIV-1 in purified progenitor cells of normal human bone marrow. Science 1988; 242:919–922.

62. Phillips NC, Skamene E, Chedid L. Correction of defective tumoricidal activity of macrophages from A/J mice by liposomal immunomodulators. Immunopharmacology 1988; 15:1–10.

63. van Rooijen N, van Nieuwmegan R. Liposomes in immunology: the immune response against antigen-containing liposomes. Immunol Commun 1977; 6:489–498.

64. Allen PM. Antigen processing at the molecular level. Immunol Today 1987; 8:270–273.

65. Richards RL, Wirtz RA, Hockmeyer WT, Alving CR. Liposomes as carriers for a malaria peptide vaccine: developmental aspects: In:Majde JA, ed. Immunopharmacology of infectious diseases: vaccine adjuvants and modulators of nonspecific resistance. New York:Alan R. Liss, 1987:171–180.

66. Ghezzi P, Dinarello Ca. IL-1 induces IL-1. J Immunol 1988; 140:4238–4244.

67. Takagi Y, Dyer CA, Curtiss LK. Platelet-enhanced apolipoprotein E production by human macrophages: a possible role in atherosclerosis. J Lipid Res 1988; 29:859–867.

68. Pepe MG, Curtiss LK. Apolipoprotein E is a biologically active constituent of the normal immunoregulatory lipoprotein, LDL-In. J Immunol 1986; 136:3716–3723.

69. Cathcart MK, Morel DW, Chisolm GM III. Monocytes and neutrophils oxidize low density lipoprotein making it cytotoxic. J Leuk Biol 1985; 38:341–350.

70. Tanabe F, Sato A, Ito M, Ishida E, Ogata M, Shigeta S. Low density lipoprotein oxidized by polymorphonuclear leukocytes inhibits natural killer cell activity. J Leuk Biol 1988; 43:204–210.

71. Mannino RJ, Gould-Fogerite S. Liposome-mediated gene transfer. Biotechniques 1988; 6:682–690.

72. Wu GY, Wu CH. Receptor-mediated gene delivery and expression *in vivo*. J Biol Chem 1988; 263:14621–14624.

73. Holt JT, Redner RL, Neinhuis AW. An oligomer complimentary to c-*myc* mRNA inhibits proliferation of HL-60 promyelocytic cells and induces differentiation. Molec Cell Biol 1988; 8:963–973.

74. Fidler IJ, Fogler WE, Kleinerman ES, Saiki I. Abrogation of species specificity for activation of tumoricidal properties in macrophages by recombinant mouse or human interferon-γ encapsulated in liposomes. J Immunol 1985; 135:4289–4296.

75. Chedid L. Muramyl peptides as possible endogenous immunopharmacological mediators. Microbiol Immunol 1983; 27:723–732.

76. Didier ES, Wheeler E, Rutherford MS, Tompkins WAF. Characterization of two highly phosphorylated cytoskeleton-associated proteins, pp58 and pp60, in tumoricidal murine peritoneal macrophages and their comparison with vimentin. Molec. Immunol 1988; 25:785–794.

77. McBride WH, Hoon DB, Jung T, Naungayan J, Nizze A, Morton DL. Cyclophosphamide-induced alterations in human monocyte functions. J Leuk Biol 1987; 42:659–666.

78. Kaushansky K, Shoemaker SG, Alfaro S, Brown C. Hematopoietic activity of granulocyte/macrophage colony-stimulating factor is dependent upon two distinct

regions of the molecule: functional analysis based upon the activities of interspecies hybrid growth factors. Proc Natl Acad Sci USA 1989; 86:1213–1217.

79. Jones EY, Stuart DI, Walker NPC. Structure of tumor necrosis factor. Nature 1989; 338:225–228.

80. Daemen T, Veninga A, Dijkstra J, Scherphof G. Differential effects of liposome-incorporation on liver macrophage activating potencies of rough lipopolysaccharide, lipid A, and muramyl dipeptide. J Immunol 1989; 142:2469–2474.

81. Shimizu J, Suda T, Yoshioka T, Kosugi A, Fujiwara H, Hamaoka T. Induction of tumor-specific *in vivo* protective immunity by immunization with tumor antigen-pulsed antigen-presenting cells. I Immunol 1989; 142:1053–1059.

82. Brown JH, Jardetzky T, Saper MA, Samradui B, Bjorkman PJ, Wiley DC. A hypothetical model of the foreign antigen binding site of class II histocompatibility molecules. Nature 1988; 332:845–850.

83. Cathcart MK, McNally AK, Morel DW, Chisolm GM, III. Superoxide anion participation in human monocyte-mediated oxidation of low-density lipoprotein and conversion of low-density lipoprotein to a cytotoxin. J Immunol 1989; 142:1963–1969.

84. Palinski W, Rosenfeld ME, Ylä-Herttuala, Gurtner GC, Socher SS, Butler SW, Parthasarathy S, Carew TE, Steinberg D, Witztum JL. Low density lipoprotein undergoes oxidatitive modification *in vivo*. Proc Natl Acad Sci USA 1989; 86:1372–1376.

7

Receptor-Dependent Targeting of Lipoproteins to Specific Cell Types of the Liver

Theo J. C. Van Berkel, J. Kar Kruijt, P. Chris De Smidt, and Martin K. Bijsterbosch
Leiden University, Leiden, The Netherlands

I. INTRODUCTION

Specific targeting of drugs can be achieved by utilizing the specific properties of the target cell. Because cell membranes are decisive for the primary interaction, the recognition marker for a drug or drug transport vehicle is obviously related to the presence for specific receptors. This chapter outlines the strategy to achieve targeting to specific cell types present in one organ, the liver. It is anticipated that the principles can be applied in a general sense to achieve delivery of drugs at other desired sites of action.

The liver is a heterogeneous tissue that contains parenchymal cells (92.5% of liver protein), endothelial cells (3.3% of liver protein), Kupffer cells (2.5% of liver protein), and fat-storing cells (1.7% of liver protein). To determine uptake of substances in a certain cell type, a cell isolation method is needed which allows a quantitative recovery of the ligand in the purified cells.

The procedure outlined in Fig. 1 prevents destruction of targeting ligands, because liver perfusion, cell separation, and cell purification are performed at a low temperature (8°C). In addition, one is able to assess recovery by comparing the radioactivity in the purified parenchymal, endothelial, and Kupffer cells (all obtained from one liver) to the radioactivity originally present in the whole rat liver. The method has been evaluated using ^{125}I-asialofetuin as the substrate and comparing quantitative autoradiographic data [1]. When asialofetuin is injected into rats, it is rapidly taken up from serum by the liver (Fig. 2). Preinjection of unlabeled asialofetuin (5 mg) slows the decay of the

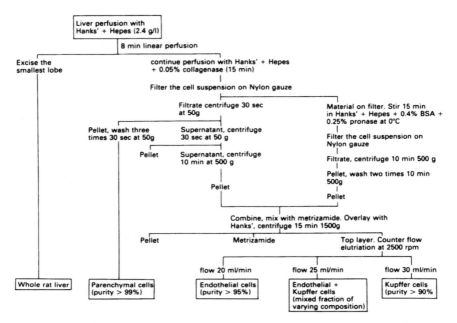

Figure 1 Procedure for the isolation of liver cells at a low temperature (8 °C).

radioactive ligand, while the apparent liver uptake is inhibited. When at 10 min after injection the different liver cell types were isolated, we found that the great majority of ^{125}I-asialofetuin had become associated with parenchymal cells. Preinjection of 5 and 25 mg of unlabeled asialofetuin inhibited the association of ^{125}I-asialofetuin with parenchymal cells by 84% and 93%, respectively, while the association with endothelial and Kupffer cells was inhibited by only 22% and 28%, respectively (Fig. 3). Hubbard et al. [1] reported, by using an autoradiographic method, that 20 mg of unlabeled asialofetuin led to inhibition of ^{125}I-asialofetuin uptake by parenchymal cells of 99% and by endothelial cells and Kupffer cells of 24%. Furthermore, preinjection of N-acetyl galactosamine blocked specifically the uptake of ^{125}I-asialofetuin by parenchymal cells, while the uptake by Kupffer and endothelial cells was specifically inhibited by preinjection of mannan (Fig. 4).

These data establish that the nonparenchymal cell preparations are free of any parenchymal cell–derived materials. Recovery calculations indicate a quantitative recovery in the isolated cells of the radioactivity originally present in the whole rat liver.

After having established the cell isolation method, the cell specificity of lipoproteins for the various liver cell types has to be investigated to verify the extent to which lipoproteins are suitable carriers for specific drug delivery in the different cell types.

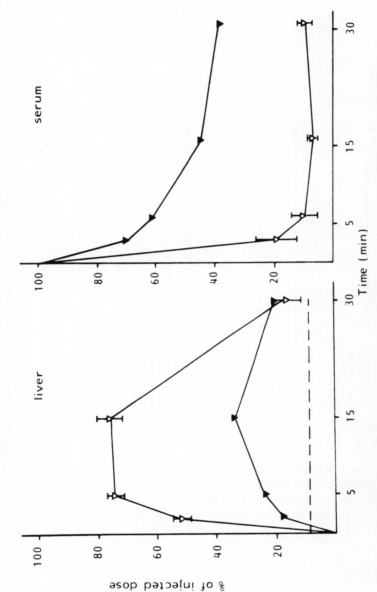

Figure 2 The effects of unlabeled asialofetuin on the serum decay and liver association of ^{125}I–asialofetuin was injected into anesthetized rats ($\triangledown$–$\triangledown$) without or with ($\blacktriangledown$–$\blacktriangledown$) preinjection of 5 mg unlabeled asialofetuin. The livers were not perfused and the dashed line represents the maximal contribution of the serum value to the liver uptake (determined with ^3H-labeled albumin).

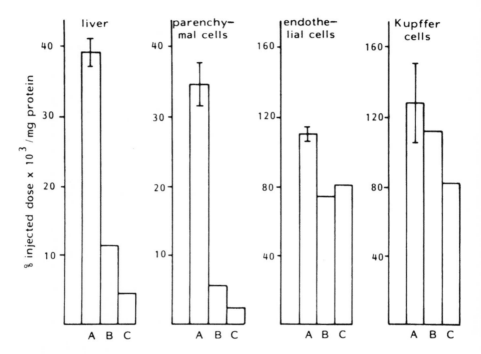

Figure 3 The effect of unlabeled asialofetuin on the association of ^{125}I-asialofetuin to parenchymal, endothelial, and Kupffer cells. Ten minutes after injection of 9 μg ^{125}I-asialofetuin the cells were isolated according to Fig. 1. A = control; B = preinjection of 5 mg asialofetuin (-1 min); C = preinjection of 25 mg asialofetuin (-1 min).

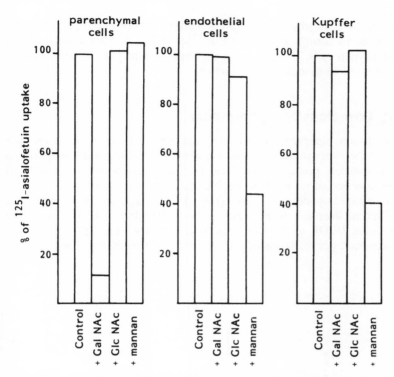

Figure 4 The effect of preinjection of GalNAc, GlcNAc, or mannan on the cell association of ^{125}I-asialofetuin to parenchymal, endothelial, and Kupffer cells.

II. RESULTS AND DISCUSSION

A. Chylomicrons

Chylomicrons are triglyceride-rich lipoproteins, synthesized by the small intestinal epithelium as vehicles to transport absorbed lipids to the plasma compartment. Electron micrographs have shown that chylomicrons are composed of a homogeneous core surrounded by a surface film. Although the protein moiety of chylomicrons is small (0.5–4% by weight), their role in the interaction with the liver is obvious. The relative amounts of apolipoprotein C-III and E are decisive for an interaction with the liver. As indicated in Fig. 5, an excess of apo C-III blocks the interaction of chylomicrons with both parenchymal and nonparenchymal liver cell types, while mixtures of apo E and C-III lead to an intermediate interaction. Addition of pure apo E leads to a clearly increased interaction of chylomicrons with both parenchymal and nonparenchymal liver cells (Fig. 6). A determination of the in vivo hepatic up-uptake of chylomicrons indicates that 65% of the injected chylomicrons are recovered in the parenchymal cells (30 min after injection) [2]. This indicates that chylomicrons form a potential vehicle to transport drugs rapidly to the liver parenchymal cells. Furthermore, incorporation of pure apolipoproteins might be utilized to modulate this function.

B. Very Low Density Lipoproteins

Very low density lipoproteins (VLDLs) are synthesized by the liver and transport lipids from the liver to extrahepatic sites. Subsequently reuptake in the liver is induced, probably modulated by the apolipoproteins E and C-III [3]. After injection of cholesterol-rich VLDL into rats a rapid serum decay is coupled to a rapid uptake in the liver (Fig. 7). Although uptake is exerted by all the different liver cell types (Fig. 8), the association of both VLDL and cholesterol-rich VLDL to Kupffer cells is respectively, only 3.5- and 1.2-fold higher (expressed per milligram of cell protein). Because Kupffer cells contain

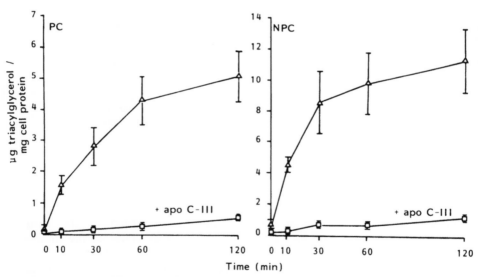

Figure 5 Time course of the cell association of chylomicrons and apo C-III enriched chylomicrons from estrogen-treated rats to parenchymal (PC) and nonparenchymal cells (NPC). The cells were incubated with 0.17 mg of triacylglycerol per milliliter. Values are means of three different chylomicron and apo C-III–enriched chylomicron preparations incubated with three different cell preparations. The results are given as a means ± SEM (indicated by the bars). The chylomicrons contained 5.9 ± 0.4 μg protein/mg triacylglycerol (△). The apo C-III–enriched chylomicrons contained an additional 12.6 ± 0.2 μg apo C-III/mg triacylglycerol (□).

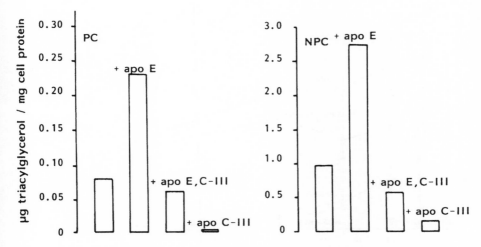

Figure 6 The cell association of chylomicrons enriched with various amounts of apo E and/or apo C-III to parenchymal (PC) and nonparenchymal cells (NPC) at 0°C. The chylomicron preparations and incubation conditions were identical with those used in Fig. 5 except that incubation was performed for 1 h at 0°C. The left (unlabeled) columns represent the cell association of the untreated chylomicrons from estrogen-treated rats.

only 2.5% of the liver protein, quantitatively the majority of VLDL and especially cholesterol ester-rich VLDL is taken up by parenchymal cells (Table 1).

C. Low Density and High Density Lipoproteins

Low density lipoproteins (LDLs) are a class of lipoproteins with a mean diameter of 23 nm, whereas the high density lipoproteins (HDLs) are smaller (about 10 nm).

To determine the fate of LDL we labeled the apolipoprotein with ^{14}C-sucrose and determined the serum decay and liver uptake [4]. Low density lipoprotein is slowly removed from serum (half-life about 4.5 h) and the major part (70–80%) is recovered in the liver. A determination of the uptake in the various liver cells at 4.5 h postinjection (Fig. 9) indicates that 71% of the total liver uptake of LDL is attributed to the Kupffer cells. When LDL is methylated, it loses its ability to interact with its specific receptor [5]. With parenchymal cells no significant difference in uptake between native LDL and methylated LDL was found, whereas it appears that only Kupffer cells internalize LDL in a receptor-dependent way. The data indicate that LDL can be used for drug transport to Kupffer cells with a relatively slow delivery rate.

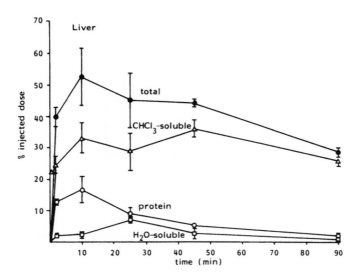

Figure 7 The amount and form of ^{125}I-CE-VLDL in liver at various times after injection. On the indicated time after injection, the liver was subjected to an 8°C perfusion with a Hanks buffer and the liver was subjected to a Folch extraction, whereafter label distribution as a percentage of the injected dose can be calculated. Values are the mean ± SEM of two experiments.

We recently synthesized a compound (**I**) in order to utilize the active receptor for asialofetuin (Fig. 2–4) as a trigger for the uptake of various types of vesicles [6]. The triantennary galactose-terminated cholesterol derivative (Tris-Gal-Chol) dissolves easily in water; upon mixing with liposomes, it is immediately incorporated into these particles. Tris-Gal-Chol addition to LDL leads to immediate incorporation [6].

In Fig. 10 the decay of LDL in serum and uptake in liver is plotted after loading the particles with varying amounts of Tris-Gal-Chol. Incorporation of Tris-Gal-Chol into LDL leads to a markedly increased removal from serum

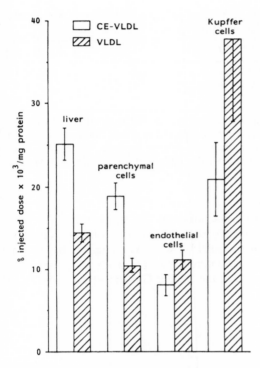

Figure 8 Association of CE-VLDL and VLDL to parenchymal, endothelial, and Kupffer cells in vivo. ^{125}I–CE-VLDL and ^{131}I-VLDL were injected into rats and 10 min after injection a liver perfusion was started. After 8 min of perfusion at 8°C the total liver association of radioactivity was determined (liver). Subsequently parenchymal, endothelial, and Kupffer cells were isolated by a low temperature procedure and the association to the isolated cells was determined. The bars represent values ± SE (N=4).

paralleled with a quantitative recovery of the label in liver (Fig. 10). Similar experiments performed with HDL apparently show a similar induction of liver uptake by Tris-Gal-Chol (Fig. 11).

The increased decay of Tris-Gal-Chol–LDL (20:13) in serum is nearly completely blocked by preinjection (1 min before the Tris-Gal-Chol–LDL) of 0.5 mmol/rat of N-acetylgalactosamine (GalNAc, Fig. 12, final concentration in the blood approximately 50 mM), while preinjection of the same dose of N-acetylglucosamine (GlcNAc) has no effect at all (not shown). Compared to GalNAc, the preinjection of 5 or 25 mg of asialofetuin influences the serum radioactivity and the liver association of Tris-Gal-Chol–LDL (20:13) only to a small extent.

Table 1 Relative Contribution of the Different Cell Types to the Total Liver Uptake of CE-VLDL and VLDL 10 Min Postinjection

Cell type	CE-VLDL(%)	VLDL (%)
Parenchymal cells	95.7 ± 0.7	88.0 ± 1.5
Endothelial cells	1.5 ± 0.3	3.4 ± 0.4
Kupffer cells	2.9 ± 0.6	8.6 ± 1.4

The amount of radioactivity per milligram of cell protein in the isolated cell fractions was multiplied with the amount of protein that each cell type contributes to total liver protein. The values are calculated as the mean of four experiments (± SE). CE-VLDL (20 μg of apolipoprotein/rat) and VLDL (20 μg of apolipoprotein/rat) were injected simultaneously in each rat.

The increased decay of Tris-Gal-Chol–HDL in the serum and liver association is also completely blocked by preinjection (1 min before Tris-Gal-Chol–HDL) of 0.5 mmol per rat of GalNAc, while GlcNAc has hardly any effect. Preinjection of 5 mg of asialofetuin per rat leads to a complete blockade of the increased liver association of Tris-Gal-chol–HDL, similar to GalNAc (Fig. 12).

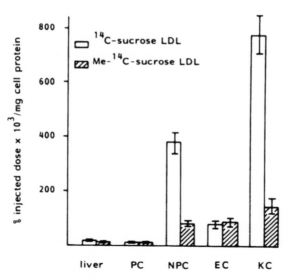

Figure 9 Distribution of ^{14}C sucrose–labeled LDL or methylated LDL between the various liver cell types at 4.5 h after injection. Screened lipoproteins were injected and 4.5 h after injection the cells were isolated. NPC, nonparenchymal cells; PC, parenchymal cells; EC, endothelial cells: KC, Kupffer cells. Values are means ± SEM for three experiments.

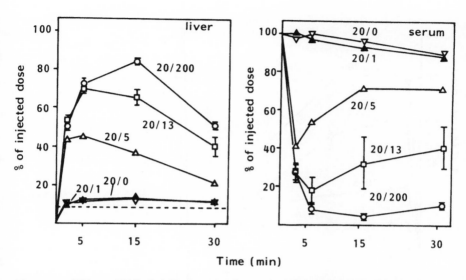

Figure 10 Effect of Tris-Gal-Chol on the liver association and serum decay of LDL. ^{125}I-LDL (20 μg of apolipoproteins) was mixed with 0 ($\triangledown$), 1 ($\blacktriangle$), 5 ($\triangle$), 13 ($\square$), or 200 (o) μg of Tris-Gal-Chol. The mixture was injected into anesthetized rats, and the liver association and serum decay were determined. When indicated, bars represent SE for three animals. The livers were not perfused, and the dashed line represents the maximal contribution of the serum value to the liver uptake (determined with ^{3}H-labeled albumin).

The complete inhibition of the Tris-Gal-chol–induced liver uptake of HDL by asialofetuin suggests that the asialoglycoprotein receptor on parenchymal cells is responsible for this uptake. At 10 min postinjection the HDL association with parenchymal cells (2.8 ± 0.1% of the injected dose x 10^4/mg cell protein, N=4) is stimulated 42- or 105-fold, respectively, by incorporation of 5 or 13 μg of Tris-Gal-Chol/20 μg of HDL apoprotein (Fig. 13). For nonparenchymal cells the association of HDL (17.5 ± 1.1% of the injected dose x 10^4/mg of cell protein, N=4) is stimulated 1.7- or 5-fold, respectively, by 5 or 13 μg of Tris-Gal-Chol incorporated into HDL. When 200 μg of Tris-Gal-Chol is mixed with 20 μg of HDL (loading to its disintegration), nonparenchymal cell uptake shows a marked further increase (Fig. 13).

The capacity of the liver uptake system of Tris-Gal-Chol–loaded HDL was studied by injection of 100 and 250 μg of HDL (mixed with Tris-Gal-Chol in the ratio of HDL protein to Tris-Gal-Chol of 20:13). Both with 100 and with 250 μg of Tris-Gal-Chol–HDL the total liver uptake (in percentage of the injected dose) was reduced by about 50% as compared with 20 μg of Tris-Gal-Chol–HDL. The percentage of the injected dose taken up by the parenchymal cells was about two times lower than with the low dose of 20 μg of Tris-Gal-

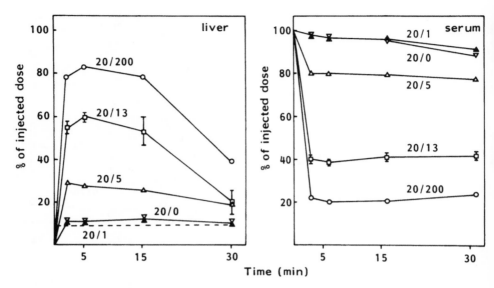

Figure 11 The effect of Tris-Gal-Chol on the liver association and serum decay of HDL. ^{125}I-HDL (20 μg of apolipoprotein) was mixed with 0 (▽), 1 (▲), 5 (△), 13 (□), or 200 (○) μg of Tris-Gal-Chol. The mixture was injected into anesthetized rats and the liver association and serum decay were determined. When indicated, bars represent SE for three animals. The livers were not perfused and the dashed line represents the maximal contribution of the serum value to the liver uptake (determined with ^{3}H-labeled albumin).

Chol–HDL, while the percentage uptake by the nonparenchymal cells was not changed.

The low extent of inhibition obtained by preinjection of asialofetuin compared to GalNAc suggests that the classical galactose (asialoglycoprotein) receptor on parenchymal cells was probably not mainly responsible for the high liver association of Tris-Gal-chol–loaded LDL. A subdivision of the liver into parenchymal and nonparenchymal cells indicates that the main effect of Tris-Gal-Chol loading of LDL is on nonparenchymal cell uptake (Fig. 14). At 10 min postinjection, the LDL association with parenchymal cells (2.5 ± 0.2% injected dose x 10^4/mg of cell protein, N=5) is stimulated 6- or 10-fold, respectively, by incorporation of 5 μg or 13 μg of Tris-Gal-Chol/20 μg of LDL apoprotein. For nonparenchymal cells, the association of LDL (131 ± 15% injected dose x 10^4/mg of cell protein, N=5) is stimulated 60- or 70-fold, respectively, by 5 μg or 13 μg of Tris-Gal-Chol incorporated into LDL. The specific radioactivity (per milligram of cell protein) in nonparenchymal cells relative to parenchymal cells is therefore 490- and 250-fold higher for these preparations (compared to a factor of 52 for native LDL).

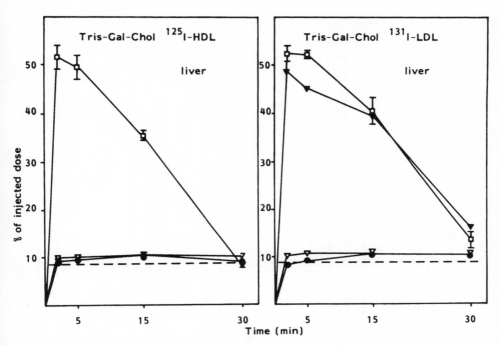

Figure 12 The effect of asialofetuin and GalNAc on the liver association of Tris-Gal-Chol–LDL. ^{125}I-HDL (20 μg of apolipoprotein) and ^{131}I-LDL (20 μg of apolipoprotein) were separately mixed with 13 μg of Tris-Gal-Chol. One minute prior to the injection of the Tris-Gal-Chol–loaded lipoprotein 5 mg asialofetuin (▼-▼) or 0.5 mmol GalNAc (•-•) was injected. ▽-▽ , Tris-Gal-Chol–loaded lipoproteins without preinjection; unloaded lipoproteins.

To study the capacity of the uptake system for Tris-Gal-Chol–loaded LDL in the different cell types, we performed similar studies with injection of 250 μg of LDL. The total liver uptake at 10 min postinjection is reduced from 41.3% to 30.1% of the injected dose when 250 μg is injected instead of 20 μg of Tris-Gal-Chol–LDL (20:13). The percentage uptake in parenchymal cells is not influenced, while the lowered uptake by nonparenchymal cells is responsible for the relative decrease in the liver's efficiency in capturing Tris-Gal-Chol–LDL.

Since the cellular distribution of Tris-Gal-Chol–HDL was seen to be so different from that of Tris-Gal-Chol–LDL, we decided to compare the cellular uptakes of the Tris-Gal-Chol–loaded lipoproteins, as well as asialofetuin, directly in one set of experiments. Furthermore, we specified the cellular uptake sites more precisely by separating endothelial from Kupffer cells. As indicated in Fig. 15, Tris-Gal-Chol stimulates the total liver uptake of LDL or HDL to a similar extent. However, the parenchymal cell uptake of Tris-Gal-

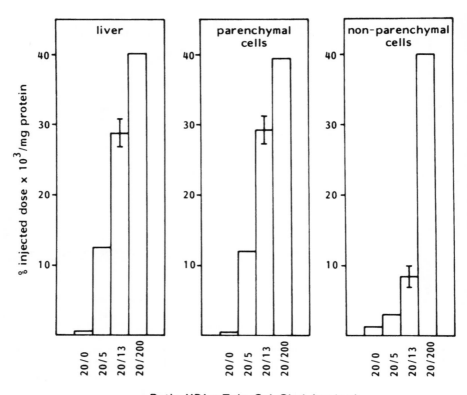

Figure 13 The effect of Tris-Gal-Chol on the association of HDL to parenchymal and nonparenchymal cells in vivo. ^{125}I-HDL (20 μg of apolipoprotein) was mixed with 0 (20/0), 5 (20/5), 13 (20/13), or 200 (20/200) μg of Tris-Gal-Chol. Ten minutes after injection a liver perfusion was started and the total liver association (after 8 min of perfusion at 8°C) and the association to the subsequently isolated parenchymal and nonparenchymal cells was determined. The bars represent values ± SE (N = 3).

Chol–HDL is much higher than that of Tris-Gal-Chol–LDL, whereas the reverse is seen with the Kupffer cells. Both for Kupffer and for endothelial cells, Tris-Gal-Chol incorporation stimulated the uptake of LDL but had only a small effect on that of HDL. Asialofetuin follows a cellular uptake pattern similar to Tris-Gal-Chol–HDL, although the cell association of Tris-Gal-Chol–HDL to endothelial and Kupffer cells is even lower than that of asialofetuin.

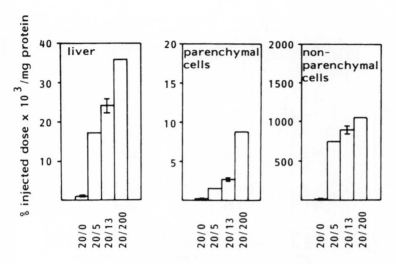

Ratio LDL : Tris-Gal-Chol (μg/μg)

Figure 14 The effect of Tris-Gal-Chol on the association of LDL to parenchymal and nonparenchymal cells in vivo. ^{125}I-LDL (20 μg of apolipoprotein) was mixed with 0 (20/0), 5 (20/5), 13 (20/13), or 200 (20/200) μg of Tris-Gal-Chol. Ten minutes after injection, a liver perfusion was started and the total liver association (after 8 min of perfusion at 8°C) and the association to the subsequently isolated (at 8°C) parenchymal and nonparenchymal cells was determined. The bars represent values ± SE (N = 3)

Table 2 Relative Contribution of the Different Liver Cell Types to the Total Liver Uptake of Asialofetuin, Tris-Gal-Chol–HDL, and Tris-Gal-Chol–LDL

Cell type	Asialofetuin (%)	Tris-Gal-Chol–HDL (%)	Tris-Gal-Chol–LDL (%)
Parenchymal cells	82.5	98.0	7.7
Endothelial cells	9.3	0.5	15.5
Kupffer cells	8.2	1.5	76.8

The amount of radioactivity per milligram of cell protein in the isolated cell fractions was multiplied with the amount of protein that each cell type contributes to total liver protein. Lipoproteins (20 μg of apolipoproteins) were mixed with 13 μg of Tris-Gal-Chol. The values are calculated from the mean of three independent experiments for each substrate.

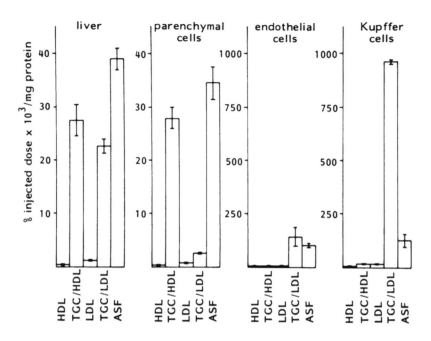

Figure 15   The effect of Tris-Gal-Chol on the association of HDL and LDL to parenchymal, endothelial, and Kupffer cells in comparison with the association of asialofetuin (ASF). ^{125}I-HDL or ^{125}I-LDL (20 μg of apolipoprotein) was mixed with 13 μg of Tris-Gal-Chol (TGC/HDL or TGC/LDL) or the equivalent amount of phosphate-buffered saline (HDL or LDL).   Ten minutes after injection of the apolipoprotein or ^{125}I-asialofetuin (9 μg), a liver perfusion was started, and the total liver association (after 8 min of perfusion at 8°C) and the association with the subsequently isolated (at 8°C) parenchymal, endothelial, and Kupffer cells was determined.   The bars represent values SE ± (N = 3).

From these data, and taking into account the amount of protein contributed by each cell type to the total liver, the percentage of the total liver uptake of asialofetuin and Tris-Gal-Chol–loaded lipoproteins for which each liver cell type is responsible can be determined (Table 2).

D. Intracellular Processing of Lipoproteins

In order to investigate the possible involvement of lysosomes in the processing of CE-rich VLDL, the liver was subjected to a subcellular fractionation. Label associated to the apolipoproteins showed a maximal association with the liver at 10 min postinjection (Fig. 16). At 45 min after intravenous injection the major part (75%) of the apolipoprotein-associated label is already released

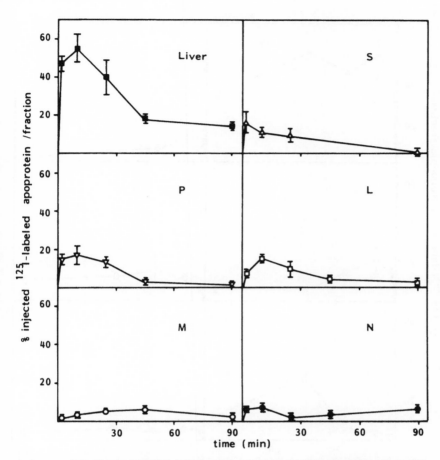

Figure 16 Distribution of ^{125}I-labeled apoprotein in subcellular fraction of the liver at various times after injection of ^{125}I–CE-VLDL. On the indicated times the liver was subjected to an 8°C perfusion with a Hanks buffer whereafter a subcellular distribution procedure was started. The subcellular fractions were subjected to a Folch extraction to determine apoprotein-associated label. Values are expressed as a percentage of the injected ^{125}I label associated with apolipoprotein of CE-VLDL that becomes associated with the indicated subcellular fraction. N, nuclear fraction; M, heavy mitochondrial fraction; L, light mitochondrial fraction containing lysosomes; P, microsomal fraction; S, final supernatant.

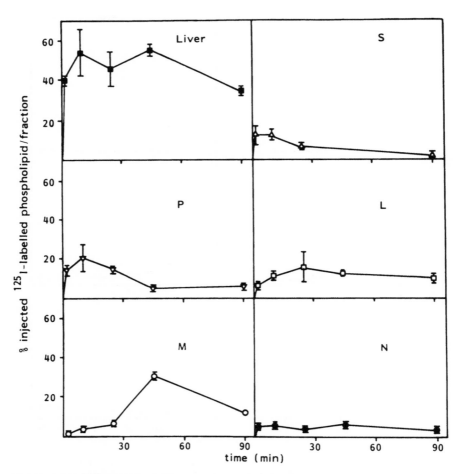

Figure 17 Distribution of ^{125}I–labeled phospholipids in subcellular fraction of the liver at various times after injection of ^{125}I–CE–VLDL. On the indicated times the liver was subjected to an 8°C perfusion with a Hanks' buffer whereafter a subcellular distribution procedure was started. The subcellular fractions were subjected to a Folch extraction to determine phospholipid-associated label. Values are expressed as a percentage of the injected ^{125}I label associated with apolipoprotein of CE-VLDL that becomes associated with the indicated subcellular fraction. N, nuclear fraction; M, heavy mitochondrial fraction; L, light mitochondrial fraction containing lysosomes; P, microsomal fraction; S, final supernatant.

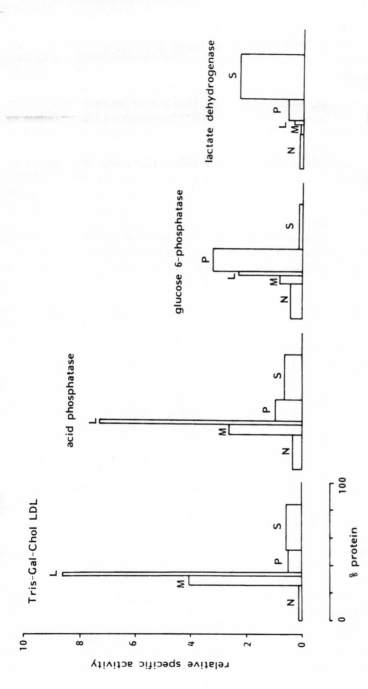

Figure 18 Distribution of ^{14}C sucrose–labeled Tris–Gal–Chol-LDL in subcellular fractions of the liver. ^{14}C Sucrose-labeled LDL (20 μg of apolipoprotein) was mixed with 13 μg of Tris–Gal–Chol. The mixture was injected into rats, and 60 min after injection, a subcellular distribution procedure was started exactly as described by De Duve et al. [7]. N, nuclear fraction; M, heavy mitochondrial fraction; L, light mitochondrial fraction; P, microsomal fraction; S, final supernatant.

from the liver. The sequence of events agrees with an initial association to the membranes, uptake in endocytotic vesicles, and lysosomal processing of the particle. However, it must be mentioned that a phospholipidlike drug may follow a dissimilar route and may end in the mitochondrial fraction (Fig. 17). The fate of Tris-Gal-Chol–loaded LDL was followed by labeling the protein moiety with ¹⁴C-sucrose. When a subcellular distribution was performed 60 min after injection of ¹⁴C-sucrose–labeled Tris-Gal-Chol–loaded LDL according to De Duve et al. [7], the radioactivity coincided with the lysosomal marker acid phosphatase (Fig. 18).

The intracellular processing of Tris-Gal-Chol–HDL was investigated by isolating hepatocytes 10 min after the in vivo injection of Tris-Gal-Chol–HDL (20:13) and incubating the cells at 37°C in vitro. Within 30 min of incubation the cells lose $45 \pm 1\%$ (N=2) of their radioactivity, of which $70 \pm 1\%$ is recovered as trichloroacetic acid soluble in the cell supernatant.

The possible involvement of lysosomes in processing Tris-Gal-Chol–HDL was investigated by pretreating the rats with either leupeptin [8] or chloroquine [9] and following the disappearance of label from liver (Fig. 19). Leupeptin

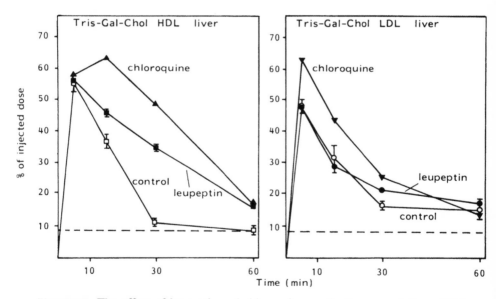

Figure 19 The effect of leupeptin and chloroquine on the liver association of Tris-Gal-Chol–HDL and Tris-Gal-Chol–LDL. ¹²⁵I-HDL (20 μg of apolipoprotein) and Tris-Gal-Chol–LDL were mixed with 13 μg of Tris-Gal-Chol. The mixtures were injected into rats which were preinjected with 5 mg of leupeptin (60 min prior to the lipoproteins) or with chloroquine (120 and 60 min prior to injection of the lipoproteins). The control was preinjected at 60 min prior to the lipoproteins with phosphate-buffered saline. The bars represent values ± SE (N = 3).

is an inhibitor of thiol proteases, while chloroquine acts as a general inhibitor of lysosomal proteolysis by increasing the lysosomal pH. In addition, chloroquine may inhibit the fusion of endocytotic vesicles or multivesicular bodies with lysosomes. Pretreatment of the rats with either leupeptin or chloroquine clearly inhibits the processing of Tris-Gal-Chol–HDL. The serum radioactivities were not influenced by the various treatments (not shown). At 30 min after injection of Tris-Gal-Chol–loaded ^{125}I-HDL in rats pretreated with leupeptin, a subcellular distribution study was performed (Fig. 20). It appears that the radioactivity is highly enriched in the lysosomal fraction. In untreated liver the lysosomal marker acid phosphatase shows a high relative activity in the L fraction. However, the pretreatment of the rats with leupeptin causes a shift in its distribution. A similar shift was found for the lysosomal markers ß-glucuronidase and cathepsin D.

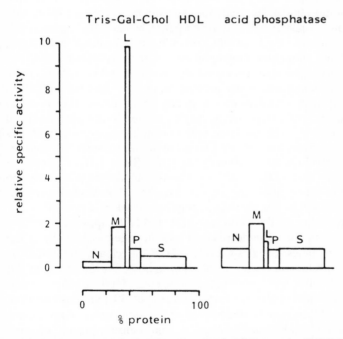

Figure 20 Distribution of Tris-Gal-Chol–HDL in subcellular fractions of the liver. ^{125}I-HDL (20 µg of apolipoprotein) was mixed with 13 µg of Tris-Gal-Chol. The mixture was injected into rats which were preinjected with 5 mg of leupeptin (60 min prior to injection of Tris-Gal-Chol–HDL). Thirty minutes after injection of Tris-Gal-Chol–HDL a subcellular distribution procedure was started exactly as described by De Duve et al. [7]. N, nuclear fraction; M, heavy mitochondrial fraction; L, light mitochondrial fraction; P, microsomal fraction; S, final supernatant.

It can be noticed that leupeptin and chloroquine are much less effective inhibitors of the catabolism of Tris-Gal-Chol–LDL than of Tris-Gal-Chol–HDL.

The Tris-Gal-Chol molecule was designed for use in directing lipoproteins to the liver parenchymal cells because it was anticipated that the galactose residues should allow an interaction with asialoglycoprotein (galactose) receptor on hepatocytes. The design of the compound was also adapted to the property of the hepatocyte receptor to bind and internalize triantennary structures with a much higher affinity and efficiency than bis- or monoglycosides. We showed that incorporation of Tris-Gal-Chol in LDL leads to an increased galactose-mediated interaction of LDL with the liver and that the nonparenchymal cells are responsible by 80–90% for the increased uptake. Incorporation of the Tris-Gal-Chol molecule into HDL leads to a dose-dependent increase in interaction with the liver similar to that with LDL, but now surprisingly 98% of the Tris-Gal-Chol–HDL is taken up by parenchymal cells. The association of Tris-Gal-Chol–HDL to total liver is blocked by asialofetuin (5 or 25 mg) and by GalNAc but not influenced by GlcNAc, so it can be concluded that Tris-Gal-Chol incorporation into HDL leads to a specific interaction of HDL with the asialoglycoprotein (galactose) receptor on parenchymal cells. Actually this interaction is more specific for Tris-Gal-Chol–HDL than for asialofetuin itself, because in accordance with autoradiographic data, some asialofetuin (10–20%) of the total liver uptake becomes associated with Kupffer and endothelial cells.

The molecular mechanism by which association of Tris-Gal-Chol with HDL and LDL leads to such a differential targeting to galactose receptors on hepatocytes or Kupffer cells, respectively, may be related to the ultrastructural receptor characterization. The galactose receptor on hepatocytes is randomly distributed on the capillary side of the plasma membrane, while the Kupffer cell receptors are preclustered in coated pits [10]. Since HDL possesses a mean size of 10 nm and LDL of 23 nm, the size of lipoprotein might be responsible for the differential fate of the Tris-Gal-Chol–loaded particles irrespective of the fact that the identical recognition marker is present.

The hepatic processing of Tris-Gal-Chol–LDL and HDL appears to involve the lysosomal compartment. The decrease in total hepatic radioactivity at the longer postinjection times of Tris-Gal-Chol–HDL is clearly inhibited by pretreatment of the rats with leupeptin or chloroquine. Furthermore, in the animals pretreated with the lysosomotropic agents, HDL apolipoprotein radioactivity accumulated in the lysosomal fraction at 30 min postinjection. Both leupeptin and chloroquine treatments led to the appearance of prominent autolysosomes with a concurrent change in the density profile of acid phosphatase [11]. We found a similar shift with ß-glucuronidase or cathepsin D. The accumulation of ^{125}I-HDL radioactivity in the fraction that does contain the highest relative activity of lysosomal enzymes in untreated rats (Fig. 20) indicates that the newly formed endocytotic vesicles and/or multivesicular bodies do not fuse with preexisting autolysosomes.

Table 3 Distribution of Radioactivity Between Liver and Serum After Intravenous Injection of Acetyl-LDL

Time after injection (min)	Radioactivity distribution[a] (% of injected dose)	
	Liver	Serum
5		
10	83.4 ± 1.7	2.2 ± 0.2
30	18.0 ± 1.2	8.4 ± 0.3

[a] Values are means of three different experiments.

The use of Tris-Gal-Chol might find application in targeting drugs, hormones, or other material of interest to specific liver cell types. The use of liposomes as transport vesicles is hampered by the difficulty in targeting these vesicles rapidly to parenchymal liver cells. The use of galactose residue–exposing liposomes is frustrated because these particles are mainly captured by the galactose receptor on Kupffer cells. The incorporation of Tris-Gal-Chol into HDL leads to a successful and rapid targeting to the asialoglycoprotein (galactose) receptor on hepatocytes. This property might be used to deliver any compound either covalently linked to the protein moiety or incorporated into the lipid core of HDL to the liver parenchymal cells.

E. Chemical Modification of Lipoproteins

The fate of the lipoproteins may also be changed upon a chemical modification of the apolipoproteins. For instance, acetylation of LDL leads to a rapid removal of LDL from serum (Table 3).

Table 4 Relative Contribution of the Different Liver Cell Types to the Total Liver Uptake of Asialofetuin, Tris-Gal-Chol–HDL, Tris-Gal-Chol–LDL, [14]C Sucrose–labeled LDL, Acetyl-LDL, Chylomicrons, and CE-Rich VLDL

Cell type	Asialo- fetuin (%)	Tris-Gal- Chol–HDL (%)	Tris-Gal- Chol–LDL (%)	[14]C Sucrose– labeled LDL (%)	Acetyl- LDL (%)	Chylo microns (%)	CE-rich VLDL (%)
Parenchymal cells	82.5	98.0	7.7	29	38	65	96
Endothelial cells	9.3	0.5	15.5	9	53	35[a]	1
Kupffer cells	8.2	1.5	76.8	62	9		3

[a] Contribution of endothelial and Kupffer cells.

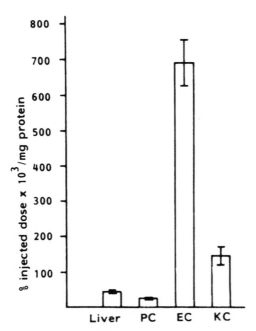

Figure 21 Distribution of acetyl-LDL between the various liver cell types at 10 min postinjection. PC, parenchymal cells; EC, endothelial cells; KC, Kupffer cells.

Isolated endothelial cells contained 5 times more acetyl-LDL per milligram of cell protein than the Kupffer cells and 31 times more than the hepatocytes (Fig. 21). This uptake is mediated by a highly active receptor (scavenger receptor) on the liver endothelial cells [12]. Morphological studies on the interaction of acetyl-LDL conjugated to 20 nm colloidal gold illustrate that the liver endothelial cells bind acetyl-LDL in coated pits, which is followed by rapid uptake [13]. Uptake proceeds through small coated vesicles and finally degradation of the apoprotein occurs in the lysosomes.

III. CONCLUDING REMARKS

The ability to rapidly introduce chylomicrons [2] and CE-rich VLDL [3] into parenchymal cells, acetyl-LDL into endothelial cells [12], Tris-Gal-Chol–LDL into Kupffer cells [14], and Tris-Gal-Chol–HDL into parenchymal cells [15] clearly indicates that in a complex tissue such as liver, successful targeting of lipoproteins to the cell of choice can be achieved (summarized in Table 4). The efficiency and specificity of these uptake processes may be modulated by individual apolipoproteins or modification of surface characteristics provided that basic knowledge (especially receptor characteristics) is available, it can be speculated that similar approaches are possible for other cell types or tissues. It will be a challenge to apply these new strategies for site-specific drug delivery in which industrial laboratories and university centers for biopharmaceutical sciences may combine practical approaches with basic cell-biological achievements.

ACKNOWLEDGMENTS

Martha Wieriks is thanked for the preparation of the manuscript and secretarial help. This research was partly supported by Grant 31.014 from the Dutch Heart Foundation.

REFERENCES

1. Hubbard AL, Wilson G, Ashwell G, Stukenbrok H. J Cell Biol 1979; 83:47–64.
2. Groot PHE, Van Berkel ThJC, Van Tol A. Metabolism 1980; 30:792–797.
3. Harkes L, Van Duijne A, Van Berkel ThJC. Eur J Biochem. ig8g; 180:241–248.
4. Harkes L, Van Berkel ThJC. Biochem J 1984; 224:21–27.
5. Weisgraber KH, Innerarity TL, Mahley RW. J Biol Chem 1978; 253:9053–9062.
6. Kempen JHM, Hoes G, Van Boom JH, Spanjer HH, De Lange J, Langendoen A, Van Berkel Th JC. J Med Chem 1984; 27:1306–1312.
7. De Duve C, Pressman BC, Gianetto R, Wattiaux R, Appelmans F. Biochem 1955; 60:604–617.
8. Dunn WA, La Badie HJ, Aronson NN. J Biol Chem 1979; 245:4191–4196.
9. Hornick CA, Jones AL, Renaud G, Hradek G, Havel RJ. Am J Physiol 1984; 246:G187–G194.
10. Kolb-Bachofen V, Schlepper-Schaefer J, Vogell W. Cell 1982; 29:859–866.
11. Furuno K, Ishikawa T, Kato K. J Biochem (Tokyo) 1982; 91:1485–1494.
12. Nagelkerke JF, Barto KP, Van Berkel ThJC. J Biol Chem 1983; 258:12221–12227.
13. Mommaas-Kienhuis AM, Nagelkerke JF, Vermeer BJ, Daems WTh, Van Berkel ThJC. Eur J Cell Biol 1985; 38:42–50.
14. Van Berkel ThJC, Kruijt JK, Spanjer HH, Nagelkerke JF, Harkes L, Kempen HJM. J Biol Chem 1985; 260:2694–2699.
15. Van Berkel ThJC, Kruijt JK, Kempen HJM. J Biol Chem 1985; 260:12203–12207.

8

Lipoproteins as Carriers for Organ-Imaging Radiopharmaceuticals

Raymond E. Counsell, Susan W. Schwendner,
Laura E. DeForge, and Mark R. DeGalan*
The University of Michigan Medical School, Ann Arbor, Michigan

Roger S. Newton
Warner-Lambert Company, Ann Arbor, Michigan

I. INTRODUCTION

Imaging procedures developed over the past several decades (e.g., computed tomography, magnetic resonance imaging, ultrasonography, radioscintigraphy) allow clinicians to assess the anatomical structure and function of internal organs without having to resort to exploratory surgery. Nuclear medicine scanning, or radioscintigraphy, has been a major contributor to this group of noninvasive imaging modalities. In this procedure, a tracer dose of a radiopharmaceutical is administered to the patient and the location of the radioactivity is determined with the aid of radiosensitive imaging devices (e.g., rectilinear scanner, scintillation camera, or emission tomography device).

For this procedure to be successful, the imaging agent or radiopharmaceutical must fulfill the following requirements:

1. The radioactivity associated with the radiopharmaceutical must readily penetrate intervening tissues in order to be detected externally. Radionuclides that decay by emission of α (e.g., radium-226) or β (e.g., carbon-14) particles are unsuitable because such radiation has poor penetrability and is largely absorbed by surrounding tissues.

Present Affiliation:
* Indiana University Medical School, Indianapolis, Indiana

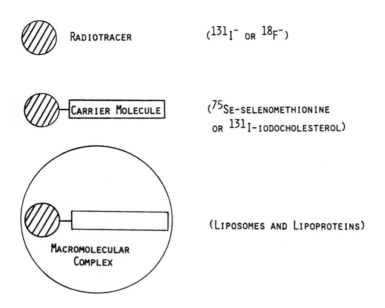

Figure 1   Strategies for the delivery of radiopharmaceuticals to specific tissues.

2. The radioactivity associated with the radiopharmaceutical should selectively localize in the organ of interest.

In rare instances, both of these requirements can be found in a single radionuclide such as the selective accumulation of radioactive iodide in the thyroid and radioactive fluoride in bone. On most occasions, however, it is necessary to link the appropriate radionuclide to a carrier molecule which will serve to direct the radionuclide to the organ of interest. Such carrier molecules can be small (e.g., ^{75}Se-selenomethionine), large (e.g., ^{99m}Tc-macroaggregated albumin), or macromolecular complexes such as liposomes (Fig. 1). Several reviews have outlined the various rationales for the selection of these carrier systems [1–3].

As discussed in previous chapters, plasma lipoproteins serve to transport plasma lipids to specific tissues. Key elements in this process are the apoproteins, found on the surface of these particles, which serve to direct them to specific cellular receptors. It is this cell-targeting aspect of lipoproteins that has called attention to their potential use in the site-specific delivery of radiopharmaceuticals.

II. CONSIDERATIONS IN THE SELECTION OF RADIOPHAR-MACEUTICALS

In general, the radionuclide for organ imaging should have the following characteristics:

1. Decay by gamma radiation alone and with an energy of about 200 kEv.
2. Physical half-life in the range of 6–12 h (i.e., long enough for radio-pharmaceutical preparation but short enough to minimize the radiation-absorbed dose to the patient).
3. Ready availability.
4. Ability to be incorporated readily into carrier molecules (i.e., chemical versatility).
5. Ability to form radiopharmaceuticals that are stable both *in vitro* and *in vivo*.

Unfortunately, no one radionuclide is currently available that fulfills all of these criteria. Indeed, only a few of the 2000 possible radionuclides are employed today for organ-imaging purposes. Among these, those of iodine and technetium-99^m are the most widely used.

Although 25 isotopic forms of iodine are known, only ^{123}I, ^{125}I, and ^{131}I are the radioactive forms used for medical purposes. Iodine-125 is not generally used for organ-imaging purposes in humans because the emitted photons are energetically too weak (Table 1) and the half-life is too long. However, these properties make ^{125}I exceptionally valuable for preliminary chemical and biological studies in small laboratory animals whereupon it can subsequently be substituted for ^{123}I or ^{131}I for studies in large animals and humans. Iodine-123 more closely fulfills the criterion for an ideal radionuclide for imaging, but, until recently, its widespread use for this purpose was hampered by its lack of availability. Iodine-131 is the most frequently employed radioisotope of iodine used in organ-imaging radiopharmaceuticals in humans. Although it is readily available and economical, the ß radiation associated with ^{131}I and its long half-life (8 days) increase the radiation dose to the patient. Despite this limitation, one important feature of ^{131}I and radioiodine in general is that it is usually readily incorporated into a variety of organic molecules either by direct iodination or by isotope exchange procedures [4]. Moreover, compared to technetium, iodine generally has only a minor effect on the physicochemical properties of the carrier molecule, making it much more useful as a metabolic tracer.

Technetium-99^m, in contrast, becomes affixed to carrier molecules by chelation. Thus the carrier molecules must have moieties that can be used for complex formation or else chelating ligands must be linked to the carrier molecule prior to labeling with technetium. Except for large molecular weight

Table 1 Commonly Used Radioisotopes of Iodine

Mass number	Principal γ-ray (mEv)	Half-life
123	0.159	13.2 h
125	0.035	60.14 days
131	0.364	8.04 days

complexes, this process normally produces marked changes in the physicochemical properties of the carrier molecule. Aside from this serious drawback, ^{99m}Tc has ideal radiation properties (monoenergetic 140 kEv gamma ray) and physical half-life (6.04 h).

Most studies dealing with lipoprotein metabolism and fate have employed radioiodinated preparations wherein the radionuclide is affixed to surface (i.e., apolipoproteins) or core (i.e., cholesteryl esters) components of the lipoproteins. To produce an apolipoprotein-labeled lipoprotein, the label is introduced by treating specific lipoprotein fractions with iodine monochloride. Most of the radioiodine becomes covalently attached to tyrosyl and histidyl residues in the apolipoproteins, but as much as 20% of the radioactivity has been known to become associated with lipid [5,6].

Although the use of ^{125}I-labeled lipoproteins has been very successful in ascertaining the *in vitro* receptor binding and metabolism of lipoproteins, this method is unreliable for assessing the metabolic contribution of various tissues *in vivo* due to the dissociation of the radiolabel from the lipoprotein. Once ^{125}I-labeled lipoproteins are internalized into the cell, they are rapidly degraded by lysosomal enzymes. This process releases free amino acids (plus label), which can readily pass through the lysosomal as well as cell membranes. Thus it is very difficult to quantitate degradation processes using apoprotein-labeled lipoproteins.

Attempts to overcome this problem have led workers to produce protein ligands that would remain in the cell once internalized. On this basis that sucrose is not hydrolyzed by intracellular enzymes of mammalian cells, Pittman and Steinberg [7] reasoned that linking radiolabeled sucrose to apolipoproteins

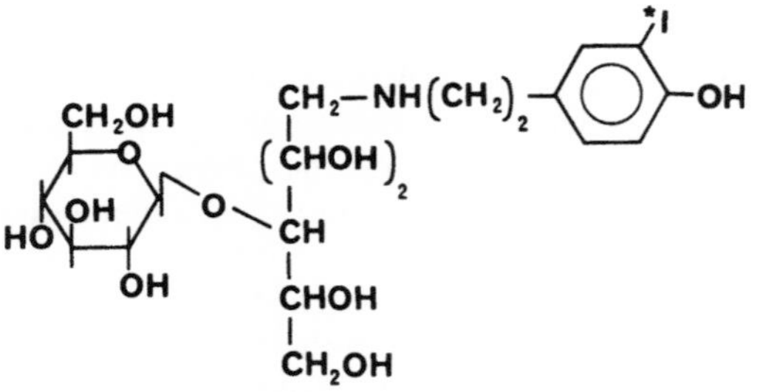

Figure 2 Radioiodinated tyramine cellobiose

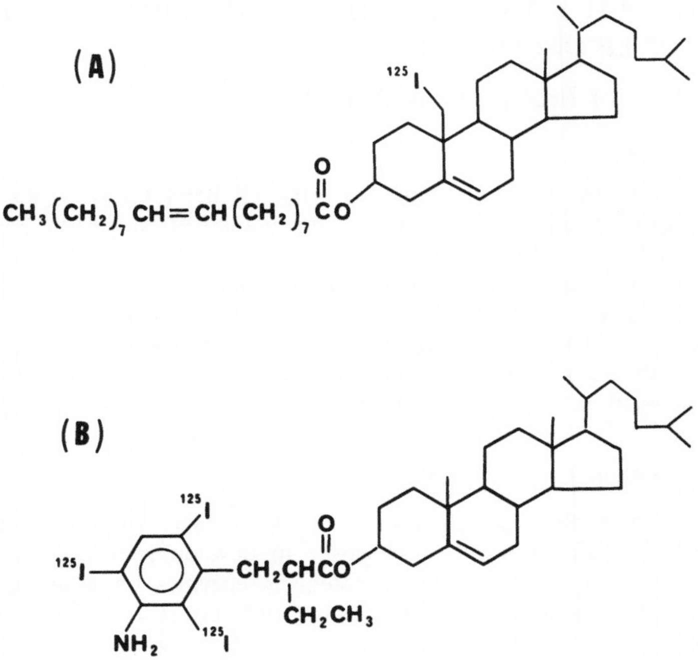

Figure 3 Radioiodinated (A) 19-iodocholesteryl oleate (^{125}I-CO) and (B) cholesteryl iopanoate (^{125}I-CI).

would provide a marker that would accumulate in cells. Using ^{14}C-labeled sucrose, they were able to verify such a hypothesis. The need to label with radioiodine led to radioiodinated tyramine cellobiose as a lipoprotein ligand (Fig. 2). The preparation, potential uses, and limitations of this ligand have been reviewed [8].

Since the apolipoproteins are a critical component in the interaction of lipoproteins with cellular receptors, others have sought ways to radiolabel the core rather than the surface elements of lipoproteins. Another potential advantage of such an approach lies in the fact that several investigators [9–11] have noted that certain cells, notably liver, adrenals, and ovary, are able to take up a disproportionately greater amount of core lipids than surface proteins—an important consideration in organ imaging. Radioiodinated 19-iodocholesteryl oleate and cholesteryl iopanoate (Fig. 3) are two such probes developed in our laboratory [12,13] that have been useful in evaluating the potential of using lipoproteins as a means of targeting imaging agents to specific tissues.

III. METHODS FOR INCORPORATING RADIOPHARMACEUTI-
CALS INTO LIPOPROTEINS

A. Modification of Protein Components

Direct Radioiodination

Iodine-125, 131, and to a lesser extent 123 have all been used to label lipoproteins for *in vitro* and *in vivo* studies. The method of choice for low density lipoprotein (LDL) involves direct iodination using iodine monochloride, as developed by McFarlane [14] and subsequently modified by Bilheimer et al. [15]. Labeling is conducted at a pH of 10 in order to minimize radioiodination of lipid components. Other iodination methods, such as those employing chloramine-T or lactoperoxidase, have been evaluated but tend to cause more oxidative protein denaturation than that produced by iodine monochloride [16].

Indirect Radioiodination

As discussed previously, the rapid metabolism of radioiodinated lipoproteins when administered *in vivo* has led investigators to develop radioiodinated markers that would not be degraded by proteolytic enzymes upon uptake within cells. The tyramine cellobiose technique evolved from such studies and involves the conjugation of tyramine with the disaccharide cellobiose [8]. The tyramine cellobiose conjugate is then radioiodinated and covalently linked to lipoproteins using cyanuric chloride. Both the secondary amino group and the phenolic hydroxyl group are attacked by this reagent and some polymerization of the radioiodinated ligand occurs. The amount of polymeric vs. nonpolymeric ligand linked to lipoproteins is unknown at this time. When this ligand is used with LDL, the reaction is best conducted at a pH of 9.5 in order to minimize labeling of the lipid moieties.

Use of the Bolton–Hunter procedure [17] has also been applied to lipoprotein studies [18,19]. This procedure is particularly appropriate for radioiodination of proteins having a minimal content of tyrosyl residues, which is the case for apo E–containing lipoproteins. In this procedure, the radioiodinated reagent [N-succinimidyl-4-hydroxy-3,5(^{125}I)-diiodophenylpropionate] acylates the lysine residues of the apolipoprotein. While it is known that acylation can markedly interfere with receptor recognition, care is taken in this procedure to ensure that no more than 0.5% of the lysine residues are modified, thus minimizing any effect on the receptor-binding properties of the lipoprotein.

Technetium Labeling

By employing the long-lived beta-emitting technetium-99, Lees and co-workers [20] were able to develop a procedure that could be used for labeling lipoproteins with γ-emitting technetium-99^m. The procedure involved adding $^{99m}TcO_4^-$ (pertechnetate) to LDL in the presence of sodium dithionite in 1 M glycine buffer at pH 9.8 (to avoid rapid decomposition of the reducing agent, dithionite). The reactants were allowed to stand at room temperature for 30 min and the labeled lipoprotein was purified by chromatography on a column of 2% agarose gel. One-third to one-half of the radioactivity was shown to be associated with the LDL peak (Fig. 4). Paper electrophoresis of the chromato-graphed product confirmed that the major amount of radioactivity (60–90%) was associated with LDL (Fig. 5). However, small areas of radioactivity occurred on either side of the LDL peak. The presence of minor components was found to vary with the time of incubation and 30 min was found to be the optimal time for LDL labeling.

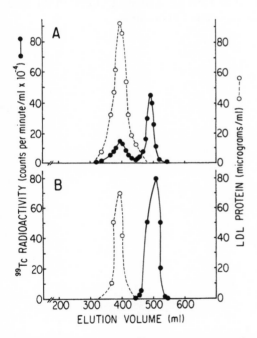

Figure 4 Elution patterns on 2% agarose gel of two representative experiments. (A) LDL was incubated with $^{99m}TcO_4^-$ and sodium dithionite (at pH 9) for 10 min and then chromatographed. (B) Native LDL and $^{99m}TcO_4^-$ were incubated together as in A, but in the absence of sodium dithionite. Solid lines represent radioactivity, dashed lines, protein. (From Ref. 20.)

B. Modification of Lipid Components

In Vivo Incorporation and Isolation

One method used to package exogenous cholesteryl esters into lipoproteins is to allow the test animal to incorporate the compound into its own lipoproteins. Blood can then be collected and the lipoprotein fractions isolated by ultracentrifugation. This method has been used in our laboratory to isolate radiolabeled rat HDL (Fig. 6) [13,21]. Donor animals are administered the radiolabeled lipid in a saline vehicle containing a small percentage of surfactant. After waiting an appropriate time, the animals are sacrificed and their blood is collected. Blood is centrifuged at low speed to obtain plasma. The density of the plasma is adjusted to 1.063–1.070 with NaCl and KBr solutions and then spun at 100,000 x g for 20 h to remove chylomicrons, very low density lipoproteins (VLDL), and LDL, leaving a high density lipoprotein (HDL)–enriched plasma. If necessary, the density can be adjusted to 1.210 for a second spin of 40 h to isolate the HDL fraction [22]. This last step frequently is omitted, however, since it was found to have no significant effect on the fate of the subsequently administered radiolabeled lipoprotein.

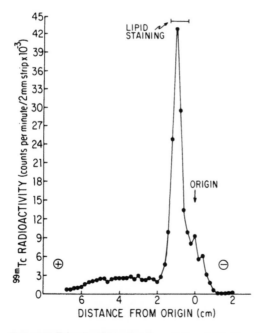

Figure 5 Distribution of radioactivity and lipid staining following paper electrophoresis of gel-filtered ^{99m}Tc-LDL. Reaction mixture had been incubated for 30 min, and about 73% of radioactivity was associated with main LDL peak. (From Ref. 20.)

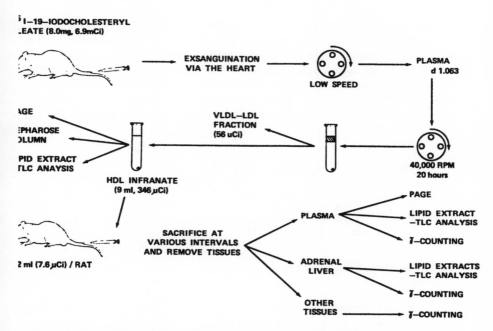

Figure 6 Method for the in vivo incorporation of radioiodinated cholesteryl ester into rat HDL.

Incorporation of 19-iodocholesteryl oleate by this procedure resulted in the appearance of approximately 5% of the dose in the HDL fraction at 30 min. Analysis of the HDL fraction by gel electrophoresis, column chromatography, lipid extraction, and thin layer chromatography (TLC) showed that the vast majority of radioactivity was associated with a single peak of activity that migrated as HDL and that this radioactivity was still associated wth cholesteryl ester. In addition, cholesteryl iopanoate has also been incorporated into rat HDL by this method [23]. In this instance, approximately 28% of the initial does was isolated in HDL-enriched plasma. Analyses as described earlier revealed similar findings, that greater than 90% of the radioactivity in the HDL band was associated with a cholesteryl ester.

In Vitro Methods

Solvent Delipidation and Reconstitution Incorporation of radiopharmaceuticals into the neutral lipid core of lipoproteins requires the use of methodologies that permit passage of the lipophilic molecule through the lipoprotein's outer hydrophilic layer of apoproteins and polar lipids. Ideally, this would result in the production of a lipoprotein with adequately high specific activity while maintaining the native characteristics of the lipoprotein.

One such procedure for labeling the lipoprotein core is the LDL reconstitution protocol outlined by Krieger et al. ([24,25]; see Chapter 3). Briefly, this procedure involves lyophilization of a mixture of 1.9 mg LDL protein and 25 mg potato starch, delipidation with ice-cold heptane, and reconstitution of the lipoprotein by incubation with the desired mixture of radiolabeled core lipids dissolved in organic solvents. The solvents are then evaporated under nitrogen, the LDL–starch mixture is dissolved in an aqueous buffer, and the LDL is

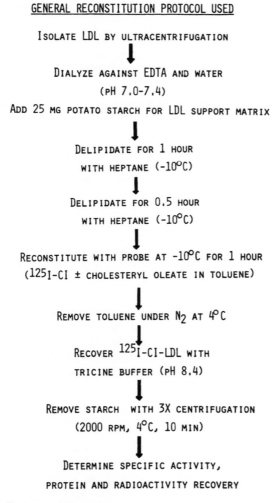

Figure 7 Method for the reconstitution of LDL with ^{125}I-cholesteryl iopanoate.

separated from the starch by means of low speed centrifugation. Krieger et al. [24] demonstrated that reconstitution of 1.9 mg LDL protein with 6 mg of exogenous cholesteryl linoleate caused no change in the phospholipid and protein content, and the particles retained the capacity to bind to cultured fibroblasts. Further, it was shown that the constitution procedure was successful for a variety of hydrophobic compounds with the structural features of long chain unsaturated fatty acyl or polyisoprenyl groups [26].

This procedure has been examined in our laboratory (Fig. 7) in an effort to reconstitute LDL with ^{125}I-cholesteryl iopanoate (^{125}I-CI), a gamma-emitting nonhydrolyzable cholesteryl ester analog that has been used experimentally both as a tracer of cholesteryl ester deposition in atherosclerotic animals and as a potential scintigraphic imaging agent [27,28]. We found that an optimal specific activity was achieved when 2 mg of a ^{125}I-CI–cholesteryl oleate (CO) mixture was used in reconstituting 1.9 mg LDL protein [29]. Protein and radioactivity recovery averaged about 35% and 30%, respectively, while the specific activity of the particles depended on both the initial specific activity of ^{125}I-CI and the ratio of ^{125}I-CI to CO used.

To assess the size distribution of ^{125}I-CI–reconstituted LDL (r[^{125}I-CI]LDL), samples were subjected to gel filtration chromatography on a 6% agarose column. The resulting elution profiles demonstrated the existence of three protein/radioactivity peaks, with the largest and smallest peaks corresponding in size to VLDL and LDL, respectively, and a third peak representing material

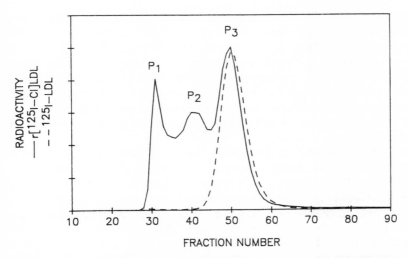

Figure 8 Elution profiles of r[^{125}I-CI]LDL and ^{125}I-LDL from a gel filtration column (6% agarose). The UV absorbance profiles (not shown) paralleled those of the radioactivity.

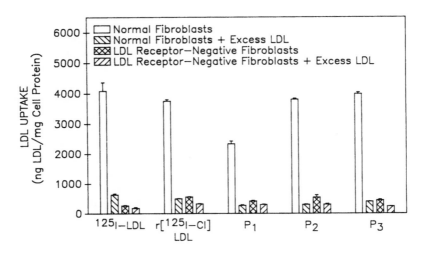

Figure 9 Uptake of r[¹²⁵I-CI]LDL, the three chromatographically separated size fractions (see Fig. 8), and ¹²⁵I-LDL by normal and LDL receptor-negative human fibroblasts. Cells were incubated with each of the lipoprotein preparations at a concentration of 10 μg/ml for 6 h either in the absence or the presence of a 20-fold excess of competing native LDL. The values represent bound, internalized, and degraded lipoprotein.

of intermediate size (Fig. 8). A comparable degree of size heterogeneity was observed upon electrophoresis of the r[¹²⁵I-CI]LDL preparations on 3% acrylamide disc gels. Although crosslinking of the free sulfhydryl groups of the apoprotein B-100 can cause extensive aggregation during certain delipidation procedures [30], electron micrographs of reconstituted LDL have shown that the larger particles appeared to have been generated by fusion of the outer lipid monolayer of the LDL particles rather than by aggregation of the protein. Similar observations were made for control preparations of LDL reconstituted with ³H-CL.

While the reconstitution procedure caused an alteration in the size distribution of the LDL, it did not appear to adversely affect the integrity of the apoprotein B-100 moiety of the particle. Analysis by 3–20% gradient SDS–polyacrylamide gel electrophoresis demonstrated that the apoprotein B-100 band of the reconstituted LDL was essentially identical to that of a native LDL control. In additional studies, the uptake of r[¹²⁵I-CI]LDL and the three size fractions (isolated by column chromatography as described earlier) was examined in normal and LDL receptor-negative human fibroblasts. The results showed that, with the exception of the P₁ fraction (largest size), uptake of the reconstituted LDL was not significantly different from that of a radioiodinated LDL control (Fig. 9). These observations indicate that r[¹²⁵I-CI]LDL appeared to be metabolized normally *in vitro*, but that, without prior fractionation, the size heterogeneity of the preparation may interfere with its use for *in vivo* experiments.

Uptake from Inert Surfaces   In this procedure the radiolabeled lipid is dissolved in a low boiling solvent such as anhydrous ether or hexane.   The solvent is then evaporated under a stream of nitrogen, which allows the lipid to adhere to the surface of the reaction vial, or other solid supports (e.g., celite, glass beads) that may be added to the vial before evaporation.   A medium containing the appropriate lipoprotein fraction is then added to the vial and the contents are incubated for a set period of time.   Such procedures have been shown to be useful for the incorporation of radiolabeled cholesterol into lipoproteins [31,32] and have been adopted for use with radioiodinated cholesterol derivatives [33,34].

Radioiodinated 19-iodocholesterol was incorporated into rat HDL by previously dispersing it on celite 545 (diatomaceous earth) in a ratio of 1:50 w/w.   This dispersion was then incubated with a solution of HDL for 24 h at 25°C in a ratio of 50 mg celite/ml HDL.   The mixture was then centrifuged and the radiolabeled HDL was separated from remaining celite by filtration (0.45 μm pore size).   This resulted in a transfer of 7.2% of the total radioactivity to the HDL filtrate.   This preparation was then analyzed by polyacrylamide gel electrophoresis according to methods described by Narayan and colleagues [35,36] in order to ascertain the integrity of the HDL and the associated radioiodinated probe.   Gel filtration on Sepharose-4B revealed two radioactive

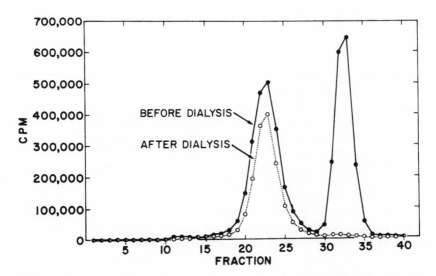

Figure 10   Sepharose-4B gel filtration analysis of rat HDL labeled with [125]I-19-iodocholesterol.

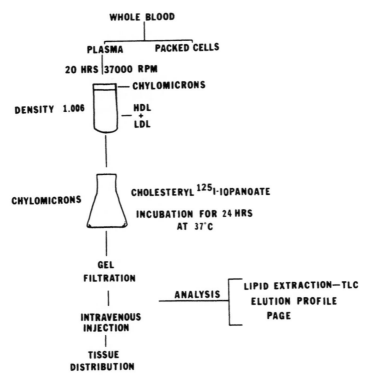

Figure 11 Chylomicron purification and incorporation of radioiodinated cholesteryl ester.

peaks, only one of which was associated with HDL. Although the second peak was not fully characterized, it was removed by dialysis (Fig. 10). Moreover, intravenous administration of this fraction to the rat showed most of the radioactivity (75% dose/g) to accumulate in the thyroid within 0.5 h. Such results underscore the necessity of careful analysis of the radiolabeled products prior to their use in *in vivo* studies. Lipid extraction of the HDL preparation followed by thin layer chromatography of the extract revealed more than 80% of the radioactivity to be radioiodinated cholesterol or radioiodinated cholesterol ester in a ratio of 4.3:1. The latter arises as a result of the presence of lecithin-cholesterol acyltransferase (LCAT) in the HDL solution.

Use of Detergents Studies by Quarfordt and Goodman [37] demonstrated that as much as 10% of the cholesteryl ester content of chylomicrons could be tagged with ^{3}H- or ^{14}C-labeled cholesteryl esters by adding an acetone solution of the latter to a suspension of washed chylomicrons. Moreover, the physical and metabolic properties of the added labeled ester were the same as those cholesterol esters incorporated into chylomicron *in vivo* [37,38].

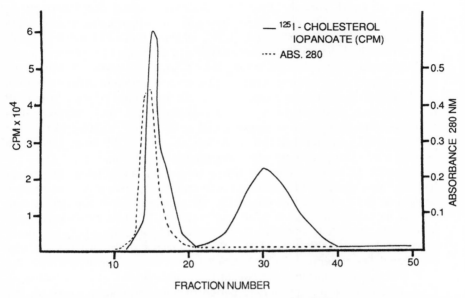

Figure 12 Gel filtration profile of incubation medium containing rat chylomicra and ^{125}I-cholesteryl iopanoate.

Such a procedure is satisfactory for the incorporation of probes that have sufficient solubility in acetone. Alternative methods are needed when this is not the case. Radioiodinated cholesteryl iopanoate posed such a problem [34]. In this case, the ester was dissolved in benzene (10 parts) to which was added a detergent, polysorbate 20 (2 parts). This mixture was vortexed and the benzene was evaporated. Physiological saline (10 parts) was then added and the mixture again was vortexed. The resulting solution was then added to a suspension of chylomicrons (25 parts) and the mixture was incubated at 37°C for 24 h (Fig. 11). The labeled chylomicrons were readily separated from unincorporated ester by gel filtration on agarose. As shown in Fig. 12, those fractions eluting in the void volume corresponded to labeled chylomicrons, whereas the second radioactive peak lacked UV absorbance at 280 nm and represented unincorporated ester. Extraction of the labeled chylomicrons and TLC analysis of the radioactivity demonstrated that it was still present as cholesteryl iopanoate. Moreover, polyacrylamide gel electrophoresis (PAGE) analysis was consistent with incorporation of the ester into chylomicrons. Following this procedure, approximately 40% of the radiolabeled probe was incorporated into the chylomicron particles. Although none of these assays provide evidence as to whether such particles are equivalent to their native counterparts, subsequent *in vivo* studies correlated closely with previous investigations using chylomicrons labeled with ^{14}C-cholesteryl oleate.

Use of Lipid Transfer Proteins   Another method for incorporating exogenous cholesteryl esters into the core of lipoproteins involves the utilization of lipid transfer protein (LTP), which is a 60–70 kilodalton acidic glycoprotein [39] that mediates the transfer and exchange of cholesteryl ester, triglyceride, and phospholipid among the lipoprotein classes. Although the lipid transfer activity present in whole plasma or lipoprotein-deficient plasma (LDP) is sufficient for many purposes, LTP can be partially purified from LDP by sequential chromatography on phenyl-Sepharose and carboxymethyl cellulose. This is readily accomplished and yields an 800- to 900-fold increase in purity [40]. In addition, this procedure separates LTP from LCAT, an enzyme which can potentially alter lipoprotein composition.

In order to radiolabel either whole plasma or an isolated lipoprotein fraction, both a source of transfer protein and a preparation of lipid vesicles containing the lipophilic radiodiagnostic are required. Lipid transfer protein sources include whole plasma, LDP, and purified LTP. Commonly used types of lipid vesicles include lipid dispersions produced by sonication [42], liposomes produced by a modification of the cholate dialysis procedure [43,44], and prelabeled lipoproteins. The radiolabeled lipid vesicles are mixed with the LTP source and lipoprotein component to be labeled, and, after incubating at $37°C$ under N_2 for an appropriate period of time (commonly 16–18 h), the desired lipoprotein fraction is reisolated by ultracentrifugation. Production of lipoproteins labeled with high specific activity can potentially be difficult since the LTP-mediated distribution of the labeled probe is determined by mass action. In this labeling procedure, it is important that the radiolabeled lipid be a substrate recognized by LTP. While this protein can mediate the transfer of a relatively wide range of lipophilic molecules, recent studies in our laboratory have shown ^{125}I-CI to be a poor substrate for LTP, with its transfer being less than 10% that observed for ^{3}H-cholesteryl oleate. Although this finding limits the methodologies that can be used to incorporate ^{125}I-CI into lipoproteins, it offers an advantage in that, after injection into animals, the radiolabel cannot readily transfer to the endogenous lipoprotein pool from the lipoprotein fraction with which it was originally associated.

IV. LIPOPROTEIN METABOLISM AND EFFECTS OF CHEMICAL MODIFICATION

The biosynthesis and metabolism of plasma lipoproteins involve complex processes that have been reviewed in previous chapters (see, e.g., Chapter 1) and other sources. Only a brief overview of the subject is presented here in order to make subsequent material more understandable.

As noted in Fig. 13, the chylomicrons (CMs) are responsible for the transport of dietary or exogenous lipids in the circulation. It is here that the CMs are acted upon by lipoprotein lipase (LPL), the lipolytic enzyme responsi-

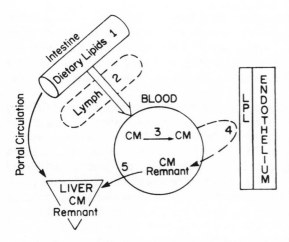

Figure 13 Sequence of biosynthesis and metabolism of chylomicrons. (1) absorption of dietary lipids; (2) transport of CM to blood via lymphatics; (3) transport of CM to extrahepatic tissues; (4) formation of CM remnant via lipoprotein lipase (LPL); (5) uptake of CM remnant by the liver.

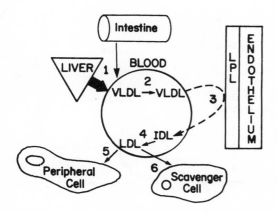

Figure 14 Sequence of biosynthesis and metabolism of VLDL and LDL. (1) biosynthesis of VLDL in liver and intestine; (2) transport of VLDL to extrahepatic tissues; (3) formation of remnant particle (IDL) via lipoprotein lipase (LDL); (4) remodeling of remnant and LDL formation; (5) high affinity uptake of LDL into peripheral cells; (6) low affinity uptake of LDL into scavenger cells.

ble for hydrolyzing triglycerides to free fatty acids. The CMs undergo restructuring during this process and acquire apolipoprotein E (apo E) along with other apolipoproteins. These smaller CMs, called CM remnants, are cleared very rapidly from the circulation ($t_{1/2} \approx 5$ min) as a result of their interaction with a specific receptor associated with hepatocytes which recognizes apo E [45–47].

In contrast to CMs, VLDLs are responsible for the transport and delivery of endogenously synthesized triglycerides and cholesteryl esters to extrahepatic tissues (Fig. 14). Synthesized primarily in the liver, VLDLs enter the circulation and become a substrate for LPL. Lipoprotein lipase utilizes VLDL much less efficiently than the larger CMs and, as a result, the plasma half-life of VLDL is much longer (1-3 h) [47]. Lipolysis reduces the size of these particles, converting them to VLDL remnants or intermediate density lipoproteins (IDLs). In humans, a substantial portion of the IDL is converted to LDL by the action of hepatic lipase, whereas in other animals such as the rat this does not occur and the IDL particles are taken up by the liver in a manner similar to that of CM remnants.

The elegant studies of Goldstein and Brown [48] have shown that extrahepatic cells acquire their cholesterol by an LDL receptor–mediated process. This process (Fig. 15) involves the binding of the apolipoproteins

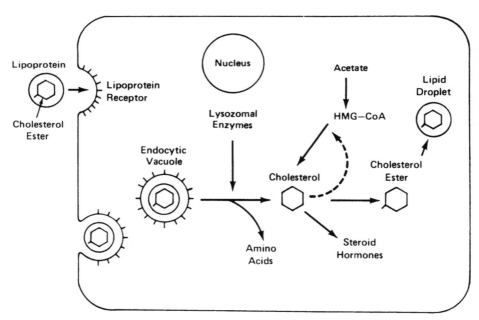

Figure 15 Receptor-mediated endocytosis of lipoprotein-cholesterol by steroid hormone-secreting cells.

present in LDL (apo B and apo E) to specific, high affinity binding sites located on the cell membrane surface (apo B, E receptor). Once binding has occurred, this LDL receptor complex is rapidly internalized by endocytosis and subsequently digested by lysosomal enzymes to liberate free cholesterol. This cholesterol not only is utilized as an important structural component for cell membrane synthesis but also serves to regulate the following intracellular processes: (a) the stimulation of the formation of cholesteryl ester, by catalyzing and serving as a substrate for acyl-CoA–cholesterol acyltransferase (ACAT); (b) the intracellular concentration of cholesterol, by suppressing ß-hydroxy-ß-methyl-glutaryl-CoA (HMG-CoA) reductase, the rate-limiting step in cholesterogenesis; and (c) the reduction of the rate of LDL receptor synthesis in order to decrease the amount of LDL cholesterol being taken up by the cell.

An additional pathway for LDL metabolism involves a lower affinity process associated with scavenger cells or macrophages of the reticuloendothelial system (Fig. 14). In humans, 33–65% of the plasma LDL is degraded by the high affinity receptor-mediated process, and the remainder is handled by the scavenger cells [49]. Overall, these degradation processes are much slower than those associated with the larger CMs and VLDLs and result in a plasma half-life of about 3–4 days for LDL [47].

In humans, HDLs have the longest plasma half-life of all the lipoproteins (5–6 days) [47]. Moreover, their biosynthesis and metabolism are the most complex. Although initially formed in the liver and intestine, they rapidly become modified in the circulation by interaction with the other lipoproteins. An important distinction of HDL is that HDL particles are the major carrier for apolipoproteins AI (apo AI) and AII (apo AII). The latter is an essential determinant for LCAT, which serves to esterify HDL cholesterol. The resulting cholesteryl esters are then transferred to VLDL and IDL via a plasma exchange protein. It has been proposed [50] that this may be a mechanism for removing cholesterol from cells for delivery to the liver in a process termed reverse cholesterol transport.

High density lipoproteins have been shown to bind to cells such as fibroblasts at specific sites that are separate from those that bind LDL [51]. Moreover, the rate of uptake of HDL by extrahepatic cells is much less than that observed for LDL, except in those species where HDL represents the major cellular source of cholesterol, such as the rat. In the rat, HDLs appear to play a role similar to that of LDLs in humans. For example, in rat steroid-secreting tissues (adrenal and gonads), rat HDLs are internalized via a specific receptor-mediated process [52] in a manner similar to that observed for LDLs in human fetal adrenals [53].

In the initial metabolic studies, radiolabeled lipoproteins were prepared by chemical procedures (usually radioiodination) that ostensibly would have only a minor effect, if any, on the metabolism of the native lipoprotein. For purposes of comparing receptor- versus nonreceptor-mediated processes, how-

Table 2 Chemical Modification of Apolipoproteins

Modification	Reagents	Reference	Remarks
Acylation[a]			
Acylation	Acetic anhydride	57	Reagents modify ε-amino of lysine residues as well as thiol- and hydroxyl-bearing residues
Acetoacetylation	Diketene	58	Largely specific for ε-amino groups
Carbamylation	Cyanate	58	
Succinylation	Succinic anhydride	59	Similar to acetylation but with a more marked increase in negative charge
Nitration[a]	Tetranitromethane	60,61	Primarily involves nitration of tyrosine residues
Schiff base formation[a]			
Malonimylation	Malondialdehyde	62	Derivatizes nearby amino functions and arginyl residues
Carbinolaminylation	Cyclohexanedione	63	Selective for arginyl residues
Reductive alkylation[b]			
Methylation	Formaldehyde + borohydride	58	Mono- or dimethylates ε-amino lysyl residues
Hydroxyethylation	Cyanoborohydride + hydroxyacetaldehyde	54	Similar to methylation but with a greater change in the hydrophilic properties of the apolipoprotein
Glucosylation	Glucose + cyanoborohydride	54	

[a] These modifications increase the negative charge associated with the apolipoprotein.
[b] These modifications have little effect on the chare associated with the apolipoprotein.

ever, it was necessary to introduce chemical modifications which prevented metabolism by the normal receptor-mediated pathway.

Chemical modifications that accomplish this goal have been reviewed [19,54] and rely upon modification of the surface apolipoproteins, specifically the lysine and arginine residues in apo B and apo E. Some of these modifications and the reagents that are employed to effect these changes are listed in Table 2. Those procedures that involve acylation or Schiff base formation of the basic amino acids cause a concomitant decrease in the positive charge of the lipoproteins, as reflected by electrophoretic mobility. Another method of increasing the negative charge of the apolipoproteins involves nitration of the tyrosyl residues with tetranitromethane, which directly affects the pKa of the phenolic hydroxyl. Increasing the anionic character of LDL by such modifications destroys the ability of these products to bind to the LDL receptor but does not interfere with their recognition by the scavenger receptor [55,56]. The scavenger receptor recognizes not only these modified lipoproteins but also a diverse group of polyanionic macromolecules such as maleyl-human albumin, polyinosinic acid, and fucoidin [55,56].

Less information is available on the alkylated lipoproteins. Although such modifications theoretically should not significantly alter the cationic character of the lipoprotein, they have been shown to interfere with the ability of LDL to bind to its receptor [54]. Methylated LDL has even been stated to go unrecognized by the scavenger receptor of cultured macrophages [54]. Moreover, both methylated and glucosylated LDL have been found to clear from the bloodstream more rapidly than cyclohexanedione-modified LDL, and this was ascribed to the more immunogenic property of the alkylated lipoproteins [54].

Clearly, the experimental use of chemically modified lipoproteins has provided us with considerable insight into the metabolic fate of lipoproteins, but much remains to be understood about the mechanisms of cellular uptake of these different macromolecules in either their native or chemically modified forms.

V. TARGETS FOR LIPOPROTEIN-MEDIATED DELIVERY OF RADIOPHARMACEUTICALS

The main feature that has drawn investigators to explore lipoproteins as drug delivery vehicles has been their ability to interact with specific cellular receptors and then be taken up into the cells by receptor-mediated endocytosis. Obviously, harnessing such a normal physiological recognition system represents an attractive process for the delivery of organ-imaging agents, provided there is (a) a means for appropriately radiolabeling the lipoproteins and (b) some selectivity between tissues in the uptake process so that anatomical and functional differences can be monitored and visualized. The following is a description of those tissues that represent the most obvious targets for such lipoprotein-mediated drug delivery, along with some of the preliminary findings.

A. Liver

Much has been learned in recent years about the role of the liver in lipoprotein uptake and clearance as well as the importance of the liver in regulating plasma cholesterol levels by using a variety of labeled lipoproteins. The liver is important because it can remove excess cholesterol from the body by biliary excretion.

Hepatocyte Uptake

The best understood pathway for lipoprotein uptake into the liver is by the LDL (apo B,E) receptor (Fig. 15) [64,65]. This is the major site of uptake of LDL, with greater than 50% of the total LDL receptors present on the hepatocyte in rabbits, hamsters, and rats [65]. This high affinity receptor has been shown to behave like the LDL receptor on the fibroblast, in that it can be regulated by metabolic conditions. The number of LDL receptors can be increased in a variety of animal models including the hyperthyroid rat and the hypolipidemic (colestipol- and lovastatin-treated) dog. Conversely, animal models of decreased LDL receptors include the hypothyroid rat, hypercholesterolemic rabbits, and the Watanabe heritable hyperlipidemic (WHHL) rabbit, which is genetically deficient in the LDL receptor [66]. Moreover, this receptor has been shown to be upregulated in the rat by administration of estrogens such as ethinyl estradiol (E-E_2) [67]. For example, rats pretreated with E-E_2 have been shown to have approximately a 20-fold increase in uptake of labeled LDL into parenchymal cells with no change in nonspecific (methylated LDL) binding. Thus the hypocholesterolemic effect of E-E_2 in the rat appears to be mediated through an increased catabolism of lipoprotein cholesterol in the liver. Studies have shown that the rat liver accounts for over 90% of the tissue accumulation of radioactivity when radiolabeled LDL is administered following E-E_2 pretreatment [68].

The other well established lipoprotein receptor in the liver is the apo E receptor [65,69]. Chylomicrons synthesized in the intestine are hydrolyzed in the peripheral tissues by LPL. This lipoprotein starts out with apo B and AI, apo C and E are acquired in the intestinal capillaries, and apo AI and C are lost in the reaction with LPL. The remaining CM remnant is different from native CM not only in apolipoprotein composition but also in its decreased content of triglyceride. Such remnants are rapidly taken up by the liver. Peripheral LPL also converts VLDL to a remnant particle (IDL) which is rapidly sequestered by the liver. A specific hepatic receptor predominantly recognizes apo E and removes these particles from circulation. Unlike the LDL receptor, this receptor is not regulated by the cholesterol content of the cell [65,70]. For example, the WHHL rabbit has normal remnant removal. This receptor is saturable and specific for apo E and is distinct from the apo B,E receptor that binds LDL.

In addition, there appears to be an HDL receptor, which, in contrast to the LDL receptor, is enhanced by cholesterol loading (HDL$_c$). This receptor binds apo AI and apo AII, and tyrosine residues are important for its recognition [65]. High density lipoprotein labeled on the AI apoprotein has been shown to be taken up by liver and kidney [71].

Nonparenchymal Uptake

Harkes and Van Berkel [72] have shown the uptake of LDL by nonparenchymal cells is by a receptor that is not affected by pretreatment with E-E$_2$. Lysine, but not arginine, is important for recognition by this receptor, whereas both lysine and arginine residues are important on parenchymal cells. This recognition site can be downregulated by the prior administration of ethyl oleate. It has been calculated that one-third of LDL clearance is by such a low affinity pathway [73].

As mentioned, chemical modification of LDL leads to uptake by scavenger cells by receptor-dependent and receptor-independent pathways. Interestingly, acetylated LDL labeled in the core with ^{14}C-cholesteryl oleate has shown that initial uptake of radioactivity by endothelial cells is followed by clearance into parenchymal cells [74].

Liver Targeting Studies

In an effort to take advantage of the known rapid clearance of chylomicron remnants by the liver, radiotracers have been incorporated into these vesicles [34]. Chylomicrons were isolated from blood drawn from rats given corn oil intragastrically (Fig. 11). The milky white chylomicron fraction was then incubated with ^{125}I-cholesteryl iopanoate for 24 h at 37°C. At the end of this time, a labeled chylomicron fraction was separated from unincorporated ester by gel filtration. The tissue distribution of this preparation versus one where the tracer was dissolved in physiological saline with the aid of polysorbate is shown in Table 3. Administration of the tracer in the chylomicron preparation more than doubled the liver uptake of radioactivity over that for physiological saline. Based on organ weights, approximately 69% of the tracer localized in the liver within 30 min following administration in chylomicrons as opposed to only 31% when given in physiological saline. This result agreed very closely with the earlier studies of Quarfordt and Goodman [37,38], who employed ^{14}C-cholesteryl oleate as the tracer for similar studies. A major difference from this latter study, however, was that cholesteryl iopanoate is resistant to hydrolysis by the liver lipases and therefore failed to be cleared at any appreciable rate from the liver (Fig. 16). Some redistribution of the tracer to

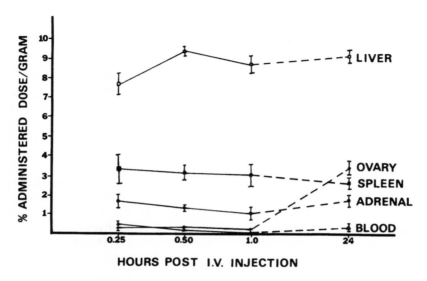

Figure 16 Tissue distribution of radioactivity following administration of chylomicra labeled with ^{125}I-cholesteryl iopanoate as a function of time.

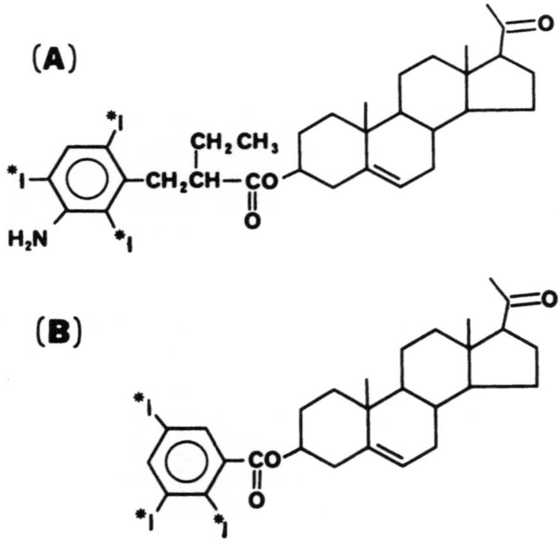

Figure 17 Structures for radioiodinated pregnenolone iopanoate (A) and pregnenolone-2,3,5-triiodobenzoate (B).

Table 3 Distribution of Radioactivity in the Rat After Intravenous Administration of ^{125}I-Cholesteryl Iopanoate in Tween/Saline Vehicle and Chylomicrons

	Percentage of administered dose/gram of tissue ± SEM			
	Tween/saline		Chylomicrons	
Tissue	0.5 hour ($N = 5$)	24 hours ($N = 5$)	0.5 hour ($N = 4$)	24 hours ($N = 4$)
Adrenal cortex	4.90 ± 0.72	11.03 ± 2.17	1.29 ± 0.13	1.92 ± 0.21
Blood	5.19 ± 0.44	0.70 ± 0.05	0.06 ± 0.01	0.23 ± 0.05
Liver	4.03 ± 0.54	6.59 ± 0.32	9.22 ± 0.25	9.15 ± 0.31
Lung	1.34 ± 0.13	0.60 ± 0.08	0.29 ± 0.06	0.25 ± 0.05
Ovary	5.16 ± 0.72	24.54 ± 2.96	0.12 ± 0.02	3.46 ± 0.77
Plasma	10.37 ± 0.86	1.34 ± 0.08	0.13 ± 0.03	0.39 ± 0.10
Spleen	0.78 ± 0.07	2.00 ± 0.13	3.14 ± 0.45	2.73 ± 0.34
Thyroid	0.87 ± 0.10	3.75 ± 0.22	0.07 ± 0.01	1.32 ± 0.32

HDL was observed to occur with time and provided an explanation for the increased radioactivity in the adrenals and ovaries noticed at 24 h.

Although radioiodinated cholesteryl iopanoate displayed exceptional liver localizing capability, its resistance to hydrolysis prevented normal clearance and diminished interest in this tracer as an imagining agent for assessment of liver function. Surprisingly, similar esters of pregnenolone and other sterols (Fig. 17) were found to undergo significant *in vivo* hydrolysis in the rat (Table 4), rabbit, and dog [12,75,76]. Moreover, significant species differences in the metabolic fate of these tracers were observed. For example, in the rat the pregnenolone esters were found to localize in the adrenal cortex in very high concentrations within 0.5 h. Unfortunately, this finding did not carry over to the rabbit or dog and thus reduced the potential of these agents as adrenal imaging agents. On the other hand, like cholesteryl iopanoate, the pregnenolone esters showed considerable liver uptake at 0.5 h but, unlike cholesteryl iopanoate, liver radioactivity was observed to clear with time. Moreover, PAGE analysis of plasma samples revealed that radioactivity rapidly became associated with all classes of lipoproteins. With time, however, this radioactivity became associated with serum albumin in a manner characteristic for the presence of the radioiodinated free acids in the circulation.

Such polyiodinated acids are used clinically as cholecystographic agents because of their ability to be rapidly cleared by the biliary tract. These events are visually observed following administration of ^{123}I-pregnenolone-2,3,5-triiodobenzoate to a dog (Fig. 18). Rapid accumulation of the tracer in the liver is followed by a time-dependent clearance into the biliary tract and gall bladder. Studies with these pregnenolone esters as hepatographic agents are continuing.

Table 4 Distribution of Radioactivity in the Rat After Intravenous Administration of ^{125}I-Pregnenolone Iopanoate and ^{125}I-Pregnenolone Triiodobenzoate in Tween/Saline Vehicle

	Percentage of Administered dose/gram of tissue ± SEM			
	Pregnenolone Iopanoate		Pregnenolone-2,3,5-triiobenzoate	
Tissue	0.5 hour ($N = 3$)	24 hours ($N = 5$)	0.5 hour ($N = 4$)	24 hours ($N = 4$)
Adrenal cortex	23.08 ± 1.58	4.26 ± 0.72	22.90 ± 1.76	7.78 ± 1.02
Blood	1.73 ± 0.10	0.04 ± <0.01	2.04 ± 0.10	0.16 ± 0.02
Liver	5.21 ± 0.50	0.39 ± 0.02	7.24 ± 0.41	1.19 ± 0.18
Ovary	2.13 ± 0.20	1.08 ± 0.12	4.78 ± 0.77	4.46 ± 0.38
Thyroid	0.94 ± 0.07	156.70 ± 24.94	1.43 ± 0.20	106.76 ± 10.91

Phospholipid-stabilized triglyceride emulsions such as Intralipid are believed to be metabolically equivalent to chylomicrons [77] and are used widely in humans for parenteral nutrition. Unlike chylomicrons, these emulsions lack both cholesterol and apoproteins. However, both *in vitro* and *in vivo* studies have demonstrated that Intralipid particles rapidly acquire apoproteins from plasma lipoproteins and are metabolized by the action of lipoprotein lipase in a manner similar to chylomicrons [78–80].

As noted, chemical modification of LDL, such as acetylation, produce vesicles that are rapidly taken up by macrophages and other cells of the reticuloendothelial system. Dresel et al. [81] labeled acetyl-LDL (Ac-LDL) with iodine-131 and compared scintigraphic images in the rate with those obtained with ^{99m}Tc-albumin was administered first as a control for liver perfusion. This was followed at 10 min by injection of ^{131}I-LDL and then at 20 min by ^{131}I–Ac-LDL. ^{99m}Tc-albumin gave rise to high levels of radioactivity in the heart and liver, the organs with the largest blood pool. The heart/liver ratio remained constant over 10 min, indicating that ^{99m}Tc-albumin was not being rapidly removed from the circulation. Although ^{131}I-LDL displayed some clearance by the liver, it was the ^{131}I–Ac-LDL that was avidly taken up by the liver. The scintiscans showed a rapid accumulation of radioligand in the liver following administration of ^{131}I–Ac-LDL. Due to the rapid removal of ^{131}I–Ac-LDL from the circulation, a significant drop in heart activity was seen shortly after administration (i.e., 20 min). At the same time, activity in the liver accumulated rapidly.

Studies evolving from the investigations of Brown and Goldstein showed that intravenously administered ^{125}I-VLDL gave rise to 32% of the radioactivity in the liver within 30 min [82]. In contrast, experiments with WHHL rabbits revealed that less than 10% of the administered radioactivity was taken up by the liver, thereby providing evidence that the uptake of ^{125}I-VLDL occurred by a receptor-mediated process. During normal catabolism, VLDL gives rise to IDL and LDL, both of which are taken up by the liver by binding to the LDL receptor. These studies prompted the subsequent labeling of VLDL with iodine-123 for scintiscanning. Scintiscans of rabbits clearly demonstrated the progressive accumulation of radioactivity in the liver and disappearance from the blood pool. As anticipated, these changes were not observed in the WHHL rabbit. While these findings were significant in their own right, the authors questioned the potential application of ^{123}I-VLDL to measure LDL receptors in humans because the clearance of VLDL in humans is much slower than rabbits and the rapid degradation of radioiodinated lipoproteins by lysosomal enzymes may preclude such a possibility.

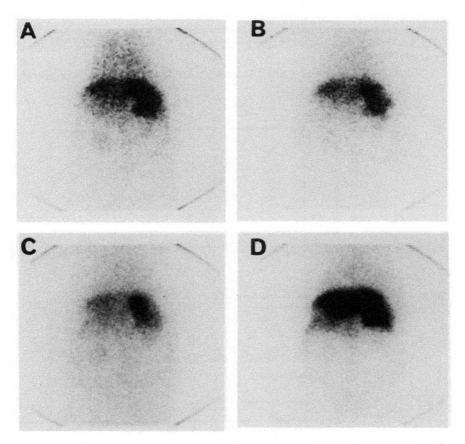

Figure 18 Gamma camera scintigraphy of ^{123}I-pregnenonlone-2,3,5-triiodobenzoate in the dog at (A) 0.5 h; (B) 1 h; (C) 3 h postadministration; (D) ^{99m}Tc-sulfur colloid.

B. Steroid Hormone-Secreting Tissues

Steroid-secreting tissues (e.g., adrenal gland and ovary) require much more cholesterol than other tissues since, in addition to the cholesterol needed for membrane synthesis, these tissues require cholesterol for steroid hormone synthesis. According to current concepts, tissues responsible for the biosynthesis of steroid hormones derive their cholesterol from three sources: (a) circulating plasma; (b) hydrolysis of intracellular cholesteryl esters; and (c) intracellular biosynthesis (Fig. 19). Most animal studies to date indicate that the adrenal cortex relies mainly on the circulating lipoproteins as the source of cholesterol [52,83]. The storage pool of cholesteryl esters varies with the activity of the tissue in hormone production. Hormone-secreting cells make up the majority of cells in the adrenal cortex, a lesser proportion of cells in the ovary, and only a small fraction of cells in the testis. Thus in the rat, for example, it is not surprising to find much higher amounts of total cholesterol (free and esterified) in the adrenal gland (40 mg/g) than in the ovary (11 mg/g) or testis (2.5 mg/g) [52].

The first evidence for a specific lipoprotein receptor pathway in adrenal

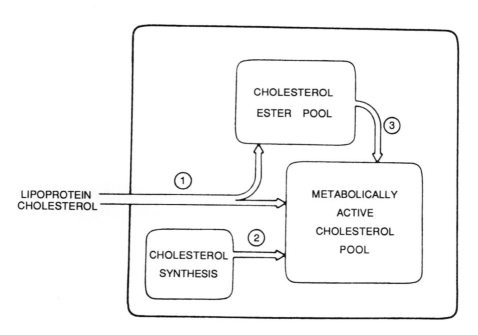

Figure 19 Cholesterol balance in adrenal and gonadal tissues.

tissue was demonstrated by incubating mouse adrenal tumor cells (Y-1 clone) in culture with ^{125}I-labeled LDL [84]. These adrenal cells bound ^{125}I-LDL at a surface binding site that showed saturability and high affinity. This lipoprotein receptor was shown to be specific for LDL by binding both human and mouse LDL, but not human or mouse HDL. Moreover, methylated LDL, which does not bind to the LDL receptor *in vitro*, was not taken up rapidly by the adrenal gland *in vivo* ref [85].

To further study this ligand–receptor specificity, mice treated with 4-aminopyrazolo-pyrimidine (4-APP) were used as a model. This drug blocks lipoprotein secretion by the liver and causes a rapid and profound drop in the plasma cholesterol level [86]. This drop in plasma cholesterol leads to an increase in the number of LDL receptors on adrenal membranes as measured *in vitro* [80] and to an increase in adrenal uptake of ^{125}I-LDL as measured *in vivo* [85,87]. Moreover, administration of rabbit antibody directed against the LDL receptor purified from bovine adrenal cortex blocks the uptake of ^{125}I-LDL, but not ^{125}I-HDL, by the adrenal gland of the 4-APP–treated mouse [88]. Similar studies have been conducted in 4-APP–treated rats. In contrast to the mouse, however, HDL was several times more effective than LDL in delivering cholesterol to the rat adrenal gland [52,89–91].

Gwynne and Hess [92] followed up on these studies by examining the binding and degradation of human and rat HDL by rat adrenocortical cells in suspension. They found that both human and rat HDL bound with equal efficiency and to a much grater degree than human LDL. Moreover, two processes of association were observed, one that was readily reversed by unlabeled HDL and a second that was not readily reversed. These studies indicated a lipoprotein specificity for HDL in these cells, but it was uncertain whether the lack of reversibility of the second process represented internalization of the HDL by the cell. Thus the question remains as to whether serum HDL may transfer cholesteryl ester during membrane binding without obligate endocytosis.

Stein and co-workers [91,93] employed ^{3}H-cholesteryl linoleyl ether (^{3}H-CLE), a nonhydrolyzable analog of cholesteryl ester, to study the metabolic fate of lipoproteins in a variety of species. Incorporation of ^{3}H-CLE into rat HDL was accomplished with the aid of rabbit cholesteryl ester transfer protein and the labeled HDL screened in donor rats prior to reinjection into test rats. Although the liver was the major site of accumulation of radioactivity at 48 h [45%], the adrenal was observed to contain about 1% of the injected dose at this time. Pretreatment of the rat with E-E$_2$ for 5 days resulted in a 20% increase in hepatic uptake, but the adrenal uptake was increased by over fivefold.

The use of this nonhydrolyzable analog of cholesteryl ester permitted a much better measure of the uptake of HDL cholesteryl ester by tissues than ^{125}I-HDL. Table 5 compares the tissue uptake of radioactivity following administration of ³H-CLE–HDL or ¹²⁵I-HDL. A much lower recovery of radioactivity was observed for the radioiodinated preparation. Such a difference suggested either that tissues take up less labeled protein than labeled cholesteryl ester (i.e., cholesteryl ester uptake is not entirely associated with a receptor-mediated endocytotic process) or that proteolytic hydrolysis and deiodination of the radioiodinated apolipoproteins contributed the decreased levels of radioiodine in tissues.

Along similar lines, studies with reconstituted lipoproteins have shown that cholesteryl esters may indeed be taken up by steroid-secreting cells by processes other than receptor-mediated endocytosis. Utilizing HDL in which apo AI was labeled with ¹²⁵I-tyramine cellobiose and the core lipids were tagged with ³H-CLE, Glass and co-workers [9,10] found the uptake of cholesteryl ether to be severalfold higher than the rates of apo AI uptake in the adrenal (seven times) and the ovary (four times). In another study, the same group noted that the selective uptake of ³H-CLE was enhanced over that of ¹²⁵I-HDL uptake in the adrenals and ovaries but not the liver of 4-APP–treated rats [93]. Similar observations made by others utilizing rat adrenal cells in culture [11] have led to the speculation that cholesteryl ester may be transferred into the cell by some intracellular protein analogous to cholesteryl ester transfer protein or that the labeled apolipoproteins return to the surface of the cell by retroendocytosis.

Receptor-dependent pathways for lipoprotein metabolism have also been investigated in the rabbit. Using ¹⁴C-sucrose–labeled LDL, Pittman and colleagues [94,95] determined the overall degradation of LDL in various organs by comparing LDL catabolism in normal and receptor-deficient mutant (WHHL) rabbits. They found that the adrenal gland was the most active

Table 5 Distribution of Radioactivity After Injection of Screened Homologous HDL into Rats

HDL	Time (hr)	Radioactivity recovered (% injected dose)	
		Liver	Adrenal
³H-CLE–HDL	6	28.3 ± 0.7	0.8 ± 0.09
	12	36.4 ± 1.4	1.1 ± 0.03
	24	38.6 ± 0.7	1.0 ± 0.02
¹²⁵I-HDL	6	8.9 ± 0.9	0.1 ± 0
	12	8.3 ± 0.7	0.1 ± 0.01
	24	6.6 ± 0.3	0.06 ± 0.01

Source: Ref. 91.

tissue in terms of its fractional catabolic rate per gram of tissue for LDL degradation. Furthermore, subsequent calculations, in which it was assumed that the only change in LDL catabolism in the receptor-deficient rabbit was the loss of receptor-mediated pathways, indicated that in the adrenal, 92% of the LDL degradation in normal animals was receptor-mediated. Using radioiodinated and ^{14}C-sucrose–labeled homologous LDL, Spady and co-workers [96,97] confirmed the adrenal gland to have the high rate of uptake in the rabbit and that 98% of this uptake was receptor-dependent.

Whereas recognition of the existence of the LDL receptor was first achieved with the aid of human skin fibroblasts [98], the direct evidence that such receptors exist in other human tissue has been slow in coming. Simpson and co-workers [53,99], however, have shown that human fetal adrenal tissue readily take up and degrade ^{125}I-LDL in a manner similar to other cells. Moreover, the number of ^{125}I-LDL binding sites in membrane fractions prepared from tissue maintained in the presence of ACTH was twice that found in membranes prepared from non–ACTH-treated tissue. They concluded that ACTH stimulates LDL uptake and degradation by fetal adrenal tissue through a mechanism that gives rise to an increase in the number of LDL receptors on the plasma membrane.

With regard to lipoprotein receptor-mediated processes in the ovary, the most information is available on the rat ovary. As was observed for the adrenal, luteinized ovaries of the rat appear to utilize exogenous cholesterol provided by HDL for purposes of steroid hormone production [100,101]. Moreover, the cell surface binding sites for the uptake of HDL are distinct from the well characterized LDL receptor [102]. Similar to the studies with adrenal cells, there is also the question of whether intact HDL enters the cell in order to deliver the cholesterol [11,103]. As noted earlier, when the apo AI of HDL was labeled with covalently linked ^{125}I-tyramine cellobiose and the core lipid was tagged with ^{3}H-cholesteryl ethers, the uptake of cholesteryl ether by the rat ovary was found to be four times that of apo AI [9,10].

Thus although the evidence for the involvement of lipoproteins in the delivery of cholesterol to the steroid-secreting tissues is convincing, it is still uncertain whether receptor-mediated endocytosis is necessary in all instances. It seems clear, however, that cholesterol derivatives are selectively taken up by these tissues and that this process is enhanced by lipoproteins.

In 1969, our laboratory reported the first synthesis of radioiodinated 19-iodocholesterol (^{125}I-C), which was to become the first noninvasive functional imaging agent of the adrenal cortex [104]. Further studies with this agent in rats revealed that 19-iodocholesterol closely mimicked cholesterol in its ability to be taken up by steroid-secreting tissues and become concentrated in an esterified form [105]. Moreover, it was shown that, unlike natural cholesteryl esters, these esters of 19–iodocholesterol were poor substrates for adrenal cholesterol esterase [106]. Thus the administered radioiodinated cholesterol appeared to be transported and taken up into tissues similar to cholesterol, but once esterified it reached a metabolic dead end in the adrenal cortex, which led to its accumulation in sufficient quantities to provide the first scintigraphic images of the adrenals in humans [107].

These findings subsequently led to an examination of the involvement of lipoproteins in the transport of radioiodinated cholesterol [33]. It was found that the radioiodinated analog rapidly became incorporated into plasma HDL upon intravenous administration to rats. Moreover, incorporation of ^{125}I-C into rat HDL prior to administration had little effect on adrenal uptake when administered to control rats but produced a greater than fourfold increase in uptake in animals treated with 4-APP (Table 6).

Electrophoresis of the plasma at 0.5 h revealed that more than half of the radioactivity in the plasma was associated with the HDL fractions and agreed with the findings of Andersen and Dietschy [52] which showed HDL to be the major lipoprotein carrier of cholesterol in the rat.

Administration of ^{125}I-C in HDL to normal rats failed to show enhanced adrenal uptake, presumably because of the considerable competition with endogenous lipoprotein (metabolic dilution). However, once the pool of circulating lipoprotein was reduced by 4-APP administration, a significant increase in adrenal uptake ensued.

Since cholesterol is transported in lipoproteins largely in its esterified form, studies were undertaken to evaluate the oleate ester of 19-iodocholesterol [13,21]. Accordingly, rat HDL was labeled *in vivo* by intravenous administration of radioiodinated 19-iodocholesteryl oleate (^{125}I-CO) and the HDL fraction subsequently was isolated by sequential ultracentrifugation. This radioiodinated HDL fraction (^{125}I-CO–HDL) was then administered to normal and hypolipidemic rats and its fate compared with that obtained for ^{125}I-CO administered in physiological saline.

Although the plasma clearance of ^{125}I-CO and ^{125}I-CO–HDL showed differences in the early distribution phase, the subsequent first order elimina-

Table 6 Uptake of Radioactivity in the Adrenal Cortex Following Intravenous Administration of ^{125}I-C in Physiological Saline Vs. Buffered HDL Solution in Untreated and 4-APP–Treated Rats

	Percentage of administered dose/gram of tissue			
	Physiologic saline[a]		Rat HDL	
	Untreated ($N = 5$)	4-APP Treated ($N = 5$)	Untreated ($N = 5$)	4-APP Treated ($N = 3$)
Mean	15.83	19.11	14.41	60.03
Range	12.21–18.27	16.18–22.21	11.80–16.48	52.75–67.23

[a] ^{125}I-C was dissolved in isotonic saline with the aid of ethanol (10%) and Tween-80 (1.6%).

tion phase was similar for the two preparations. This was not surprising since within 0.5 h after administration, ^{125}I-CO becomes associated with plasma lipoproteins, and the greatest percentage (67%) is associated with HDL. Moreover, at 1 h and later time periods following the administration of ^{125}I-CO or ^{125}I-CO–HDL, there was little difference in the distribution of radioactivity among the various plasma lipoproteins. In addition, the plasma half-life of 10.0 h for ^{125}I-CO–HDL corresponded to the value of 10.5 h found by Roheim et al. [108] and that of 10.6 h noted by Sigurdsson et al. [109] with [131]I- and ^{125}I-labeled HDL, respectively. The close agreement of these clearance values provides additional evidence for the intimate association of the ^{125}I-CO with the circulating HDL.

The most striking difference between ^{125}I-CO and ^{125}I-CO–HDL was in the amount of radioactivity taken up and retained by steroid-secreting tissues, the adrenal cortex and ovaries. Moreover, the enhanced uptake of ^{125}I-CO by adrenal cortex and ovary (Fig. 20) when it was administered as ^{125}I-CO–HDL agreed with earlier studies by others [9,10] with tritiated cholesteryl ether.

Similar to the earlier observation with iodocholesterol, the modest increase of uptake of radioactivity by target tissues when ^{125}I-CO–HDL was administered to normal rats could be explained by metabolic dilution with endogenous HDL. However, when rats were made hypolipidemic with 4-APP or E-E$_2$ prior to administration of ^{125}I-CO or ^{125}I-CO–HDL, the appearance of radioactivity in the adrenal cortex and ovary at 0.5 h was dramatically increased over that observed for normal rats (Fig. 20). Moreover, in the 4-APP–treated rats, the concentration of radioactivity in the adrenal cortex was more than fivefold greater following administration of ^{125}I-CO–HDL than was observed for ^{125}I-CO. In contrast to the adrenal cortex, the ovaries were less responsive to hypolipidemia, but prior incorporation of ^{125}I-CO into HDL did cause a slight increase in uptake of radioactivity by the ovary following pretreatment with 4-APP, and this enhancement was even more pronounced in rats made hypolipidemic with E-E$_2$. These results were consistent with the work of Strauss et al. [100], who showed that the uptake of rat ^{125}I-labeled HDL was greatest for the adrenals at 15 min following intravenous administration to immature 4-APP–treated rats. This uptake of radioactivity by the adrenals was markedly reduced by coadministration of human HDL.

At 24 h following the administration of ^{125}I-CO–HDL to normal rats, the percentage of administered dose per gram of tissue was 9.5 and 16.1 for the adrenal cortex and ovary, respectively. This retention of radioactivity in the steroid-secreting tissues following administration of ^{125}I-CO–HDL represents the major distinction from results obtained with radioiodinated HDL (^{125}I-labeled HDL) preparations. Radioactivity localizing in the adrenal gland was found to decline rapidly within the first hour following intravenous administration of ^{125}I-labeled HDL to mice [110]. This lack of retention of radioactivity by the adrenal was explained by the rapid degradation of ^{125}I-labeled HDL by proteolytic enzymes. Radioiodinated cholesteryl esters, on the other hand, are not subject to the same rapid metabolic fate and are stored in steroid-secreting tissues. The latter is particularly true for esters of 19-iodocholesterol since they have been shown to be poor substrates for rat adrenal cholesteryl ester hydrolase [106]. Consequently, it was reasonable for radioactivity to persist in adrenal and ovarian tissue for a prolonged period of time following administration of ^{125}I-CO–HDL.

Additional evidence supporting a lipoprotein receptor-mediated process for the enhanced uptake of ^{125}I-CO–HDL in steroid-secreting tissues came from studies with acetylated lipoproteins [21]. Acetylation of ^{125}I-CO–HDL with

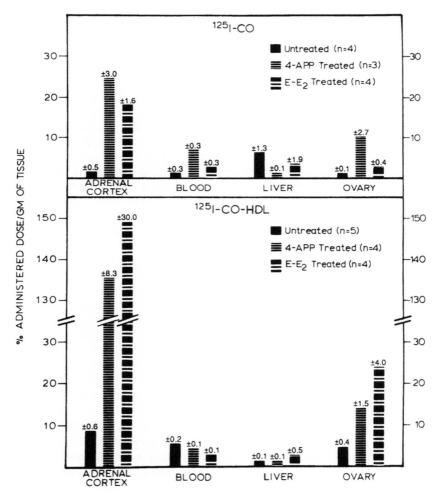

Figure 20 Concentration of radioactivity in tissues at 0.5 h following intravenous administration of ^{125}I-CO or ^{125}I-CI–HDL in normal and hypolipidemic rats.

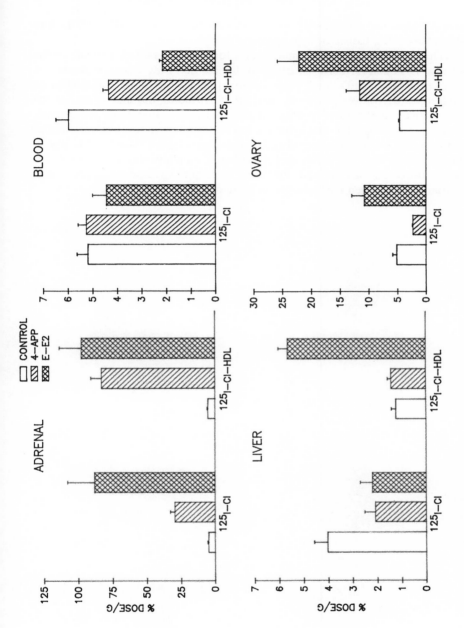

Figure 21 Concentration of radioactivity in tissues at 0.5 h following intravenous administration of ^{125}I–CI and ^{125}I–CI–HDL in normal and hypolipidemic rats.

acetic anhydride significantly decreased the uptake of ^{125}I-CO–HDL into steroid-secreting tissues of hypolipidemic rats.

The possibility of utilizing lipoproteins for targeting radiodiagnostics to the adrenal has been further examined with the aid of ^{125}I-cholesteryl iopanoate, a nonhydrolyzable cholesteryl ester analog [12,111]. Unlike radioiodinated cholesterol derivatives previously used for such studies, this analog carries the radioiodine on the ester portion of the molecule rather than the steroid. Despite this change, the tissue distribution of this tracer in rats was very similar to that noted earlier for radioiodinated cholesteryl oleate (Table 7). Moreover, incorporation into rat HDL and subsequent administration to 4-APP- or estrogen-treated rats showed the same dramatic increase in adrenal and ovarian uptake when compared to untreated animals (Fig. 21).

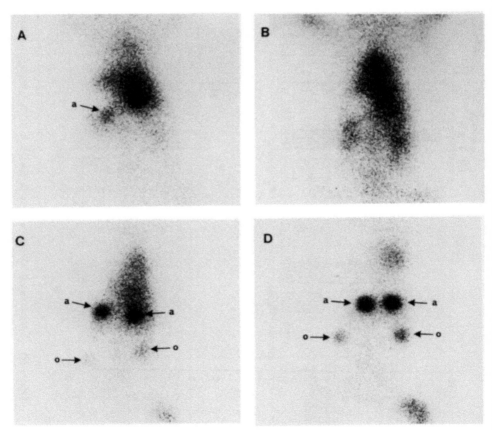

Figure 22 Gamma camera scintigraphy of 4-APP–treated rats at 0.5 h following administration of (A) ^{125}I–NP-59; (B) ^{125}I-CI; (C) ^{125}I-CI–HDL; (D) ^{125}I-CI–HDL and surgical removal of the liver. a, adrenal; o, ovary.

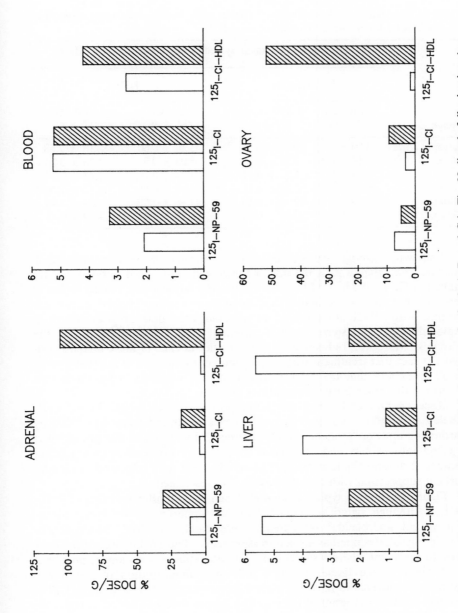

Figure 23 Concentration of radioactivity in tissues of animals A, B, and C in Fig. 22 directly following imaging (approximately 1 h following administration of radipharmaceutical) in comparison to normolipidemic controls (blank).

Table 7 Distribution of Radioactivity 0.5 Hour after Intravenous Administration of ^{125}I-Cholesteryl Iopanoate in Tween/Saline or HDL

	Percentage of administered dose/gram of tissue ± SEM			
	Tween/Saline		HDL	
Tissue	Untreated ($N = 5$)	4-APP-treated ($N = 5$)	Untreated ($N = 4$)	4-APP–treated ($N = 5$)
Adrenal cortex	4.90 ± 0.72	29.47 ± 3.26	5.30 ± 0.60	83.01 ± 7.68
Blood	5.19 ± 0.44	5.24 ± 0.34	5.98 ± 0.51	4.35 ± 0.23
Liver	4.04 ± 0.54	2.07 ± 0.44	1.24 ± 0.18	1.43 ± 0.14
Ovary	5.16 ± 0.72	2.28 ± 0.11	4.66 ± 0.16	11.52 ± 2.36
Plasma	10.37 ± 0.86	12.10 ± 0.53	11.86 ± 1.32	9.69 ± 0.49
Thyroid	0.87 ± 0.10	0.34 ± 0.02	0.47 ± 0.06	1.19 ± 0.73

Follow-up scintigraphy studies were performed in female rats 0.5 h postinjection. Figure 22 is the scan obtained with ^{125}I-CI incorporated into rat HDL (^{125}I-CI–HDL) in 4-APP–treated rats. The adrenals are readily visible in this scan and much more readily delineated than in scans obtained in a similar manner with ^{125}I-CI administered in physiological saline or with NP-59, a radioiodinated cholesterol analog currently used clinically to image the adrenals. Moreover, radioanalyses of the tissues in each of these animals verified the fact that treatment with 4-APP caused an increase in adrenal uptake in all cases and that the HDL-mediated uptake was over three times that noted for the other tracer preparations (Fig. 23).

Recently, Lees and co-workers [112] reported their results with human LDL labeled with ^{99m}Tc. Labeled lipoprotein was separated from free ^{99m}Tc by chromatography on Sephadex G-50. Between 3 and 5 mCi of this product was injected into rabbits and abdominal images were obtained with a gamma camera at different time periods. The rabbits were then subjected to adrenal cortical suppression with dexamethasone and rescanned.

Figure 24 shows a typical scintigram taken at 21 h after injection of ^{99m}Tc-LDL. Although significant radioactivity was apparent in the liver, spleen, kidneys, gut, and bladder, the adrenals were readily visualized. Moreover, as seen in Fig. 24, dexamethasone suppression reduced adrenal uptake of radioactivity to the point where visualization by scintigraphy was no longer possible.

Analysis of the tissues at 24 h following administration of the tracer revealed the adrenal to have the highest concentration of radioactivity (1.13% of inject-ed dose per gram of tissue). The adrenals were followed in order by spleen (0.77%), liver (0.2%), and kidney (0.06%). These adrenal values were very similar to those reported earlier for ^{125}I-CI in the female rabbit (1.15%) [75]. However, ^{125}I-CI administered in isotonic saline produced higher levels

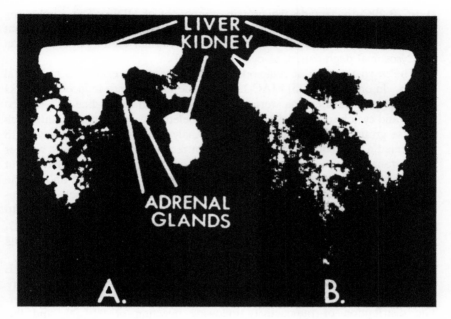

Figure 24 Anterior scintigrams of the abdomen of a rabbit obtained 21 h after injection of ^{99m}Tc-LDL (A) prior to adrenal cortical suppression and (B) after a 3 day course of treatment with dexamethasone (2 mg) injected intramuscularly per day. (From Ref. 113.)

in the liver (0.50) than ^{99m}Tc-LDL. It will be interesting to observe what improvement in adrenal/liver ratio may result by incorporating radioiodinated CI into LDL.

C. Tumors

The fact that cells, especially rapidly dividing ones, require cholesterol for normal growth and function (e.g., synthesis of cell membranes) has led a number of investigators to analyze the interaction of lipoproteins with neoplastic cells. Gal and co-workers [113,114] studied a number of different types of cancer cells in monolayer culture. They found neoplastic cells to metabolize LDL at higher rates than nonneoplastic cells. In epidermoid cervical carcinoma, for example, LDL was metabolized at a rate 15–20 times greater than that of normal cells, such as fetal adrenal tissue in organ culture. Similar studies with radioiodinated HDL showed human endometrial carcinoma cells

(HEC-B-296) to degrade the protein component of HDL. This catabolism, however, was found to be the result of the action of a proteolytic enzyme that was present on the external surface of the cells, rather than by internalization and lysosomal degradation, as occurs with LDL.

The *in vivo* uptake of radioiodinated LDL has been examined by several groups. In one study [115] MAC-13 soft tissue tumors were induced in NMRI mice and parotid adenoma were induced in CFLP mice. Iodine-125–labeled native and cyclohexanedione-modified ^{131}I-labeled LDL were injected into each of these groups as well as control mice. Analysis of tissue radioactivity at 18 h revealed a high concentration of counts in receptor-rich tissues such as liver and adrenal. In each instance, the tumor was moderately radioactive, but when account was made for its mass it was found to be second only to the liver in activity. Efforts to improve the tumor selectivity showed that pro-administration of sodium taurocholate and hydrocortisone sodium succinate to downregulate receptor-mediated uptake in the liver and adrenals, respectively, caused a reduction in uptake of radioactivity by these tissues without affecting tumor activity.

In another study [116], ^{125}I-LDL and ^{131}I-cyclohexanedione–modified LDL were administered to BALB/c mice bearing M-2 fibrosarcoma. Animals were sacrificed at 24 h and tumor uptake was found to be 2.5 times that of liver. The distribution of radioactivity following injection of ^{125}I-LDL and ^{131}I-cyclohexanedione–modified preparation indicated that both receptor-dependent and receptor-independent pathways were involved in the uptake of LDL by the tumor.

Several laboratories have noted higher LDL receptor activity in leukemia cells over normal white blood cells [117,118]. Cells from patients with acute myelogenous leukemia showed markedly higher degradation rates when compared with mononuclear cells and granulocytes from peripheral blood of healthy individuals. Furthermore, leukemia cells from patients with monocytic or myelomonocytic leukemia were observed to have the highest degradation rates. In most cases, there was little difference in the LDL receptor activity between leukemia cells isolated from bone marrow. Ginsberg and co-workers [119] also noted that patients with myeloproliferative disease catabolize both native LDL and cyclohexanedione-modified LDL at an increased rate and suggested that receptor-independent pathways for lipoprotein catabolism also deserve consideration in such patients.

Not all tumors show an increased capacity to catabolize lipoproteins over that of surrounding normal cells. Renal carcinoma implanted in isogeneic Wistar-Lewis rats, for example, has a cholesterol ester content 14 times greater than normal kidney at 12 weeks, but, unlike normal kidney, failed to acquire its cholesterol by a LDL receptor–mediated process. Instead, growth of this tumor was sustained largely by *de novo* cholesterol biosynthesis by the tumor and to a lesser extent by uptake of lipoprotein by a receptor-independent process (Fig. 25) [120].

Similarly, studies with rat hepatoma (HTC No. 7288C) *in vivo* have shown such tumors to have considerably fewer receptors for chylomicron remnants than normal rat liver [121]. E-E$_2$ treatment increased the number of lipoprotein receptors in both liver (1.1-fold) and tumor (1.2- to 1.6-fold) membranes, but hypercholesterolemia did not produce any significant changes in remnant binding to either liver or hepatoma membranes. Administration of apo E-rich lipoproteins of various sizes (chylomicron remnants, VLDL, IDL, and HDL$_c$) showed none to be taken up to any significant extent by this rat hepatoma cell line. Since it is known that hepatoma cells *in vitro* can bind, internalize, and degrade such lipoproteins, it is believed that, in addition to reduced lipoprotein receptors, limited access of the lipoproteins to the hepatoma *in vivo* may contribute further to the decreased delivery of lipoprotein cholesterol to these tumors.

The potential of utilizing lipoproteins to image tumors has also been of keen interest to our laboratory [122]. To assess the possibility of delivering γ-emitting probes to tumors via a lipoprotein receptor–mediated process, we once again employed the labeled nonhydrolyzable cholesteryl ester ^{125}I-CI. As noted previously, the uptake of this probe by rat liver and adrenal could be significantly enhanced by prior incorporation into chylomicron remnants [34] and HDL [111], respectively. The question was whether a similar response could be demonstrated for tumors.

To address this question, rats bearing the Walker-256 sarcoma in the thigh were employed as the tumor model. As seen in Fig. 26, this tumor accumu-

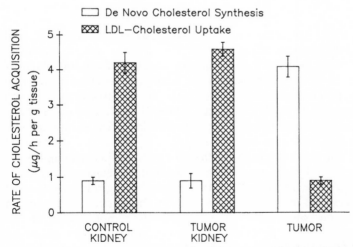

Figure 25 The rates of cholesterol acquisition by kidney of control rats and kidney and tumor of rats bearing a renal cell carcinoma. (Adapted from Ref. 120.)

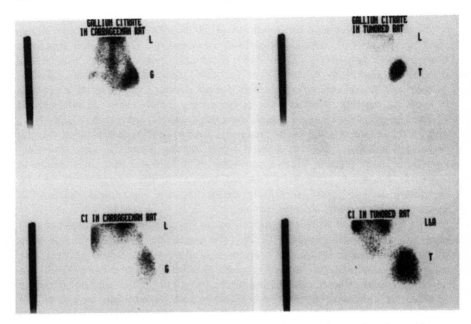

Figure 26 Comparison of gamma camera scintiscans of rats bearing either a carrageenan-induced granuloma or Walker-256 sarcoma in the left thigh and following administration of either gallium-67 citrate (top) or ^{125}I-CI (bottom). A, adrenal; G, granuloma; L, liver; T, tumor.

lates ^{125}I-CI to about the same extent as gallium-67 citrate, a widely used tumor-imaging radiopharmaceutical. One of the disadvantages of gallium-67 citrate in the clinic is its inability to distinguish neoplastic from inflammatory lesions. Figure 26 clearly demonstrates the ability of gallium-67 citrate to concentrate in an inflammatory lesion such as a carrageenan-induced granuloma present in the thigh of the rat. While ^{125}I-CI showed some ability to localize in this inflammatory lesion, its concentration in the granuloma was visibly less than that present in the tumor.

Quite unlike our previous results for the liver and steroid-secreting tissues, the uptake of ^{125}I-CI by the rat tumors could not be enhanced by prior incorporation into rat HDL (Fig. 27). As a matter of fact, administration of ^{125}I-CI in HDL resulted in a decreased uptake by the tumor. Moreover, in stark contrast to our observations for the adrenals, tumor uptake of ^{125}I-CI–HDL was unaffected by pretreating the animals with estrogen, indicating a lack of tumor responsiveness to hypocholesterolemia. At this point, there is no obvious explanation for the ability of ^{125}I-CI to accumulate in the Walker-256 tumor.

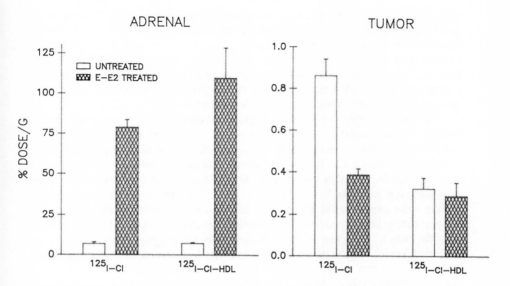

Figure 27 Concentration of radioactivity in tissues at 0.5 h following intravenous administration of ^{125}I-CI and ^{125}I-CI–HDL in untreated and estrogen-treated Walker-256 tumored rats.

D. Atheroma

Atherosclerosis is an extremely complex disease in terms of both its multifactoral etiology and the structure and cellular composition of the lesions themselves. As a result of its causal linkage to myocardial infarction, stroke, angina pectoris, and other forms of cardiovascular disease, it remains a leading cause of death in the United States [123].

The genesis of atherosclerosis can be simplistically viewed in terms of two interrelated processes: the abnormal proliferation of arterial smooth muscle cells and the over accumulation of lipoprotein-derived lipid. Although the proliferative component is certainly a critical aspect in the progression of the disease, emphasis in this discussion is placed on arterial accumulation of lipid to provide a basis for later discussion of incorporation of lipoprotein-associated radiopharmaceuticals into progressing lesions. Here, a general description of the anatomy of the normal artery wall and the alterations that occur in the different stages of lesion progression is followed by a brief survey of the major arterial cell types and their capacity for receptor-mediated uptake of various lipoproteins. Finally, arterial uptake of lipoproteins by nonreceptor-mediated mechanisms is considered.

The arterial wall can be divided into three major regions: the intima, the media, and the adventitia. The intima consists of a single layer of endothelial cells lining the luminal surface of the artery. It serves the extremely important function of limiting the rates of uptake of blood-borne macromolecules. While lipoproteins can cross normal intima by either receptor-mediated endocytosis or transport through plasmalemmal vesicles [124], disruption of the endothelium can enhance the influx of lipoproteins into the artery [125] and serve as a stimulus for platelet aggregation with the concomitant release of smooth muscle cell mitogens such as platelet-derived growth factor [126].

The most peripheral region of intima is defined by the internal elastic lamina, a fenestrated layer of elastic fibers. The next layer, the media, is composed of smooth muscle cells, collagen, elastic fibers, and proteoglycans. The smooth muscle cell is very important in lesion formation. Its movement from the media, through the internal elastic lamina, and into the intima is an essential step in early lesion formation [127]. Once in the intima, smooth muscle cells can increase the size of the lesion by actively proliferating and by generating large quantities of connective tissue matrix [128]. In addition, smooth muscle cells are able to accumulate large cytoplasmic droplets of cholesteryl ester, taking on an appearance that has earned them the appropriate name of foam cells. Although they are capable of *de novo* cholesterol synthesis, most cholesterol found in smooth muscle cells, and in atheroma in general, is deposited by other mechanisms [129]. The outer edge of the media is defined by the exterior elastic lamina, and the outermost layer of the artery is the adventitia, comprised primarily of fibroblasts, connective tissue, smooth muscle cells, and fat.

Although atherosclerotic lesions are extremely varied in terms of size, cellular composition, extracellular configuration, and anatomic location, they can be divided into three broad categories: early lesions, advanced lesions, and complicated lesions.

The most common type of early lesion is the fatty streak. These lesions are grossly visible, flat or slightly raised above the artery wall, and appear yellowish due to the high concentration of lipid, predominantly cholesteryl ester [130]. The majority of cells found within fatty streaks are monocyte-derived macrophage foam cells and smooth muscle cells [131]. Fatty streaks are extremely common lesions; they are found in aortas of nearly all children by the age of 10 and can increase with age until they cover nearly 50% of the aorta [132]. Although it is still questionable whether the fatty streak is a precursor to the advanced lesion, certain evidence suggests that this may be the case in the coronary vasculature [133].

Advanced lesions are distinguishable from early lesions by the presence of large amounts of connective tissue and by the protrusion of the lesion into the lumen of the vessel. Elastin, collagen, and proteoglycans can combine to form a "cap" on the lumen side of the lesion, covering an often necrotic center consisting of cell debris, and intracellular and extracellular cholesterol and cholesteryl ester. The advanced lesion does not form with the same frequency as fatty streaks and usually appears later in life [134].

The third general category of atheroma, the complicated lesion, is formed when the characteristics of the advanced lesion degenerate to further compromise the integrity of the vessel. The most common alterations are lesion calcification, cellular necrosis, and hemorrhage into the vessel wall, all of which diminish the strength and elasticity of the vessel. These changes predispose the vessel to more sudden events that commonly produce the morbidity associated with atherosclerosis, such as rupture of the plaque, hemorrhage into the body cavity, or thrombosis [135].

All the major cell types in atherosclerotic lesions possess cell surface receptors capable of mediating the binding and internalization of various lipoproteins. The receptor types of major importance include the LDL receptor, the acetylated LDL or scavenger receptor, and the ß-VLDL receptor. The receptor for native LDL is present on endothelial and smooth muscle cells [136] and on monocyte-derived macrophages [137]. As in other tissues, this receptor is sensitive to downregulation as a result of cellular cholesterol loading. The second major receptor is the acetyl-LDL or scavenger receptor, which is present on endothelial cells [138] and macrophages [139]. Unlike the situation for the native LDL receptor, the scavenger receptor is expressed regardless of the cholesterol loading state of the cell, thereby permitting massive cholesterol overloading to form foam cells [139]. This receptor was first described using LDL that was chemically modified on the epsilon amino groups of the apoprotein so that the net negative charge of the particle was increased; such chemical modifications include carbamylation, malondialdehyde modification, acetoacetylation, and, most commonly, acetylation (see Table 2).

A great deal of effort has been centered around the search for a biological equivalent of acetylated LDL and several such possibilities now exist. A primary candidate is LDL modified by endothelial or smooth muscle cells [140]. This appears to occur through free radical–induced peroxidation and results in a particle with increased density, increased electrophoretic mobility, and decreased amino group reactivity [141–143]. In addition, it was shown recently that charge-modified VLDL can also serve as a ligand for the scavenger receptor [144].

Finally, receptors also exist for ß-VLDL, a lipoprotein mainly produced in hypercholesterolemic animals which floats in the density range of VLDL but has the beta electrophoretic mobility of LDL. These receptors occur predominantly on endothelial cells [145] and macrophages [146], and their expression is poorly regulated in response to cholesterol loading. More recent studies, however, have demonstrated that the ß-VLDL binding site does not represent a distinct receptor type since it is immunologically related to the LDL receptor [147,148].

In addition to the receptor-mediated uptake of lipoproteins, receptor-independent means of uptake also exist. In addition to the high affinity, low capacity mechanism for LDL uptake via the receptor pathway, a low affinity, high capacity system also exists which becomes increasingly important with elevated serum cholesterol levels [96]. Large quantities of cholesterol can be cleared from the plasma and accumulate in cells via this mechanism. In addition, lipoproteins can become trapped in the extracellular matrix of the arterial wall; this has been shown to be particularly relevant with regard to damaged and reendothelialized regions of the vasculature, where large quantities of negatively charged proteoglycans in the neointima can trap circulating lipoproteins [149,150].

Because atherosclerosis is typically asymptomatic until it becomes quite severe, the ability to assess the status of the disease in patients at risk would be of great clinical importance. Ideally, such a procedure would allow early detection of lesions and thereby permit dietary or pharmacological intervention before the disease progressed to a stage associated with an increased risk of heart attack or stroke.

Several different approaches have been taken toward noninvasive diagnosis of atherosclerosis. While angiography remains the standard to which many of these techniques are compared [151], other methodologies include plethysmography [152], Doppler and ultrasound techniques [152,153], magnetic resonance imaging [154], computed tomography [155], and scintigraphic detection of ^{111}In-labeled platelets [156]. Another potential approach, which is the focus of the following discussion, is based on the finding that accumulation of LDL in the arterial wall is increased in areas involved in lesions [157,158]. Use of this differential uptake of LDL as an approach to lesion imaging is a valid rationale considering the critical role played by LDL in atherogenesis [159]. In addition, an imaging modality such as this would theoretically permit the visualization of lesions at stages earlier than are detectable by angiography since stenosis of the vessel would not be required for lesion detection.

In an effort to determine the suitability of this approach, Lees et al. [160] performed a study in which 100 μCi of autologous ^{125}I-LDL was injected into three patients with carotid atherosclerosis demonstrated by angiography and one hypercholesterolemic patient without carotid disease. Two days after injection, blood pool activity had decreased and focal uptake of ^{125}I-LDL into the arterial wall was detectable by gamma camera scintigraphy in two of the three patients with carotid disease (Fig. 28). This uptake corresponded to the sites of the lesions seen on the angiograms with considerably lower quantities of radioactivity accumulating in the nondiseased areas. In the third atherosclerotic patient, a suggestion of localized ^{125}I-LDL uptake was observed, while no focal LDL uptake was seen in the control subject.

In addition, Lees *et al.* developed methodology for labeling LDL with [99m]Tc [20]. This radioisotope offers a distinct advantage over ^{125}I in that it emits a higher energy, allowing better spatial resolution of the images. Moreover, the radiolabel does not readily become dissociated from the lipoprotein *in vivo*. In two different studies, ^{99m}Tc-LDL was injected into atherosclerotic rabbits with some success in lesion detection [20,161]. Studies have also been done in humans using LDL labeled with [125]I [162]. Of 17 patients studied, carotid artery lesions were detected in 10 and femoral artery lesions in 6.

An alternative means of tracing lipoprotein uptake into the artery wall involves the use of lipoproteins containing radiolabeled cholesteryl esters or ethers in the inner lipid core. Our own efforts have involved the use of ^{125}I-CI in a cholesterol-fed, New Zealand white rabbit model of atherosclerosis. Although the compound was injected in a Tween/physiological saline vehicle, it was found to become very rapidly associated with plasma lipoproteins, predominantly ß-VLDL in the cholesterol-fed rabbit [27,28]. Because the ester linkage of ^{125}I-CI is resistant to hydrolysis and further metabolism, the radioactivity that accumulates in the lesions of the animals is representative of total cholesteryl ester uptake.

Preliminary studies were performed to ensure that ^{125}I-CI accumulated preferentially in atheroma relative to normal arterial tissue. In these studies, ^{125}I-CI was solubilized in Tween/isotonic saline and injected into cholesterol-

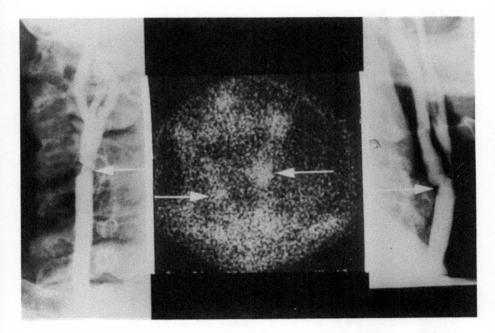

Figure 28 Comparison of bilateral carotid angiograms with a 2-day scintigram. Thyroid gland is visible at bottom of scintigram, and the blood pool in cerebral venous sinuses at top. The isolated right common carotid lesion (left side of scintigram) sequesters ^{125}I-LDL, while extensive disease of left common and internal carotids, which extends much higher in the neck, accumulates far more radiolabeled lipoprotein (right side of scintigram). (From Ref. 161.)

fed rabbits that had been subjected to balloon catheter deendothelialization of the aorta six weeks earlier. Animals were sacrificed 48 h later and autoradiography was performed on the isolated aortas (Fig. 29). The autoradiograms showed distinct regions of radioactivity which closely corresponded to the pattern of lesion development seen upon oil red O staining of the aortas. In contrast, the normal vessels contained little or no radioactivity or oil red O

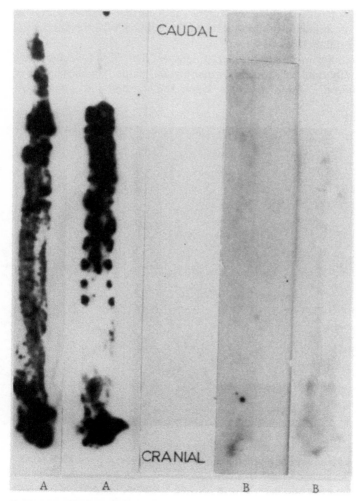

Figure 29 En-face autoradiographs of whole aortas of either atherosclerotic (A) or normal (B) rabbits injected with ^{125}I-CI and sacrificed 48 h later. The probe was injected 6 weeks after denudation of the aorta in atherosclerotic rabbits.

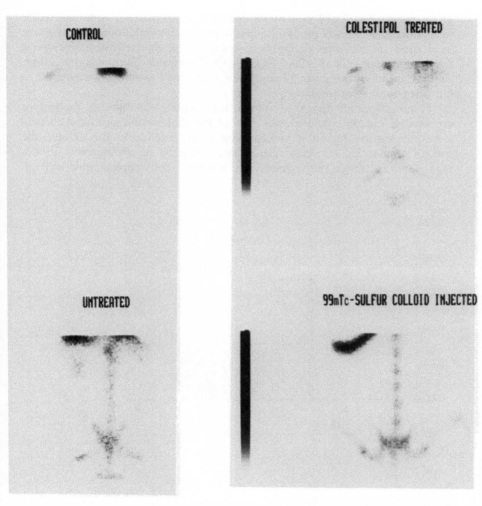

Figure 30 Representative scintigraphic scans with ^{125}I-CI of animals in group N (control), group C (untreated), group CP (colestipol treated), and a group C animal injected with ^{99m}Tc-sulfur colloid.

staining. These results indicated that the probe preferentially accumulated in the lesioned, lipid-containing areas of the vessels.

Based on the favorable results of this preliminary investigation, further studies were performed in an effort to image the aorta of intact animals via gamma camera scintigraphy. In one such study [163], New Zealand white rabbits were made atherosclerotic by feeding a diet containing 2% cholesterol for 15 weeks. While one group of animals remained untreated, two other groups were treated with either colestipol (1% w/w in diet) or clofibrate (0.3% w/w in diet). A control group of animals was fed a normal diet. Starting one week after the initiation of the diets, animals were injected twice weekly via the marginal ear vein 10 μCi of ^{125}I-CI again dissolved in Tween/isotonic saline. The last injection was given 13 days prior to sacrifice to allow for clearance of the compound from the blood.

The scintigraphic images of animals representative of three of the groups are seen in Fig. 30. Whereas the scan of the cholesterol-fed animal showed a distinct vertical region of radioactivity in the midline of the animal, the control animal fed a normal diet showed virtually no radioactivity in this region, and the image of the animal that was fed a diet supplemented with co-

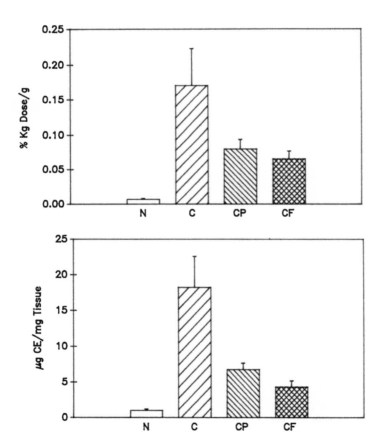

Figure 31 Uptake of ^{125}I-CI and accumulation of cholesteryl ester in the abdominal-thoracic region of the aorta of normal rabbits (N), cholesterol-fed rabbits (C), and cholesterol-fed rabbits treated with either colestipol (CP) or clofibrate (CF).

lestipol demonstrated the existence of an intermediate amount of radioactivity in the same position. Similar results were observed in the group treated with clofibrate. However, by implanting capsules containing $Na^{125}I$ into the aorta of an eviscerated animal, it was found that the lowest level of radioactivity that was detectable on the equipment used was tenfold higher than the average amount of radioactivity present in the aortas of the cholesterol-fed group of animals. One of the cholesterol-fed animals was also injected with ^{99m}Tc-sulfur

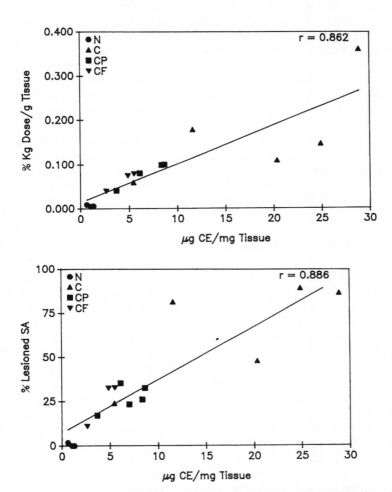

Figure 32 Correlation between radioactivity and cholesteryl ester accumulation (upper panel) and between percentage of lesioned surface area (SA) and cholesteryl ester content (lower panel) in the abdominal-thoracic aortas of rabbits that were normal (N), cholesterol-fed (C), or cholesterol-fed and treated with colestipol (CP) or clofibrate (CF).

colloid, which localizes in the bone marrow, and rescanned (Fig. 30). The similarity of this scan to the ^{125}I scans suggested that much of the apparent radioactivity was localized in the bone marrow.

Regardless of the generalized accumulation of ^{125}I-CI, the probe still served as a good tracer of cholesteryl ester deposition in the aorta. When the aortas were removed from the animals upon sacrifice and analyzed, it was found that both the radioactivity content and the cholesteryl ester content of the abdominal/thoracic region of the aorta were lowest for the group fed normal chow, highest for the cholesterol-fed group, and intermediate for the two drug-treated groups (Fig. 31). Moreover, when the esterified cholesterol content of the abdominal/thoracic aorta was plotted against both the accumulation of radioactivity and the percentage of lesioned surface area of the same aortic region (Fig. 32), the correlation between the two sets of variables was highly significant. The similarity of these two parameters in their ability to provide an index of the cholesteryl ester content of the abdominal/thoracic aorta indicates that ^{125}I-CI is a good tracer for cholesteryl ester deposition.

The goal of future studies is therefore to more selectively target the ^{125}I-CI to atheromatous lesions. The approach will involve both incorporation of the probe into a single atherogenic lipoprotein fraction prior to injection and induction of a less severe hyperlipidemia in order to reduce the metabolic dilution of the ^{125}I-CI–labeled lipoprotein.

VI. CONCLUDING REMARKS

At this time, the clinical potential of radiolabeled lipoproteins remains an uncertainty. Studies performed to date have done more to indicate the limitation in the use of these macromolecules as delivery vehicles than to extol their promise. Technetium-99[m] would be an ideal nuclide for labeling lipoproteins and this approach has been followed by several laboratories. While some promising results have been achieved with such preparations, we still have no information on whether such labeling modifies the normal receptor-mediated mechanism for lipoprotein processing. Labeling with radioiodine, on the other hand, does not compromise receptor recognition *in vitro*, but the rapid metabolism of these produces limits their *in vivo* utility. Radioiodinated nonmetabolizable lipids have been incorporated into lipoproteins, but it is still too early to judge their clinical potential.

Among the tissues for imaging, the most advanced studies are those with the liver and adrenals, both known to be well endowed with lipoprotein receptors. Obviously, the ability to ascertain the lipoprotein receptor responsiveness of these tissues by noninvasive imaging would have wide clinical application. Lipoprotein-mediated uptake of radioactivity by tumors is not nearly as well defined, nor does it appear that the tumor lipoprotein receptors are under the same degree of homeostatic regulation as the liver and adrenal.

Nonetheless, the accumulation of radioiodinated cholesteryl esters by an animal tumor justifies further efforts in this area. Finally, the use of lipoproteins for the *in vivo* visualization of atherosclerotic lesions appears to offer the greatest challenge. Work to date offers little encouragement for this approach in humans. These studies, however, suggest that radioiodinated lipoproteins may be useful in animal models of atherosclerosis to assess the efficacy of drugs on the *in vivo* progression or regression of atherosclerotic lesions.

VII. ADDENDUM

Since completion of this chapter, several notable advances in methodology permitting imaging of atherosclerotic lesions and the biodistribution of LDL have been reported. These include a study using LDL labeled with ^{123}I-tyramine cellobiose, a tracer that residualizes in tissues [164]. The superiority of this preparation over ^{123}I-LDL, which was degraded and lost from tissues, was readily apparent from scans obtained 24 h postinjection. In addition, further studies utilizing ^{99m}Tc-LDL have been reported [165,166]. Vallabhajosula *et al.* [165] compared the plasma clearance rates, tissue distribution, and scintigraphic scans of normal and hypercholesterolemic rabbits injected with ^{99m}Tc-LDL, ^{131}I-tyramine cellobiose–LDL, and ^{131}I-LDL. The ^{99m}Tc-LDL preparation had the combined advantage of clearing from the plasma similarly to the ^{131}I-tyramine cellobiose–LDL. The ^{99m}Tc-LDL preparation was also used in studies involving imaging of human atherosclerotic lesions. Focal accumulation of radiolabel was observed in surgical specimens, and some positive images were obtained, particularly in "active" lesions with an abundance of cells and a hemorrhagic or necrotic component.

ACKNOWLEDGMENTS

The authors wish to express their appreciation for the technical assistance provided by Ms. Edie Quenby in the preparation of the chapter and to Mrs. Denise Watkins for typing the chapter. Those studies in the chapter which outline some of our own research were sponsored in part by the National Cancer Institute (CA-08349) and the American Heart Association, Michigan Chapter.

REFERENCES

1.   Counsell RE, Ice RE. The design of organ-imaging radiopharmaceuticals. In: Areins EJ, ed. *Drug design*, vol. 6. New York:Academic Press, 1975:171–259.

2. Counsell RE, Korn N. Biochemical and pharmacological rationale in radiotracer design. In:Colombetti LG, ed. *Principles of radiopharmacology*, vol. I, Boca Raton, FL:CRC Press, 1979:189–250.

3. Fritzberg AR, ed. *Radiopharmaceuticals: progress and chemical perspectives*, vol. 1–2. Boca Roton, FL:CRC Press, 1986.

4. Seevers RH, Counsell RE. Radioiodination techniques for small organic molecules. *Chem Rev* 1982; 82:575–590.

5. Eisenberg S, Windmueller HG, Levy RI. Metabolic fate of rat and human lipoprotein apoproteins in the rat. *J Lipid Res* 1973; 14:446–458.

6. Sigurdsson G, Noel S, Havel RJ. Quantification of the hepatic contribution to the catabolism of high density lipoproteins in rats. *J Lipid Res* 1979; 20:316–324.

7. Pittman RC, Steinberg D. A new approach for assessing cumulative lysosomal degradation of proteins or other macromolecules. *Biochem Biophys Res Commun* 1978; 81:1254–1259.

8. Pittman RC, Taylor CA, Jr. Methods for assessment of tissue sites of lipoprotein degradation. *Methods Enzymol* 1986; 129 (part B):612–628.

9. Glass C, Pittman RC, Weinstein DB, Steinberg D. Dissociation of tissue uptake of cholesteryl ester from that of apoprotein A-I of rat plasma high density lipoproteins: selective delivery of cholesteryl ester to the liver, adrenal, and gonad. *Proc Natl Acad Sci USA* 1983; 80:5435–5439.

10. Glass C, Pittman RC, Given M, Steinberg D. Uptake of high-density lipoprotein-associated apoprotein A-I and cholesterol esters by 16 tissues of the rat *in vivo* and by adrenal cells and hepatocytes *in vitro*. *J Biol Chem* 1985; 260:744–750.

11. Leitersdorf E, Israeli A, Stein D, Eisenberg S, Stein Y. The role of apopolipro-teins of HDL in the selective uptake of cholesteryl linoleyl ether by cultured rat and bovine adrenal cells. *Biochim Biophys Acta* 1986; 878:320–329.

12. Seevers RH, Grosiak MP, Weichert JP, Schwendner SW, Longino MA, Counsell RE. Potential tumor- and organ-imaging agents. 23. Sterol esters of iopanoic acid. *J Med Chem* 1982; 25:1500–1503.

13. Counsell RE, Korn N, Pohland RC, Schwendner SW, Seevers RH. Fate of intravenously administered high-density lipoprotein labeled with radioiodinated cholesteryl oleate in normal and hypolipidemic rats. *Biochim Biophys Acta* 1983; 750:497–503.

14. McFarlane AS. Efficient trace-labelling of proteins with iodine. *Nature* 1958; 182:53.

15. Bilheimer DW, Eisenberg S, Levy RI. The metabolism of very low density lipoproteins. I. Preliminary *in vitro* and *in vivo* observations. *Biochim Biophys Acta* 1972; 260:212–221.

16. Schaeffer EJ, Ordovas JM. Metabolism of apolipoproteins A-I, A-II and A-III. *Methods Enzymol* 1986; 129 (part B):420–443.

17. Bolton AE, Hunter WM. The labelling of proteins to high specific radioactivi-ties by conjugation to ^{125}I-containing acylating agents. *Biochem J* 1973; 133:529–539.

18. Mahley RW, Innerarity TL, Weisgraber KH, Oh SY. Altered metabolism (*in vivo* and *in vitro*) of plasma lipoproteins after selective chemical modification of lysine residues of the apoproteins. *J Clin Invest* 1979; 64:743–750.

19. Innerarity TL, Pitas RE, Mahley RW. Lipoprotein–receptor interactions. *Methods Enzymol* 1986; 129 (part B):542–565.

20. Lees RS, Garabedian HD, Lees AM, Schumacher DJ, Miller A, Isaacsohn JL, Derksen A, Strauss HW. Technetium-99ᵐ low density lipoproteins: preparation and biodistribution. *J Nucl Med* 1985; 26:1056–1062.

21. Pohland RC, Counsell RE. The role of high density lipoproteins in the biodistribution of two radioiodinated probes in the rat. *Toxicol Appl Pharmacol* 1985; 77:47–57.

22. Havel RJ, Eder HA, Bradgow JH. The distribution and chemical composition of ultracentrifugally separated lipoproteins in human serum. *J Clin Invest* 1955; 34:1345–1353.

23. Schwendner SW, McConnell DS, Counsell RE. Uptake of ¹²⁵I-cholesteryl iopanoate labelled HDL by Walker 256 sarcoma in the rat. *FASB J* 1988; 2:A781.

24. Krieger M, Brown MS, Faust JR, Goldstein JL. Replacement of endogenous cholesteryl esters of low density lipoprotein with exogenous cholesteryl linoleate. Reconstitution of a biologically active lipoprotein particle. *J Biol Chem* 1978; 253:4093–4101.

25. Krieger M, McPhaul MJ, Goldstein JL, Brown MS. Replacement of neutral lipids of low density lipoprotein with esters of long chain unsaturated fatty acids. *J Biol Chem* 1979; 254:3845–3853.

26. Krieger M, Smith LC, Anderson RGW, Goldstein JL, Kao YJ, Pownall HJ, Gotto AM Jr, Brown MS. Reconstituted low density lipoprotein: a vehicle for the delivery of hydrophobic fluorescent probes to cells. *J Supramol Struct* 1979; 10:467–478.

27. DeGalan MR, Schwendner SW, Weichert JP, Counsell RE. Radioiodinated cholesteryl iopanoate as a potential probe for the *in vivo* visualization of atherosclerotic lesions in animals. *Pharmaceut Res* 1986; 3:52–55.

28. DeGalan MR, Schwendner Sw, Gross MD, Counsell RE. Assessment of the antiatherogenic efficacy of a hypolipidemic drug by scintigraphy. In:Billinghurst MW, ed. *Current applications in radiopharmacology.* Proceedings of the Fourth International Symposium on Radiopharmacology. New York:Pergamon Press, 1985:109–117.

29. DeForge LE, DeGalan MR, Newton RS, Counsell RE. Use of radioiodinated cholesteryl iopanoate in the reconstitution of biological active LDL. In preparation.

30. Cardin AD, Witt KR, Barnhart CL, Jackson RL. Sulfhydryl chemistry and solubility properties of human plasma apolipoprotein B. *Biochemistry* 1982; 21:4503–4511.

31. Avigan J. A method for incorporating cholesterol and other lipids into serum lipoproteins *in vitro*. *J Biol Chem* 1958; 234:787–790.

32. Jonas A, Hesterberg LK, Drengler SM. Incorporation of excess cholesterol by high density serum lipoproteins. *Biochem Biophys Acta* 1978; 528:47–57.

33. Counsell RE, Schappa LW, Korn N, Huler RJ. Tissue distribution of high-density lipoprotein labeled with radioiodinated cholesterol. *J Nucl Med* 1980; 21:852–858.

34. Damle NS, Seevers RH, Schwendner SW, Counsell RE. Potential tumor- or organ-imaging agents. XXIV. Chylomicron remnants as carriers for hepatographic agents. *J Pharma Sci* 1983; 72:898–901.

35. Narayan KA. Electrophoretic methods for the separation of serum lipoproteins. In:Perkins EG, ed. *Analysis of lipids and lipoproteins* Champaign, IL: American Oil Chemists Society 1975:225–249.

36. Narayan KA, Creinin HL, Kummerow FA. Disc electrophoresis of rat plasma lipoproteins. *J Lipid Res* 1966; 7:150–157.

37. Quarfordt AH, Goodman SD. Incorporation of labeled cholesterol esters in chylomicrons *in vivo*. *J Lipid Res* 1966; 7:708–710.

38. Quarfordt SH, Goodman DS. Metabolism of doubly-labeled chylomicron cholesteryl esters in the rat. *J Lipid Res* 1967; 8:264–273.

39. Tall AR. Plasma lipid transfer proteins. *J Lipid Res* 1986; 27:361–367.

40. Pattnaik NM, Montes A, Hughes LB, Zilversmit DB. Cholesteryl ester exchange protein in human plasma. Isolation and characterization. *Biochim Biophys Acta* 1978; 530:428–438.

41. Morton RE, Zilversmit DB. Inter-relationship of lipids transferred by the lipid-transfer protein isolated from human lipoprotein-deficient plasma. *J Biol Chem* 1983; 258:11751–11757.

42. Morton RE, Zilversmit DB. A plasma inhibitor of triglyceride and cholesteryl ester transfer activities. *J Biol Chem* 1981; 256:11992–11995.

43. Morton RE, Zilversmit DB. Purification and characterization of lipid transfer protein(s) from human lipoprotein-deficient plasma. *J Lipid Res* 1982; 23:1058–1067.

44. Brunner J, Skrabal P, Hauser H. Single bilayer vesicles prepared without sonication. Physico-chemical properties. *Biochim Biophys Acta* 1976; 455:322–331.

45. Windler E, Chao Y, Havel RJ. Regulation of the hepatic uptake of triglyceride-rich lipoproteins in the rat. *J Biol Chem* 1980; 255:8303–8307.

46. Grundy SM, Mok HYI. Chylomicron clearance in normal and hyperlipidemic man. *Metabolism* 1976; 25:1225–1239.

47. Brown MS, Kovanen PT, Goldstein JL. Regulation of plasma cholesterol by lipoprotein receptors. *Science* 1981; 212:628–635.

48. Brown MS, Goldstein JL. A receptor-mediated pathway for cholesterol homeostasis (Nobel Lecture). *Angew Chem Int Ed Engl* 1986; 25:583–602 and references cited therein.

49. Goldstein JL, Brown MS. Atherosclerosis: the low-density lipoprotein receptor hypothesis. *Metabolism* 1977; 26:1257–1275.

50. Glomset JA. The plasma lecithin:cholesterol acyltransferase reaction. *J Lipid Res* 1968; 9:155–167.

51. Wu J, Butler J, Bailey JM. Lipid metabolism in cultured cells. XVIII. Comparative uptake of low density and high density lipoproteins by normal, hypercholesterolemic and tumor virus-transformed human fibroblasts. *J Lipid Res* 1979; 20:472–480.

52. Andersen JM, Dietschy JM. Relative importance of high and low density lipoproteins in the regulation of cholesterol synthesis in the adrenal gland, ovary, and testis of the rat. *J Biol Chem* 1978; 253:9024–9032.

53. Carr BR, Parker CR Jr, MacDonald PC, Simpson ER. Metabolism of high density lipoprotein by human fetal adrenal tissue. *Endocrinology* 1987; 107:1849–1854.

54. Shepherd J, Packard CJ. Receptor-independent low-density lipoprotein catabolism. *Methods Enzymol* 1986; 129 (part B):566–590.

55. Brown MJ, Goldstein JL. Lipoprotein metabolism in the macrophage. *Ann Rev Biochem* 1983; 52:223–261.

56. Goldstein JL, Ho YK, Basu SK, Brown MS. Binding site on macrophages that mediates uptake and degradation of acetylated low density lipoprotein, producing massive deposition. *Proc Natl Acad Sci USA* 1979; 76:333–337.

57. Basu SK, Goldstein JL, Anderson RGW, Brown MS. Degradation of cationized low density lipoprotein and regulation of cholesterol metabolism in homozygous familial hypercholesterolemic fibroblasts. *Proc Natl Acad Sci USA* 1976; 73:3178–3182.

58. Weisgraber KH, Innerarity TL, Mahley RW. Role of the lysine residues of plasma lipoproteins in high affinity binding to cell surface receptors on human fibroblasts. *J Biol Chem* 1978; 253:9053–9062.

59. Haberland ME, Olch CL, Fogelman AM. Role of lysines in mediating interaction of modified low density lipoproteins with the scavenger receptor of human monocyte macrophages. *J Biol Chem* 1984; 259:11305–11311.

60. Chacko GK. Modification of human high density lipoprotein (HDL$_3$) with tetronitromethane and the effect on its binding to isolated rat liver plasma membranes. *J Lipid Res* 1985; 26:745–754.

61. Nestler JE, Chacko GK, Strauss JF III. Stimulation of rat ovarian cell steroidogenesis by high density lipoprotein modified with tetronitromethane. *J Biol Chem* 1985; 260:7316–7321.

62. Fogelman AM, Shechter I, Seager J, Hokom M, Child JS, Edwards PA. Malondialdehyde alteration of low density lipoproteins leads to cholesteryl ester accumulation in human monocyte-macrophages. *Proc Natl Acad Sci USA* 1980; 77:2214–2218.

63. Mahley RW, Innerarity TL, Pitas RE, Weisgraber KH, Brown JH, Gross E. Inhibition of lipoprotein binding to cell surface receptors of fibroblasts following selective modification of arginyl residues in arginine-rich and B apoproteins. *J Biol Chem* 1977; 252:7279–7287.

64. Havel RJ. Functional activities of hepatic lipoprotein receptors. *Ann Rev Physiol* 1986; 48:119–134.

65. Hoeg JM, Brewer HB. Human lipoprotein metabolism and the liver. *Prog Liver Dis* 1986; 8:51–64.

66. Kovanen PT. Regulation of plasma cholesterol by hepatic low-density lipoprotein receptors. *Am Heart J* 1987; 113:464–469.

67. Rudling J. Role of the liver for the receptor-mediated catabolism of low-density lipoprotein in the 17 -ethinylestradiol-treated rat. *Biochim Biophys Acta* 1987; 919:175–180.

68. Kovanen PT, Brown MS, Goldstein JL. Increased binding of low density lipoprotein to liver membranes from rats treated with 17 -ethinyl estradiol. *J Biol Chem* 1979; 254:11367–11373.

69. Hui DY, Innerarity TL, Mahley RW. Lipoprotein binding to canine hepatic membranes: metabolically distinct apo-E and apo-B,E receptors. *J Biol Chem* 1981; 256:5646–5655.

70. Mahley RW, Innerarity TL. Lipoprotein receptors and cholesterol homeostasis. *Biophys Acta* 1983; 737:197–222.

71. van't Hooft FM, Van Tol A. The sites of degradation of purified rat low density lipoprotein and high density lipoprotein in the rat. *Biochim Biophys Acta* 1985; 836:344–353.

72. Harkes L, Van Berkel TJC. *In vivo* characteristics of a specific recognition site for LDL on nonparenchymal rat liver cells which differs from the 17 ethinyl estradiol-induced LDL receptor on parenchymal liver cells. *Biochim Biophys Acta* 1984; 794:340–347.

73. Cooper AD. Role of the liver in the degradation of lipoproteins. *Gastroenterology* 1985; 88:192–205.

74. Nagelkerke JF, Van Berkel TJC. Rapid transport of fatty acids from rat liver endothelial to parenchymal cells after uptake of cholesteryl ester-labeled acetylated LDL. *Biochim Biophys Acta* 1986; 875:593–598.

75. Longino MA, Glazer GM, Weichert JP, Groziak MP, Schwendner SW, Counsell RE. Esters of iopanoic acid as liver-specific CT contrast agents: biodistribution and CT evaluation. *J Comput Assist Tomog* 1981; 5:843–846.

76. Van Dort M, Schwendner SW, Skinner RWS, Gross MD, Counsell RE. Potential tumor or organ-imaging agents. 24. Radioiodinated pregnenolone esters. *Steroids* 1984; 44:85–93.

77. Carlson LA, Hallberg D. Studies on the elimination of exogenous lipids from the blood stream. The kinetics of the elimination of a fat emulsion and of chylomicrons in the dog after single injection. *Acta Physiol Scand* 1963; 59:52–61.

78. Weinberg RB, Scanu AM. In vitro reciprocal exchange of apoproteins and nonpolar lipids between human high density lipoproteins and an artificial triglyceride–phospholipid emulsion (Intralipid). *Atherosclerosis* 1982; 44:141–152.

79. Robinson SF, Quarfordt SH. Apoproteins in association with Intralipid incubations in rat and human plasma. *Lipids* 1979; 14:343–349.

80. Carlson LA. Studies on the fat emulsion Intralipid I. Association of serum proteins to Intralipid triglyceride particles. *Scand J Clin Lab Invest* 1980; 40:139–144.

81. Dresel HA, Fredrich EA, Waldherr R, Via DP, Schettler G. Imaging and visualization of the acetyl-LDL receptor of the rat liver. In:Crepaldi G, Tiengo A, Baggio G, eds. *Diabetes, obesity and hyperlipidemias*, vol. 3. Amsterdam: 1985:69–76.

82. Huettinger M, Corbett JR, Schneider WJ, Willerson JL, Brown MS, Goldstein JL. Imaging of hepatic low density lipoprotein receptors by radionuclide scintiscanning *in vivo*. *Proc Natl Acad Sci USA* 1984; 81:7599–7603.

83. Balasubramaniam S, Goldstein JL, Faust FR, Brunschede GY, Brown MS. Lipoprotein-mediated regulation of 3-hydroxy-3-methylglutaryl coenzyme A reductase activity and cholesteryl ester metabolism in the adrenal gland of the rat. *J Biol Chem* 1977; 252:1771–1779.

84. Faust JR, Goldstein JL, Brown MS. Receptor-mediated uptake of low density lipoprotein and utilization of its cholesterol for steroid synthesis in cultured mouse adrenal cells. *J Biol Chem* 1977; 252:4861–4871.

85. Brown MS, Kovanen PT, Goldstein FL. Evolution of the LDL receptor concept—from cultured cells to intact animals. *Ann NY Acad Sci* 1980; 348:48–68.

86. Henderson JF. Studies on fatty liver induction by 4-aminopyrazolopyrimidine. *J Lipid Res* 1963; 4:68–74.

87. Kovanen PT, Goldstein JL, Chappell DA, Brown MS. Regulation of low density lipoprotein receptors by adrenocorticotropin in the adrenal gland of mice and rats *in vivo*. *J Biol Chem* 1980 225:5591–5598.

88. Kita T, Beisiegel U, Goldstein JL, Schneider WJ, Brown MS. Antibody against low density lipoprotein receptor blocks uptake of low density lipoprotein (but not high density lipoprotein) by the adrenal gland of the mouse *in vivo*. *J Biol Chem* 1981; 256:4701–4703.

89. Anderson JM, Dietschy JM. Regulation of sterol synthesis in adrenal gland of the rat by both high and low density human plasma lipoproteins. *Biochem Biophys Res Commun* 1976; 72:880–885.

90. Gwynne JT, Mahaffee D, Brewer HB Jr, Ney RL. Adrenal cholesterol uptake from plasma lipoproteins: regulation by corticotropin. *Proc Natl Acad Sci USA* 1976; 73:4329–4333.

91. Stein Y, Dabach Y, Hollander G, Halperin G, Stein O. Metabolism of HDL-cholesteryl ester in the rat studied with a nonhydrolyzable analog, cholesteryl linoleyl ether. *Biochim Biophys Acta* 1983; 752:98–105.

92. Gwynne JT, Hess B. The role of high density lipoproteins in rat adrenal cholesterol metabolism and steroidogenesis. *J Biol Chem* 1980; 255:10875–10883.

93. Halperin G, Stein O, Stein Y. Synthesis of ether analogs of lipoprotein lipids and their biological applications. *Methods Enzymol* 1986; 129 (part B):816–848.

94. Glass C, Pittman RC, Given M, Steinberg D. Uptake of high density lipoprotein-associated apoprotein A-1 and cholesterol esters by 16 tissues of the *in vivo* and by adrenal cells and hepatocytes *in vitro*. *J Biol Chem* 1985; 260:774–750.

95. Pittman RC, Carew TE, Attie AD, Witstum JL, Watanabe Y, Steinberg D. Receptor-dependent and receptor-independent degradation of low density lipoprotein in normal rabbits and in receptor-deficient mutant rabbits. *J Biol Chem* 1982; 257:7994–8000.

96. Spady OK, Huettinger M, Bilheimer DW, Dietschy JM. Role of receptor-independent low density lipoprotein transport in the maintenance of tissue cholesterol balance in the normal and WHHL rabbit. *J Lipid Res* 1987; 28:32–41.

97. Spady DK, Dietschy JM. Rates of cholesterol synthesis and low density lipoprotein uptake in the adrenal glands of the rat, hamster and rabbit *in vivo*. *Biochim Biophys Acta* 1985; 836:167–175.

98. Goldstein JL, Brown MS. Binding and degradation of low density lipoproteins by cultured human fibroblasts: comparison of cells from a normal subject and from a patient with homozygous familial hypercholesterolemia. *J Biol Chem* 1973; 249:5153–5162.

99. Ohashi M, Carr BR, Simpson ER. Effects of adrenocorticotropic hormone on low density lipoprotein receptors of human fetal adrenal tissue. *Endocrinology* 1981; 108:1237–1242.

100. Strauss JF III, MacGregor LC, Gwynne JT. Uptake of high density lipoprotein by rat ovaries *in vivo*, and dispersed ovarian cells *in vitro*, direct correlation of high density lipoprotein uptake with steroidogenic activity. *J Steroid Biochem* 1982; 16:525–531.

101. Christie MH, Gwynne JT, Strauss JF III. Binding of human high density lipoproteins to membranes of internalized rat ovaries. *J Steroid Biochem* 1981; 14:671–678.

102. Gwynne JT, Strauss JF III. The role of lipoproteins in steroidogenesis and cholesterol metabolism in steroidogenic glands. *Endocr Rev* 1982; 3:299–329.

103. Reaven E, Chen YI, Spicher M, Azhar S. Morphologic evidence that high density lipoproteins are not internalized by steroid-producing cells during *in situ* organ perfusion. *J Clin Invest* 1984; 74:1384–1397.

104. Counsell RE, Ranade VV, Blair RJ, Beierwaltes WH, Weinhold PA. Tumor localizing agents. IX. Radioiodinated cholesterol. *Steroids* 1970; 15:317–328.

105. Korn N, Nordblom G, Floyd E, Counsell RE. Potential organ or tumor imaging agents. XX. Ovarian imaging with 19-radioiodinated cholesterol. *J Pharma Sci* 1980; 69:1014–1017.

106. Nordblom GD, Schappa LW, Floyd EE, Langdon RB, Counsell RE. A comparison of cholesteryl oleate and 19-iodocholesteryl oleate as substrates for adrenal cholesterol esterase. *J Steroid Biochem* 1980; 13:463–466.

107. Blair RJ, Beierwaltes WH, Lieberman LM, Boyd CM, Counsell RE, Weinhold PA, Varma VM. Radiolabeled cholesterol as an adrenal scanning agent. *J Nucl Med* 1970; 12:176–182.

108. Roheim PS, Rachmilewitz D, Stein O, Stein Y. Metabolism of iodinated high density lipoproteins in the rat. *Biochim Biophys Acta* 1971; 248:315–329.

109. Sigurdsson G, Noel S, Havel RJ. Quantification of the hepatic contribution to the catabolism of high density lipoproteins in rats. *J Lipid Res* 1979; 20:316–324.

110. Kovanen PT, Schneider WJ, Hillman GM, Goldstein JL, Brown MS. Separate mechanisms for the uptake of high and low density lipoproteins by the mouse adrenal glands *in vivo*. *J Biol Chem* 1979; 254:5498–5505.

111. Counsell RE, Schwendner SW, Gross MD, Longino MA, McConnell DS. Lipoprotein incorporation enhances uptake of radioiodinated cholesteryl ester into steroid hormone-secreting tissues. *J Nucl Med* 1989; 30:1088–1094

112. Isaacsohn JL, Lees AM, Lees RS, Strauss HW, Barlai-Kovach M, Moore T. Adrenal imaging with technetium-99m-labelled low density lipoproteins. *Metabolism* 1986; 35:364–366.

113. Gal D, MacDonald PC, Porter JC, Simpson ER. Cholesterol metabolism in cancer cells in monolayer culture. III. Low density lipoprotein metabolism. *Int J Cancer* 1981; 28:315–319.

114. Gal D, Ohashi M, MacDonald PC, Buchsbaum HJ, Simpson ER. Low-density lipoprotein as a potential vehicle for chemotherapeutic agents and radionucleotides in the management of gynecologic neoplasms. *Am J Obstet Gynecol* 1981; 139:877–885.

115. Hynds SA, Welsh J, Stewart JM, Jack A, Soukop M, McArdle CS, Calman KC, Packard CJ, Shepherd J. Low-density lipoprotein metabolism in mice with soft tissue tumors. *Biochim Biophys Acta* 1984; 795:589–595.

116. Norata G, Cant G, Ricci L, Nicolin A, Trezzi E, Catapano AL. *In vivo* assimilation of low density lipoprotein by a fibrosarcoma tumor line in mice. *Cancer Lett* 1984; 25:203–208.

117. Ho YK, Smith G, Brown MJ, Goldstein JL. Low density lipoprotein (LDL) receptor activity in human myelogenous leukemia cells. *Blood* 1978; 52:1099–1114.

118. Vitols S, Gahrton G, Ort A, Peterson C. Elevated low density lipoprotein receptor activity in leukemia cells with monocytic differentiation. *Blood* 1984; 63:1186–1193.

119. Ginsberg H, Goldberg IJ, Wang-Iverson P, Gitler E, Le NA, Gilbert HS, Virgil Brown W. Increased catabolism of native and cyclohexanedione-modified low density lipoprotein in subjects with myeloproliferative diseases. *Arteriosclerosis* 1983; 3:233–241.

120. Clayman RV, Bilhartz LE, Buja LM, Spady DK, Dietschy JM. Renal cell carcinoma in the Wistar-Lewis rat: a model for studying the mechanism of cholesterol acquisition by a tumor *in vivo*. *Cancer Res* 1986; 46:2958–2963.

121. Barnard GF, Erickson SK, Cooper AD. Regulation of lipoprotein receptors on rat hepatomas *in vivo*. *Biochim Biophys Acta* 1986; 879:301–303.

122. Unpublished results.

123. Solberg LA, Strong JP. Risk factors and atherosclerotic lesions. A review of autopsy studies. *Arteriosclerosis* 1983; 3:187–198.

124. Gimbrone MA Jr. Vascular endothelium and atherosclerosis. In:Moore S, ed. *Vascular injury and atherosclerosis*. New York:Marcel Dekker, 1981:25–52.

125. Schwartz SM. Role of endothelial integrity in atherosclerosis. *Artery* 1980; 8:305–314.

126. St. Clair RW. Pathogenesis of the atherosclerotic lesion: current concepts of cellular and biochemical events. In:Cox RH, Tulenko TN, eds. *Recent advances in arterial disease: atherosclerosis, hypertension, and vasospasm*. New York:Alan R. Liss, 1986:1–29.

127. Nilsson J. Smooth muscle cells in the atherosclerotic process. *Acta Med Scand (suppl)* 1986; 715:25–31.

128. Ross R, Klebanoff SJ. The smooth muscle cell. I. *In vivo* synthesis of connective tissue proteins. *J Cell Biol* 1971; 50:159–171.

129. Davignon J, Marcel YL. Lipid metabolism of the lesion and interaction of lipoproteins with cells. In:Moore S, ed. *Vascular injury and atherosclerosis*. New York:Marcel Dekker, 1981:175–207.

130. Lofland HB, Clarkson TB. The bi-directional transfer of cholesterol in normal aorta, fatty streaks and atheromatous plaques. *Proc Sco Exp Biol Med* 1970; 133:1–8.
131. Ross R. The pathogenesis of atherosclerosis—an update. *N Engl J Med* 1986; 314:488–500.
132. Haust MD. The natural history of human atherosclerotic lesions. In:Moore S, ed. *Vascular injury and atherosclerosis*. New York:Marcel Dekker, 1981:1–23.
133. McGill HC Jr. Persistent problems in the pathogenesis of atherosclerosis. *Arteriosclerosis* 1984; 4:443–451.
134. Wissler RW. Progression and regression of advanced atherosclerosis as studied by quantitative methods. In:Born GRV, Catapano AL, Paoletti R, eds. *Factors in the formation and regression of the atherosclerotic plaque.* New York:Plenum Press, 1982:59–78.
135. Constantinides P. Atherosclerosis—a general survey and synthesis. *Surv Synth Path Res* 1984; 3:477–498.
136. Newton RS. Modulation of hepatic and extrahepatic LDL receptors: involvement in the progression of atherosclerosis. *Drug Dev Res* 1985; 6:141–154.
137. Fogelman AM, Haberland ME, Seager J, Hokom M, Edwards PA. Factors regulating the activities of the low density lipoprotein receptor and the scavenger receptor on human monocyte-macrophages. *J Lipid Res* 1981; 22:1131–1141.
138. Voyta JC, Via DP, Butterfield CE, Zetter BR. Identification and isolation of endothelial cells based on their increased uptake of acetylated-low density lipoprotein. *J Cell Biol* 1984; 99:2034–2040.
139. Brown MS, Basu SK, Falck JR, Ho YK, Goldstein JL. The scavenger cell pathway for lipoprotein degradation: specificity of the binding site that mediates the uptake of negatively-charged LDL by macrophages. *J Supramol Struct* 1980; 13:67–81.
140. Henriksen T, Mahoney EM, Steinberg D. Enhanced macrophage degradation of biologically modified low density lipoprotein. *Arteriosclerosis* 1983; 3:149–159.
141. Steinbrecher UP, Witztum JL, Parthasarathy S, Steinberg D. Decrease in reactive amino groups during oxidation or endothelial cell modification of LDL. Correlation with changes in receptor-mediated catabolsim. *Arteriosclerosis* 1987; 7:135–143.
142. Parthasarathy S, Fong LG, Otero D, Steinberg D. Recognition of solubilized apoproteins from delipidated, oxidized low density lipoprotein (LDL) by the acetyl-LDL receptor. *Proc Natl Acad Sci USA* 1987:84:537–540.
143. Steinbrecher UP. Oxidation of human low density lipoprotein results in derivatization of lysine residues of apoprotein B by lipid peroxide decomposition products. *J Biol Chem* 1987; 8:3603–3608.
144. Mazzone T, Lopez C, Bergstraesser L. Modification of very low density lipoproteins leads to macrophage scavenger receptor uptake and cholesteryl ester deposition. *Arteriosclerosis* 1987; 7:191–196.
145. Baker DP, Van Lenten BJ, Fogelman AM, Edwards PA, Kean C, Berliner JA. LDL, scavenger, and β-VLDL receptors on aortic endothelial cell. *Arteriosclerosis* 1984; 4:248–255.

146. Pitas RE, Innerarity TL, Mahley RW. Foam cells in explants of atherosclerotic rabbit aortas have receptors for ß-very low density lipoproteins and modified low density lipoproteins. *Arteriosclerosis* 1983; 3:2–12.

147. Koo C, Wernette-Hammond ME, Innerarity TL. Uptake of canine ß-very low density lipoproteins by mouse peritoneal macrophages is mediated by a low density lipoprotein receptor. *J Biol Chem* 1986; 261:11194–11201.

148. Ellsworth JL, Kraemer FB, Cooper AD. Transport of ß-very low density lipoproteins and chylomicron remnants by macrophages is mediated by the low density lipoprotein receptor pathway. *J Biol Chem* 1987; 262:2316–2325.

149. Camejo G, Ponce E, Lopez F, Starosta R, Hurt E, Romano M. Partial structure of the active moiety of a lipoprotein complexing proteoglycan from human aorta. *Atherosclerosis* 1983; 49:241–254.

150. Srinivasan SR, Vijayagopal P, Dalferes ER Jr, Abbate B, Radhakrishnamurthy B, Berenson GS. Dynamics of lipoprotein–glycosaminoglycan interactions in the atherosclerotic rabbit aorta *in vivo. Biochim Biophys Acta* 1984; 793:157–168.

151. Brown BG, Bolson EL, Dodge HT. Arteriographic assessment of coronary atherosclerosis. Review of current methods, their limitations, and clinical applications. *Arteriosclerosis* 1982; 2:2–15.

152. Lees RS, Myers GS. Noninvasive diagnosis of arterial disease. *Adv Int Med* 1982; 27:475–509.

153. Blankenhorn DH, Rooney JA, Curry PJ. Noninvasive assessment of atherosclerosis. *Prog Cardiovasc Dis* 1983; 26:295–307.

154. Herfkens RJ, Higgins CB, Hricak H, Lipton MJ, Croods LE, Sheldon PE, Kaufman L. Nuclear magnetic resonance imaging of atherosclerotic disease. *Radiology* 1983; 148:161–166.

155. Limpert JD, Vogelzang RL, Yao JST. Computed tomography of aortoiliac atherosclerosis. *J Vasc Surg* 1987; 5:814–819.

156. Sinzinger H, Fitscha P. Scintigraphic detection of femoral artery atherosclerosis with 111-indium-labeled autologous platelets. *VASA* 1984; 13:350–353.

157. Roberts AB, Lees AM, Lees RS, Strauss HW, Fallon JT, Taveras J, Kopiwoda S. Selective accumulation of low density lipoproteins in damaged arterial wall. *J Lipid Res* 1983; 24:1160–1167.

158. Poledne R, Reinis Z, Lojda Z, Hanus K, Cihova Z. The inflow rate of low density lipoprotein cholesterol to the arterial wall in experimental atherosclerosis. *Physiol Bohemoslov (ABBR)* 1986; 35:313–318.

159. Rudel LL, Parks JS, Johnson FL, Babiak J. Low density lipoproteins in atherosclerosis. *J Lipid Res* 1986; 27:465–474.

160. Lees RS, Lees AM, Strauss HW. External imaging of human atherosclerosis. *J Nucl Med* 1983; 24:154–156.

161. Vallabhajosula S, Goldsmith SJ, Ginsberg HN, Badimon JJ, Le N-A, Paidi M, Fuster V. Tc-99m labeled lipoproteins: new agents for studying lipoprotein metabolism and non-invasive imaging of atherosclerotic lesions. *Nuklearmedizin(suppl)*1986; 22:606–608.

162. Sinzinger H, Bergmann H, Kaliman J, Angelberger P. Imaging of human atherosclerotic lesions using ^{123}I-low-density lipoprotein. *Eur J Nucl Med* 1986; 12:291–292.

163. DeForge LE, Schwendner SW, DeGalan MR, McConnell DS, Counsell RE. Noninvasive assessment of lipid disposition in treated and untreated atherosclerotic rabbits. *Pharm Res.* 1989; 6:1011–1016.

164. Moerlein SM, Dalal KB, Ebbe SN, Yano Y, Budinger TF. Residualizing and non-residualizing analogues of low-density lipoprotein as iodine-123 radiopharmaceuticals for imaging LDL catabolism. *Nuc Med Biol* 1988; 15:141–149.

165. Vallabhajosula S, Paidi M, Badimon JJ, Le N-A, Goldsmith SJ, Fuster V, Ginsberg HN. Radiotracers for low density lipoprotein biodistribution studies in vivo: technetium-99m low density lipoprotein versus radioiodinated low density lipoprotein preparations. *J Nucl Med* 1988; 29:1237–1245.

166. Lees AM, Lees RS, Schoen FJ, Isaacsohn JL, Fischman AJ, McKusick KA, Strauss HW. Imaging human atherosclerosis with 99mTc-labeled low density lipoproteins. *Arteriosclerosis* 1988; 8:461–470.

9

In Vivo Application of Lipoproteins as Drug Carriers: Pharmacological Evaluation of Sterylglucoside–Lipoprotein Complexes

Makoto Sugiyama and Junzo Seki
Nippon Shinyaku Co., Ltd., Kyoto, Japan

I. INTRODUCTION

Most biopharmacy and pharmacology texts note the importance of plasma proteins, such as albumin and γ_1-acid glycoprotein, in the ultimate disposition of drug molecules. Recent knowledge of the physiology of plasma lipoproteins indicates that they are an important factor not only in drug disposition but also in the pharmacological activity of a drug [1]. This chapter summarizes the in vivo potency of lipoproteins as carriers of pharmacological agents and as possible vehicles in drug delivery.

Plasma lipoproteins are well known as transport vehicles of water-insoluble lipids, such as cholesterol and triglycerides, in the blood circulation. One might well expect lipophilic drugs to combine with plasma lipoproteins. If a lipophilic compound joins with plasma lipoproteins and if the complex enters cells by an endocytotic process, the amount of the compound entering cells may vary widely in different tissues because each lipoprotein has a different metabolic fate [2,3], as discussed in detail in Chapter 1. Numerous lipophilic compounds, as well as cholesterol and triglycerides, combine with lipoproteins. Among these compounds are certain vitamins, hormones, toxins, pesticides, and therapeutic agents [4–9].

Many lipophilic substances when given orally are absorbed via the intestinal lymphatics in association with chylomicrons. Chylomicrons are primary plasma lipoproteins for the transport of dietary or exogenous lipophilic compounds in the circulatory system. For example, vitamin A is esterified in the mucosal cells during absorption. The retinyl esters associate with chylomicrons and are taken up by the liver with chylomicron remnants, which are metabolites of chylomicrons [10-12]. Furthermore, very low density lipoprotein (VLDL) and low density lipoprotein (LDL), which are responsible for the transport and delivery of endogenous lipids from the liver to extrahepatic tissues, also transport highly lipophilic compounds. For example, coenzyme Q_{10} is a highly lipophilic compound used in the treatment of congestive heart failure and angina pectoris. Intravenously injected coenzyme Q_{10} is initially taken up by the liver and subsequently transported from the liver to extrahepatic tissues through the medium of lipoproteins secreted by the liver [13]. Both VLDL and LDL are responsible for the transport of coenzyme Q_{10} to extrahepatic tissues.

Probucol, a drug widely used in the treatment of hypercholesterolemia, also combines with plasma lipoproteins following oral administration [14]. It has been shown that probucol when combined with plasma LDL alters structure and metabolism of LDL [15] and the rate and mode of interaction of lipids and apoproteins in surface-component transfer during lipolysis [16]. Probucol also prevents oxidative changes of LDL [17]. Cyclosporin A, a cyclic endecapeptide and immunosuppressant drug, combines with lipoproteins at more than two-thirds of the plasma-bound drug [18]. All lipoproteins have a nonsaturable, low affinity, and high capacity uptake for cyclosporin A [19]. High density lipoprotein (HDL) reportedly shows the highest affinity for cyclosporin A, followed by LDL then VLDL with the lowest affinity [20]. The maintenance of cyclosporin A blood levels within the optimal therapeutic range is complicated by large individual differences, which may reflect variations in lipoprotein patterns. Lipoprotein-bound cyclosporin A also interacts with lymphocytes.

Plasma lipoproteins show characteristic metabolic fates in vivo (refer to Chapter 1), so they are considered to be possible carriers in a site-specific drug delivery system, which may well have great therapeutic possibilities. Pioneering studies [21-27] have investigated in vitro the possibility of LDL as a carrier of pharmacological agents, such as cytotoxic agents, but surprisingly few studies have reported the carrier potential of lipoproteins in vivo. Recently we showed pharmacologically, for the first time, the carrier potential of lipoproteins in vivo [1]. The pharmacological effects of a sterylglucoside are closely related to lipoprotein metabolism. We confirmed that the pharmacological effects are mediated only by the metabolic processes of lower density lipoproteins; that is, the hemostatic effect in mice and the inhibitory effect on vascular permeability in rats were observed only after the intravenous administration of complexes of the sterylglucoside with lower density lipoproteins, but not with HDL. Human lipoproteins also showed the same carrier potentials. These results were supported by studies of the disposition of the complexes. This chapter discusses the problems concerned with the in vivo application of lipoproteins in drug delivery.

II. IN VIVO APPLICATION OF LIPOPROTEINS AS DRUG CARRIERS

This section briefly reviews the possible utility of plasma lipoproteins for the site-specific delivery of diagnostic and pharmacological agents.   Several researchers have employed a variety of strategies to confirm the carrier potential of lipoproteins in vivo.

As described in earlier chapters, most tissues receive their cholesterol from these circulating lipoprotein macromolecules via a specific receptor-mediated process, especially tissues responsible for the biosynthesis of steroid hormones deriving their cholesterol mainly from the plasma pool. In humans LDL is the major cholesterol carrier, whereas in rats and dogs HDL serves this function. Also LDL plays a key role in atherogenesis.

As described by Peterson et al. in Chapter 5, replicating cells, such as neoplastic cells, continuously require large amounts of cholesterol for the synthesis of cell membranes. Many types of cancer cells metabolize LDL at higher rates than nonneoplastic cells, and patients with leukemia commonly have hypocholesterolemia. Leukemic blood and bone marrow cells from most patients with acute myeloblastic leukemia have higher LDL receptor activities than do white blood cells and bone marrow cells from healthy subjects [28,29]. The LDL receptor activity shows an inverse correlation with plasma cholesterol concentration [30]. Consequently, hypocholesterolemia in leukemia and other neoplastic disorders may be due to increased LDL receptor activity followed by high uptake of plasma cholesterol by the malignant cells. Moreover, the content of LDL receptors in tumor cells from patients with primary breast cancers showed an inverse correlation with their survival time [31].

A. Site-Specific Delivery of Lipoproteins

The concept of site-specific in vivo delivery by lipoproteins has been particularly rewarding in the field of organ imaging, where only tracer doses of a radiodiagnostic are required to obtain the desired information.

Counsell and colleagues ([32]; Chapter 8) describe the use of plasma lipoproteins for site-specific delivery of diagnostic agents in a series of studies on potential organ- or tumor-imaging agents. For example, iopanoic acid, an established cholecystographic agent, when esterified with cholesterol can be incorporated into chylomicron remnants [33].   In whole animal tissue distribution studies, the intravenous injection of cholesteryl iopanoate labeled with iodine-125 in physiological saline resulted in the appearance of about 31%

of the dose in the liver at 0.5 h, but prior incorporation of [^{125}I]cholesteryl iopanoate into chylomicrons resulted in an almost threefold (87%) increase in the liver accumulation of radioactivity in the same time period. Similarly, the incorporation of 19-radioiodinated cholesteryl oleate into rat HDL, a major cholesterol carrier in rats, produced a fivefold increase in adrenal uptake and an almost fourfold enhancement of ovarian accumulation of radioactivity at 0.5 h following intravenous injection ([34]; Chapter 8).

Adrenal uptake is further supported by the work of Isaacsohn et al. [35] in which the delivery was assessed using ^{99m}Tc-labeled LDL injected into rabbits to obtain external images of the adrenal glands and to evaluate adrenal cortical function. Radioactivity in the adrenal glands after dexamethasone suppression was significantly less than in the control rabbits. The extent of the decrease in radioactivity was a function of the duration of dexamethasone treatment.

Finally, Lees and colleagues [36,37] observed that the focal uptake by atherosclerotic lesions of [^{125}I]LDL and [^{99m}Tc]LDL can be demonstrated by external imaging. Radiolabeled LDL was localized under the regrowing endothelial edge of the rabbit aorta after balloon deendothelialization, and carotid atherosclerosis could be imaged externally with the gamma scintillation camera. Therefore, LDL is considered to be a useful diagnostic tool for detecting atherosclerotic disease because early lesions of atherosclerosis encroach on the vessel wall, rather than the lumen, and are thus difficult to detect by angiography. Therefore, as a preliminary, lipoproteins can be considered possible drug carriers to tumor cells and tissues which have significant receptor activity such as the liver, adrenals, ovaries, and perhaps atherosclerotic lesions.

The majority of studies using lipoproteins (mainly LDL) as drug carriers have been carried out as in vitro experiments with tumor cells [21–27]. We too have demonstrated the accumulation of lipoproteins by tumor cells in vitro as seen in Fig. 1. Human LDL, HDL, and lipid nanospheres (lipid microemulsions) were examined. The lipid nanospheres consisted of egg yolk phosphatidylcholine, triolein, and [^{3}H]cholesteryl linoleate (5:4:1, w/w/w) in 0.15 M NaCl, 0.1 mM EDTA (pH 7.5), and were prepared by treatment with ultrasound for 60 min at 4°C. The particle size was analyzed by a photon correlation spectrometer; the mean diameter was 19.1 nm, as seen in Fig. 2. The lipid nanosphere was considered to be an excellent model of LDL without apo B, a protein-free analog of LDL, because it has a diameter similar to that of native LDL. Both LDL and HDL were labeled with [^{3}H]cholesteryl linoleate by the procedure of Roberts et al. [39]. P388 leukemic cell, Ehrlich ascites carcinoma, S-180 sarcoma, and Meth-A fibrosarcoma were preincubated for 48 h with RPMI 1640 containing 10% human lipoprotein-deficient serum ($d>1.21$ fraction) in a CO_2 incubator, then lipoproteins or lipid nanospheres were added. After incubation for 24 h, cells were collected by centrifugation and washed three times with buffered saline. The radioactivity was counted to de-

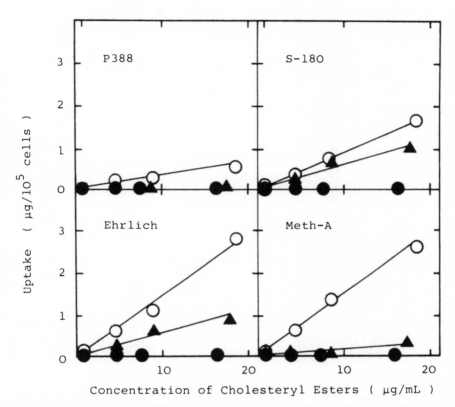

Figure 1 Uptake by cells of lipoproteins and lipid nanospheres. Cells were incubated with RPMI 1640 containing 10% human LPDS for 48 h followed by the addition of lipoproteins or microemulsion containing [³H]cholesteryl linoleate to the culture medium. Cells used in this experiment are indicated. Vehicles are human LDL (○), HDL (▲), and the lipid nanospheres (●) consisting egg yolk phosphatidylcholine, triolein, and cholesteryl linoleate (5:4:1, w/w/w). Each point shows the mean of three experiments.

termine the incorporation of [³H]cholesteryl linoleate into the cells. All of the cells examined showed much more LDL than HDL accumulation, as seen in Fig. 1. Surprisingly, these tumor cells did not take in lipid nanospheres [38]. Most in vitro studies, including our own, have provided strong evidence that rapidly dividing tumor cells are rich in LDL receptors [21–31].

Uptake of LDL by soft tissue tumors illustrates the in vivo assimilation of lipoproteins [40]. Hynds et al. [40] studied control and MAC 13 tumor-bearing NMRI mice injected with tracer doses of ¹²⁵I-labeled native and cyclohexanedione-modified ¹³¹I-labeled LDL. The radioactivities assimilated by various tissues

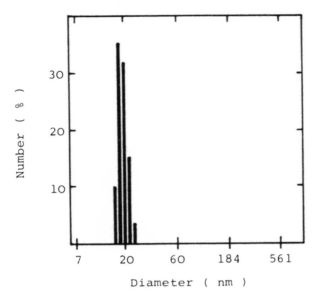

Figure 2 Size distribution of the lipid nanospheres, a protein-free analog of LDL, determined by a photon correlation spectrophotometer.

were measured and related with plasma activity at 18 h postinjection. All tissues expressed LDL receptors, but in tumor-inoculated animals neoplastic lesions were second only to liver in their net assimilation of LDL. Pretreatment of the mice with taurocholate and hydrocortisone reduced LDL receptor activities in the liver and the adrenal gland, respectively, without concomitant change in the activity of the tumor. Norata et al. [41] also showed assimilation of [125I]LDL by MS-2 fibrosarcoma in BALB/c mice. The in vivo data clearly indicated that the MS-2 tumor was, among the tissues tested, the most active in assimilating LDL by both pathways in the receptor- and non–receptor-mediated catabolism of LDL. These findings show the usefulness of lipoproteins as site-specific delivery vehicles in cancer diagnosis.

B. Pharmacological Evaluation of Sterylglucoside–Lipoprotein Complexes

Although it would be very interesting to use lipoproteins in vivo as carriers of pharmacological agents, the in vivo usefulness of lipoproteins has not been demonstrated. The pharmacological evaluation of a drug carrier is most important in the development of a drug delivery system.

Recently we showed for the first time the in vivo effectiveness of lipoproteins as carriers of the pharmacologically active agent ß-sitosteryl-ß-D-glucopyranoside, not only by studies on drug disposition but also by pharmacological tests [1].   We reported that the pharmacological effects of the sterylglucoside were closely related to lipoprotein metabolism and were observed only after the intravenous administration of complexes of the sterylglucoside with lower density lipoproteins.   This section summarizes our work with sterylglucoside–lipoprotein interactions and discusses the therapeutic possibilities of lipoproteins as carriers of pharmacological agents.

Sterylglucosides

ß-Sitosteryl-ß-D-glucopyranoside (SG), an abundant plant sterylglucoside, has some pharmacological activities.   For example, SG inhibits pulmonary and cutaneous vascular fragility in mice, rats, and guinea pigs; bleeding by cutting the tail end in mice; hemorrhage induced by snake venom in rats; increased capillary permeability by histamine, hyaluronidase, and lecithinase A in rats; and increased cerebral vascular permeability or edema induced by the bilateral carotid occlusion in SHR.   It has no influence on prothrombin time, plasma recalcification time, and euglobulin lysis time [42].   Further, SG prevents ulcer formation in mice under restrained and water immersion conditions [43].   It has been reported that a mixture containing SG as the major constituent has antitumor activity against P388 leukemia (T/C = 130% at 15 µg/kg) [44,45].

In vivo Association with Lipoproteins: Pharmacological Activity

Water-insoluble lipids, such as cholesterol and triglycerides, are absorbed from the intestine by forming chylomicrons. Solubility of SG in water was estimated to be less than 10 ng/mL.   Because SG is also considered to be a water-insoluble compound, the lymphatic absorption of SG was evaluated after the oral administration of [22,23-³H]ß-sitosteryl-ß-D-glucopyranoside ([³H]SG) in thoracic duct cannulated rats. As noted in Table 1, a limited extent of the orally administered [³H]SG was recovered in the lymph.   The recoveries increased when 1 mL of condensed milk was fed orally to facilitate chylomicron formation 30 min before the administration of [³H]SG.   This lymphatic absorption was almost the same as the absorption efficiency (2.1%) of SG, which was estimated by comparing the urinary excretion ratios of the radioactivity after the oral and intravenous administrations of [³H]SG to intact animals (Table 2). The lymph contained metabolites of SG, as seen in Table 1.   The major metabolite in the lymph was identified as ß-sitosterol.

Figure 3 shows the primary metabolic pathway of SG in mice, rats, rabbits, and dogs.   There were also ß-sitosteryl esters in the plasma after the administration of SG.   The recoveries of total radioactivity and [³H]SG were

Table 1 Lymphatic Absorption of [³H]SG After Oral
Administration[a]

	Percentage of dose	
Condensed milk	[³H]Radioactivity	[³H]SG
With[b]	2.78 ± 0.05[c]	1.29 ± 0.05
Without	1.59 ± 0.29	0.61 ± 0.11

[a] Lymph was collected for 48 h following oral administration of [³H]SG (5 mg/kg) in thoracic duct–cannulated rats.
[b] Condensed milk (1 mL) was administered orally 30 min before the administration of [³H]SG.
[c] Each value is expressed as cumulative absorption (mean ± SEM, N = 3).

83% and 76% of the dose, respectively, and 92% of the radioactivity in the feces was recovered as the parent compound (SG). Therefore, β-sitosterol in the lymph resulted from a metabolic conversion, hydrolysis of SG, in the absorption process in the intestinal cells, not in the intestinal lumen. In addition, it was found that most of the [³H]SG in the lymph was associated with chylomicrons when assessed by the centrifugal flotation technique. Therefore, it was concluded that SG is absorbed to a limited extent from the gastrointestinal tract and that SG associated with chylomicrons in the lymph enters into the circulatory system.

Because SG cannot be present in plasma as a free form due to its water insolubility, it was considered to be associated with a constituent of the plasma. Therefore, gel filtration of the plasma containing [³H]SG was carried out to identify the binding constituents for SG in the plasma. Figure 4 illustrates the elution pattern of plasma incubated with [³H]SG for 60 min in vitro. More than 88% of the applied [³H]SG was recovered in the eluate. [³H]SG was

Table 2 Excretion of Radioactivity Following Oral and
Intravenous Administration of [³H]SG to Rats

		Recovery (% of dose)[a]	
Route	Dose (mg/kg)	Urine	Feces
Intravenous	0.1	14.1 ± 1.15	249.92 ± 1.15
Oral	5.0	0.316	85.08

[a] Urine and feces were collected for 12 days after the administration of [³H]SG.

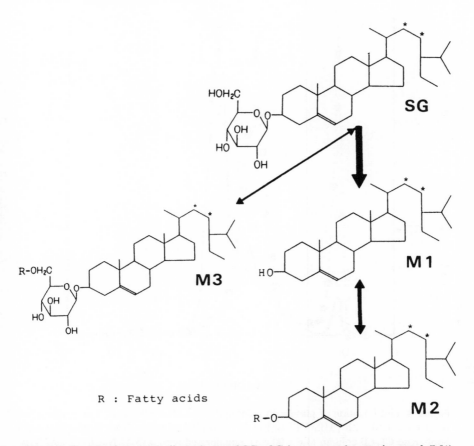

Figure 3 Prospective metabolic pathway of SG. SG is converted to ß-sitosterol (M1) by hydrolysis and to SG esters (M3) by acylation. ß-Sitosterol also receives acylation to form ß-sitosterol esters (M2). These acylated metabolites, M2 and M3, are hydrolized to M1 and SG, respectively. Glycosilation of ß-sitosterol (M1) was not found in the experimental animals; i.e., forming M1 from SG in an irreversible step. M1 and M2 were major metabolites, but M3 was minor. R, Fatty acids.

found in the VLDL/LDL fraction (26%) and the HDL fraction (68%), whereas a small amount eluted in the albumin fraction. In addition, more than 95% of SG in plasma samples was recovered in the plasma lipoproteins when assessed by ultracentrifugal flotation. These results indicate that SG associates with the lipoproteins in plasma.

Figure 5 shows the relationship between the time course of hemostatic time and the concentration of [³H]SG in serum and its subfractions after oral

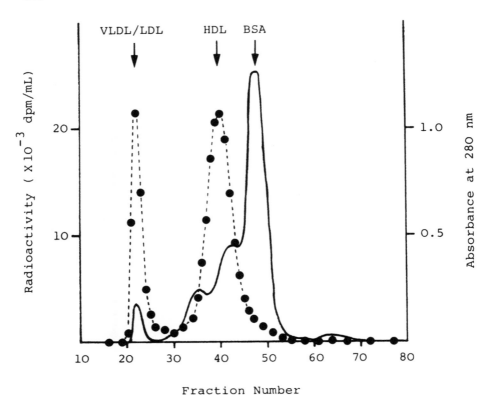

Figure 4 Gel filtration of plasma containing [³H]SG. Rat plasma (0.5 mL) which has been incubated with [³H]SG (5 μg) at 37°C for 60 min was chromatographed on a Sepharose CL-6B column (1.6 x 90 cm), equilibrated with 0.15 M NaCl, 0.2 mM Tris-HCl, pH 7.1. Three-milliliter fractions were collected, and the absorbance at 280 nm (solid curve) was monitored. One-milliliter aliquotes were taken for scintillation counting (●). The arrows indicate the elution peaks of each lipoprotein isolated from rat plasma and bovine serum albumin (BSA) chromatographed under the same conditions.

administration. The hemostatic time was determined in male ddY mice by spectrophotometry [1]. Briefly, their tails were cut about 1 cm from the end and dipped in gently stirred water which was recirculated in a spectrophotometer by a pump. The hemostatic time was defined as the time when the absorbance at 413 nm no longer increased. The heparin–manganese method was used for conventional fractionation of the serum lipoproteins [1]. This method precipitates only lower density lipoproteins. The pharmacological effect of SG did not correspond with the serum concentration of the drug.

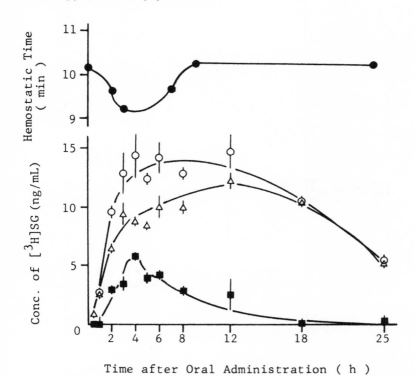

Time after Oral Administration (h)

Figure 5 Time courses of hemostatic time and concentration of [³H]SG in serum and its subfractions after oral administration of [³H]SG. The hemostatic time (•) was determined by the spectrophotometric measurement of hemorrhage from the tail after the oral administration of SG to mice (2 mg/kg, N = 5). Serum was obtained from rats after the oral administration of [³H]SG (5 mg/kg), and 1 ml of the serum was treated with heparin−manganese to fractionate the lipoproteins. Each symbol shows the concentration of [³H]SG in the serum (○), in the S (supernatant) fraction (Δ), and in the P (precipitate) fraction (■) (mean ± SEM, N = 3).

The amount of [³H]SG in the serum and the S (supernatant) fraction also remained high 18 h following oral administration. However, the concentration of [³H]SG in the P (precipitate) fraction decreased rapidly, and this time profile was closely related to the time course of the hemostatic effect, which appeared at 2 h and disappeared 10 h following the oral administration. Therefore, it was considered that the hemostatic effect depends on the concentration of SG only in the P fraction consisting of the lower density lipoproteins (CM, VLDL, IDL, LDL). This observation suggests that SG needs to be associated with the lower density lipoproteins in order to exhibit pharmacological effects.

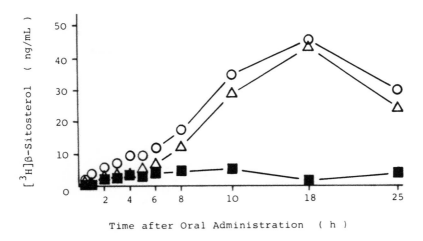

Figure 6 Time courses of concentration of [³H]ß-sitosterol in serum and its subfractions after oral administration of [³H]SG to rats. Symbols are defined in Fig. 5.

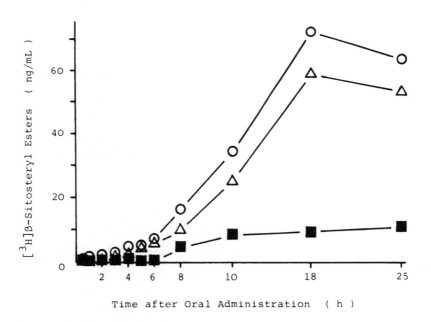

Figure 7 Time courses of concentration of [³H]ß-sitosteryl esters in serum and its subfractions after oral administration of [³H]SG to rats. Symbols are defined in Fig. 5.

There were no active metabolites of SG, because the time course of the concentration of the metabolites did not correspond with that of the pharmacological effect, as seen in Figs. 6 and 7. Pharmacological tests showed that these compounds have no pharmacological activities at the same dose. The metabolites of SG associated with plasma lipoproteins [46] and their disposition have also been described [47–51].

Hemostatic and Vascular Permeability Effects

Pharmacological potency was determined after the intravenous administration of SG complexed with each lipoprotein to confirm the hypothesis that only SG complexed with lower density lipoproteins has a pharmacological potency. Incorporation of [^{3}H]SG into the lipoproteins isolated from rat plasma by preparative ultracentrifugal flotation was carried out on the surface of Celite by a modification of the procedure described by Avigan [52]. The determination of radioactivity in the treated lipoproteins revealed that [^{3}H]SG was incorporated, to some extent, into each lipoprotein (Table 3). The uptake of SG into lipoproteins depended on the incubation temperature and incubation time, as seen in Fig. 8. There were no significant differences between the treated and the nontreated lipoproteins, as demonstrated by disc electrophoresis and gel filtration. In addition, negligible radioactivity was found in the solution without protein, and no metabolite of [^{3}H]SG was formed during these incubations.

The SG–lipoprotein complexes thus obtained were administered intravenously to mice, and the hemostatic time was determined. As seen in Fig. 9, SG in the conventional vehicle, 2% HCO-60 (hydrogenated castor oil polyethylene glycol ether) had a hemostatic effect 2 h following the administration of 25 μg/kg, whereas SG complexed with HDL and the bottom fraction (BF, the final infranatant of the last ultracentrifugation and lipoprotein-deficient serum, $d \geq$ 1.21 g/mL) had no hemostatic effect. Complexed with lower density lipoproteins such as chylomicrons (CMs), VLDL, IDL, and LDL, SG showed the same hemostatic time as with the conventional vehicle, even at a dose of 0.23 μg/kg. There were no significant differences between the effects of the lower density lipoproteins. Specifically, CMs, VLDL, IDL, and LDL used as drug carriers decreased significantly the dose of SG needed to exhibit a hemostatic effect.

It was also found that SG complexed with the lower density lipoproteins showed its maximum effect as early as 30 min after administration. When injected with the conventional vehicle, SG needed at least 90 min to show a hemostatic effect. Thus SG complexed with lower density lipoproteins had a more rapid pharmacological effect than SG injected with a conventional vehicle. These results suggest that SG with lower density lipoproteins is delivered more efficiently and rapidly to their site of action. However, when SG is injected with a conventional vehicle, only part of the SG seems to be

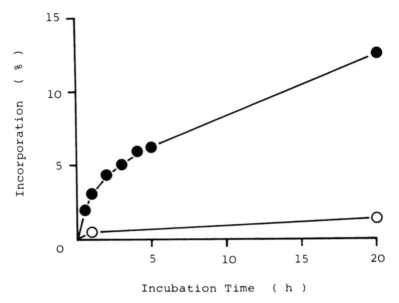

Figure 8 Incorporation of [³H]SG into serum lipoproteins. [³H]SG (500 μg) was immobilized on the surface of Celite (50 mg) and incubated with rat serum (2 mL) at 37°C (●) and 4°C (○).

Table 3 Incorporation of [³H]SG into Rat Lipoproteins in Vitro[a]

Vehicle	[³H]SG (μg/mg of protein)
CM	300.9[b]
VLDL	151.9
IDL	126.1
LDL	116.4
HDL	12.1
BF[c]	0.8

[a] [³H]SG (100 μg) was immobilized on the surface of Celite (50 mg) and then incubated with each lipoprotein solution (2 mL) at 37°C for 20 h.
[b] Each value represents the mean concentration of [³H]SG in the supernatant after triplicate incubations.
[c] The final 9 mL of the last centrifugation is the bottom fraction (BF). $d \geq 1.21$ g/mL.

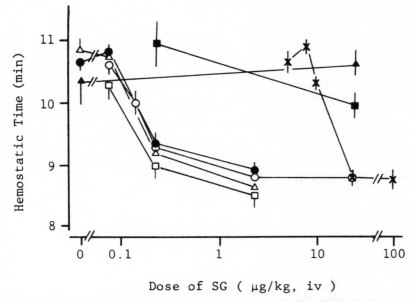

Figure 9 Hemostatic activity of SG after administration of SG lipoprotein complex. The complex prepared in vitro was administered intravenously to mice, and the hemostatic time was determined 1 h following the injection of SG complexed with CM (o), VLDL (□), IDL (Δ), LDL (•), HDL (■) and BF (▲), and determined at 2 h after the injection of SG solubilized with 2% HCO-60 (X) (mean ± SEM, N = 5).

incorporated in lower density lipoproteins in vivo. Most of the SG is probably trapped in HDL and tissues, and its pharmacological activities are masked. The incorporation of SG into lower density lipoproteins occurs in the circulatory system or the liver. The incorporation process may be one reason for the time lag in the appearance of the pharmacological effect of SG with the conventional vehicle.

As shown by inhibition of plasma protein leakage from the capillaries when determined by Evans blue leakage, SG inhibits vascular permeability. Vascular permeability, or the amount of the leakage after the intracutaneous injection of histamine, was inhibited by the preadministration of SG complexed with the lower density lipoproteins ($d < 1.063$ g/mL) as shown in Table 4. However, SG incorporated into the higher density lipoproteins ($d \geq 1.063$ g/mL) failed to affect vascular permeability. These results show that SG complexed with the lower density lipoproteins not only exhibits a hemostatic effect but also decreases vascular permeability.

Table 4 Inhibitory Effect of SG–Lipoprotein Complexes on Vascular Permeability[a]

Vehicle	Dose (μg/kg)	Vascular permeability
Control	–	1.00 ± 0.18[b]
$d < 1.063$[c]	42	0.49 ± 0.15[d]
$d \geq 1.063$	50	0.85 ± 0.16

[a] The SG–lipoprotein complex, prepared in vitro, was administered intravenously to rats. The vascular permeability was evaluated 60 min following injection of SG by the determination of Evans blue leakage.
[b] Each value shows the permeability relative to that of control rats (mean ± SEM, N = 5).
[c] The plasma proteins were fractionated into two classes at density of 1.063 g/mL.
[d] Significantly different from the control (P < 0.05).

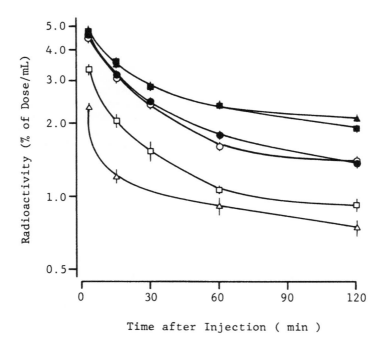

Figure 10 Time course of concentration of radioactivity in blood after intravenous administration of [³H]SG–lipoprotein complex in rats (dose: [³H]SG = 9 μg/kg). Each symbol shows the total radioactivity in the blood after the injection of [³H]SG complexed with CM (o), VLDL (□), IDL (△), LDL (•), HDL (■), and BF (▲) (mean ± SEM, N = 3).

Figure 10 shows the time course of the radioactivity in the blood after the intravenous administration of [³H]SG complexed with various lipoproteins. [³H]SG complexed with VLDL and IDL disappeared rapidly from the circulatory system, but [³H]SG complexed with HDL and BF was retained in the circulatory system in relatively high concentrations.   Therefore, the SG–lower density lipoprotein complexes are rapidly and efficiently carried from the circulatory system into those target cells that are the action sites of SG. These properties are essential for a drug carrier in the site-specific drug delivery system.

Transfer/Interchange of Drug Between Lipoprotein Classes

The plasma half-lives of the SG–lipoprotein complexes did not agree with those of the various lipoproteins (generally, CM, <5 min; VLDL, 1–3 h; LDL, 3–4 days; HDL, 5–6 days). It has been reported that drug-induced alterations in lipoprotein structure might alter the metabolism of the lipoproteins [15,16,53,54]. There may also be an interchange of [³H]SG among the various lipoprotein classes.   Considerable redistribution of [³H]SG among the lipoprotein classes is suggested in Table 5.   After the injection of the [³H]SG–lipoprotein complex, the distribution of the radioactivity in the serum lipoprotein classes in the serum was determined by an Airfuge microultracentrifugation system (Beckman). At 30 min following administration, the radioactivity in the lipoprotein fractions was different from the radioactivity in the injected lipoproteins. When the [³H]SG–CM complex was administered, the radioactivity in the $d < 1.006$ g/mL fraction was only 2.7%. This result is apparently related to the fact that CMs have the fastest clearance rate. Distributions of radioactivity in the $1.006 \leq d < 1.063$ g/mL fraction were higher after the injection of [³H]SG complexed with the lower density lipo-

Table 5   Distribution of Radioactivity in Serum Lipoprotein Fractions After Intravenous Administration of [³H]SG–Lipoprotein Complex[a]

Vehicle	$d < 1.006$ (%)	$1.006 \leq d < 1.063$ (%)	$d \geq 1.063$ (%)
CM	2.7 ± 2.1	68.4 ± 3.4	28.9 ± 1.3
VLDL	6.4 ± 3.4	64.6 ± 2.8	29.0 ± 6.8
IDL	18.3 ± 1.6	39.7 ± 1.0	42.0 ± 0.8
LDL	–	68.1 ± 6.4	33.1 ± 2.8
HDL	6.5 ± 4.1	42.7 ± 3.9	50.8 ± 0.9
BF	–	40.7 ± 1.5	60.2 ± 1.2

[a] The [³H]SG–lipoprotein complex prepared in vitro was administered intravenously to rats. Thirty minutes following the injections, serum was obtained from the rats and ultracentrifuged to determine the distribution of the radioactivity in the lipoprotein classes.
[b] Each value represents the mean ± SEM of three animals.

proteins than after that with HDL or BF.   However, the distribution of
radioactivity in this fraction after the administration of the [³H]SG–IDL
complex was relatively low. This result corresponds with the fact that IDLs are
rapidly eliminated from the circulatory system in rats.

The transfer and/or interchange of SG between lipoproteins may occur as has
been reported for α-tocopherol [54]. We have confirmed the transfer of SG
in vitro by incubating the [³H]SG–CM complex with normal rat serum at 37°C,
and then fractionating the lipoproteins by the heparin–manganese method.
The recovery of [³H]SG in the HDL fraction was increased with the incubation
time, as seen in Fig. 11. Table 6 and Fig. 12 also show the in vitro transfer/
interchange of SG and cholesteryl esters between lipoproteins. A tracer dose
of [³H]SG or [³H]cholesteryl linoleate was incorporated into fractionated human
lipoproteins by the procedure of Roberts et al. [39] and purified by ultracentrifugation at appropriate densities.   Nonlabeled lipoprotein (acceptor) was
incubated with the labeled lipoprotein (donor) in 0.15 M NaCl, 0.1 mM EDTA,
pH 7.5 at 37°C, and then the lipoproteins were fractionated by micro/ultracen-
trifugal flotation (TL-100, Beckman, 100,000 rpm, 2 h) at appropriate density.
The top (40%) was removed after the ultracentrifugation and the infranatant

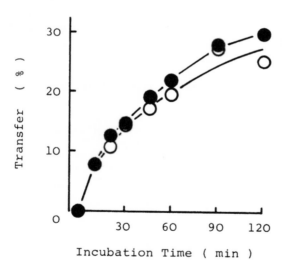

Figure 11 In vitro transfer of [³H]SG from chylomicrons to HDL. [³H]SG–CM
complex was incubated with normal rat serum (○) and posthaperin plasma (●).
Lipoproteins were fractionated by heparin–manganese method after incubation. Each
point represents the mean of triplicate incubations.

Table 6 In Vitro Transfer/Interchange of SG and Cholesteryl Linoleate Between Lipoproteins[a]

		Transfer (%)[b]					
	Donor:	VLDL		LDL		HDL	
Experimental condition	Acceptor:	LDL	HDL	VLDL	HDL	VLDL	LDL
[³H]SG							
37°C		34.3	36.8	2.8	28.1	2.1	35.1
		± 2.5	± 2.5	± 3.8	± 1.3	± 3.5	± 6.6
25°C		13.0	15.2	2.7	14.0	1.7	23.5
		± 0.7	± 1.5	± 4.0	± 0.8	± 1.0	± 1.1
4°C		2.4	4.8	<0.1	6.1	0.3	6.3
		± 1.9	± 1.5		± 0.9	± 8.7	± 2.7
37°C +LPDS[c]		41.0	42.6	2.0	27.6	8.1	32.4
		± 3.8	± 3.2	± 2.6	± 1.8	± 1.0	± 0.9
[³H]Cholesteryl linoleate							
37°C		<0.1	<0.1	16.1	3.1	<0.1	1.3
				± 4.2	± 0.6		± 7.6
4°C		<0.1	<0.1	9.6	0.7	<0.1	2.7
				± 3.3	± 1.0	.	± 7.1

[a] [³H]SG or [³H]cholesteryl linoleate was incorporated into human lipoproteins. The labeled lipoprotein (donor) was incubated with the different type of non-labeled lipoprotein (acceptor) for 60 min in saline (pH 7.5). The final concentration of each lipoprotein corresponds with the plasma concentration. Next the lipoproteins were fractionated by ultracentrifugation and the infranatant was applied for scintillation counting.
[b] Each value shows the mean ± SD (N = 3).
[c] Lipoprotein-deficient serum (50%).

was examined for scintillation counting. [³H]SG was gradually transferred between LDL and HDL in conditions without lipoprotein-deficient serum (LPDS). Lipoprotein-deficient serum contains some lipid transfer proteins that promote the exchange or net transfer of cholesteryl esters, triglycerides, and phospholipids among plasma lipoproteins [55–59]. The transfer/interchange of SG and cholesteryl linoleate depended on the incubation temperature and was suppressed at 4°C. Very low density lipoproteins did not receive [³H]SG from the other lipoproteins. On the other hand, the transfer/interchange of [³H]cholesteryl linoleate was extremely low in the lipoproteins under these conditions. From these results, it is clear that SG moves readily among lipoproteins in plasma.

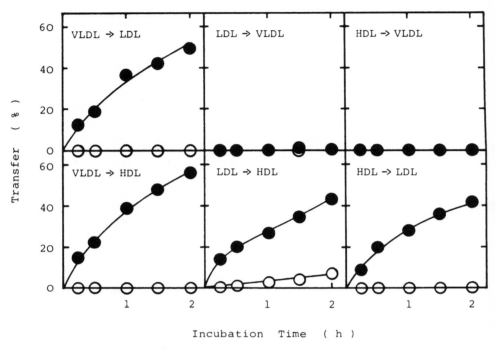

Figure 12 In vitro transfer/interchange of SG (●) and cholesteryl linoleate (○) between lipoproteins. Experimental conditions were the same as for Table 6.

Tissue Distribution and Interspecies Studies

In tissue distribution studies, the adrenal gland uptake of [³H]SG complexed with the various lipoproteins after injection differed markedly, as shown in Table 7. The lowest and highest uptake by the adrenal gland were observed with IDL and HDL, respectively. The amount of uptake by the liver was less than that by the adrenal gland. It is well established that rat HDL is internalized into the adrenal gland via a specific receptor-mediated process in a manner similar to that observed for LDL in the human fetal adrenal glands [60]. It was concluded that these results were caused by the nature of plasma lipoproteins: each lipoprotein has a different metabolic fate, as described in Chapter 1.

Each of the lower density lipoproteins is a precursor or a product of lipoprotein metabolism. As lipolysis occurs, VLDL decreases, leading to the

Table 7 Concentration of [³H]SG in the Liver and the Adrenal Glands After Intravenous Administration of [³H]SG Lipoprotein Complex[a]

Vehicle	Liver (% of dose/g)		Adrenal gland (% of dose/g)	
	[³H]Radioactivity	[³H]SG	[³H]Radioactivity	[³H]SG
CM	1.26 ± 0.51[b]	0.88 ± 0.25	0.81 ± 0.08	0.61 ± 0.06
VLDL	1.83 ± 0.06	1.31 ± 0.58	0.64 ± 0.03	0.41 ± 0.01
IDL	1.32 ± 0.03	0.91 ± 0.02	0.20 ± 0.02	0.13 ± 0.01
LDL	1.31 ± 0.03	0.93 ± 0.03	0.64 ± 0.03	0.49 ± 0.02
HDL	1.36 ± 0.05	0.92 ± 0.03	4.03 ± 0.42	3.25 ± 0.38
BF	1.25 ± 0.02	0.97 ± 0.01	1.26 ± 0.12	1.64 ± 0.10

[a] The [³H]SG–lipoprotein complex prepared in vitro was administered intravenously to rats. Thirty minutes after the injection, the liver and the adrenal glands were removed and homogenized to determine the concentration of total radioactivity and [³H]SG.
[b] Each value represents the mean ± SEM of three animals.

formation of IDL, which serves as the precursor of LDL, the major carrier of cholesterol to extrahepatic tissues in humans. The metabolic alteration of lipoproteins occurs rapidly in vivo. It is still unclear whether all of the lower density lipoproteins have a direct carrier potential to elicit the pharmacological effects of SG. The major finding in this study is that the pharmacological effects of SG are closely related to lipoprotein metabolism. It was confirmed that the pharmacological effects of SG are mediated only by the disposition process of the lower density lipoproteins. This will be useful in pharmacological studies of the action site of SG. For example, since only a small amount of the SG–lower density lipoprotein complex is taken up by the adrenal gland, the adrenal gland is not considered to be the site of action for SG. In fact, SG had a hemostatic effect even in adrenalectomized mice. The site of action is discussed later.

Lipoproteins obtained from healthy humans and mice were used as drug carriers to examine the species differences among the lipoproteins. Plasma from various species, including humans, contained a small amount of endogenous SG, and it was associated specifically with the lipoproteins in the plasma. The plasma used in this experiment, obtained from mice that had been fed a diet with 0.015% SG for one week, and its concentration was 507 ng/mL of SG. The human plasma level of SG was 61.4 ng/mL.

Figure 13 shows the hemostatic effect of SG which was associated endogenously with these plasma lipoproteins. The hemostatic effect was observed following the administration of SG associated not only with mouse

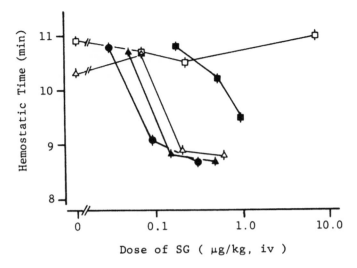

Figure 13 Hemostatic activity following intravenous administration of SG complexed with mouse and human lipoproteins. The complexes were obtained from mouse and human plasma which contained endogenous SG. Each symbol shows the hemostatic time that was determined 1 h following the injection of SG complexed with $d < 1.063$ g/mL (△) and $d \geq 1.063$ g/mL (□) fractions of mouse plasma and VLDL (●), LDL (▲), and HDL (■) fractions in human plasma (mean ± SEM, N = 5).

lower density lipoproteins ($d < 1.063$ g/mL) but also with human lower density lipoproteins (VLDL and LDL). Although SG associated with human HDL had some hemostatic potency, it may have been due to contamination of the HDL fraction with LDL, because human plasma contains more LDL than HDL [61]. It was concluded that any lower density lipoprotein had carrier potential regardless of its origin. No evident difference in the carrier potential was noticed between homologous and heterologous lipoproteins. Human lipoproteins are metabolized in rats in a manner analogous to the metabolism of the corresponding rat lipoproteins [62]. The incorporation procedures of SG into lipoproteins, both in vitro and in vivo, also had no relation to the efficacy. This may be the result of SG partitioning into the same thermodynamically stable region of the lipoproteins. The glucose residue of SG may reside in the surface of lipoproteins, and sitosteryl group may play a role as an anchor in the association of SG with lipoproteins. This concept is supported by the studies of SG associated with liposomes. Egg yolk phosphatidylcholine liposomes were prepared with various amounts of SG in 0.3 M glucose solution. Liposomes containing SG (PC/SG = 7:2) showed a greater uptake of glucose than cholesterol-loaded liposomes (PC/Ch = 7:2), as shown in Table 8. This result shows that an aqueous layer or the inner space of the

Table 8 Influence of Lipid Composition on Encapsulation Efficiency of Glucose into Liposomes

Lipid composition[a]	Encapsulation efficiency (%)
PC	1.03[b]
PC + SG (7:1)[c]	1.17
PC + SG (7:2)	1.50
PC + cholesterol (7:2)	1.31

[a] The liposomes were prepared in 0.3 M glucose solution using a probe-type sonicator. Each 1 mL of liposome suspension contained 15 mg of egg yolk phosphatidylcholine. The encapsulation efficiency was determined by an enzymatic determination of glucose after gel filtration of the liposomes.
[b] Each value represents the mean of three experiments.
[c] Molar ratio.

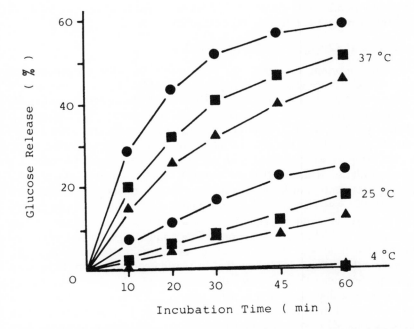

Figure 14 Effect of SG on glucose release from egg yolk phosphatidylcholine liposomes. (●), PC liposome; (■), PC/SG = 7:3 (mol/mol); (▲), PC/cholesterol = 7:2 (mol/mol). Liposomes were incuabted at 37, 25, and 4°C as indicated.

Table 9 Plasma Concentration of Endogenous SG

	Plasma concentration (ng/mL)
Mouse (ddY, Male)	194.6 ± 21.5 (N = 4)
Rat (Wistar, male)	138.0 ± 37.7 (N = 4)
Rat[a] (Wistar, male)	8.9 ± 1.2 (N = 4)
Rabbit (JW-NIBS, male)	12.5 ± 5.5 (N = 18)
Dog (beagle, male)	103.8 ± 32.3 (N = 10)
Human (Japanese, male)	64.3 ± 21.8 (N = 15)

[a] The rats were fed on an SG-free diet for a week.
Each value represents the mean ± SD.

liposomes was expanded by the addition of SG. Since SG is located in surface domains of lipoproteins, it can easily move between lipoprotein particles, as discussed elsewhere, and may influence various lipoprotein properties, such as receptor recognition. However, it has not yet been confirmed that SG changes the properties of lipoproteins. As seen in Fig. 14, SG also inhibited glucose release from liposomes. The data show that SG has a stabilizing effect on the liposome membranes. However, it does not seem to affect the nature of lipoproteins, having less effect on them than cholesterol.

Exogenous and Endogenous Sterylglucosides

It was also important to clarify the relationship between exogenous and endogenous SG. There was more endogenous SG within the circulatory system than the amount of SG administered. Table 9 shows the concentration of SG in the plasma of various normal animals and healthy humans. The endogenous SG obviously comes from the diet because many foodstuffs and animal feeds contain SG (Table 10). The amount of SG in these samples became larger when the samples were treated with methanolic potassium hydroxide for alkaline hydrolysis. It means that the samples contains not only SG but also SG esters (esterification of hydroxy groups of glucose residue). The esterification (SG esters) was also confirmed as a minor metabolite of SG in experimental animals, as seen in Fig. 3. Foods derived from plants contain sterylglucosides in higher concentrations. Cow's milk, eggs, corned beef, and salmon contain little SG. Feeds for experimental animals also include SG. Pharmacokinetic studies of SG show that its plasma concentrations in humans and animals fed normal diets every day correspond with the same way as during multiple dosing. In addition, there are negligible amounts of SG in the plasma of rats and mice fed an SG-free diet for a week. These results show that endogenous SG obviously comes from the diet and that ß-sitosterol does not undergo glucosylation in vivo. The distribution of endogenous SG among the lipoproteins in the plasma is shown in Table 11. There are species differ-

Table 10 Concentration of SG in Animal Feeds and Foodstuffs

	Concentration of SG (μg/g)
Animal feeds	
Rat (F-2)	107.3
Rat (CA-1)	114.3 (163.8)[a]
Dog (CD-1)	69.4
Rabbit (NRT-1S)	66.0
Foodstuffs	
Rice	30.1 (105.0)
Wheat flour	63.7 (107.9)
Soybean flour	214.9 (427.7)
Soybean paste	154.8 (173.9)
Soybean juice	26.8 (36.3)
Sesame	470.7
Peanut	163.1 (289.1)
Cabbage	34.5
Radish	16.6 (29.4)
Onion	35.7 (49.4)
Potato	8.9 (48.0)
Cow's milk	0.8 (0.1)
Hen's egg	− (0.8)
Corned beef	12.0 (21.9)
Salmon	− (0.1)

[a] The values in parentheses show SG concentration after treatment of the samples with methanolic KOH at 65°C for 30 min.

Table 11 Distribution of Endogenous SG Among Lipoproteins in Plasma

Species	P fraction[a] (%)	S fraction[a] (%)	VLDL[b] (%)	LDL[b] (%)	HDL[b] (%)
Mouse	34.6	65.4	−	−	67.5
Rat	30.7	69.0	−	−	−
Dog	20.4	87.9	4.2	19.5	67.5
Human	83.5	21.0	10.0	58.4	22.2

[a] Lipoproteins in the plasma were fractionated by the heparin–manganese method.
[b] Lipoproteins in the plasma were isolated by preparative ultracentrifugal flotation. Each value shows the percentage of SG in the subfraction against the plasma concentration of SG.

rences, and the distribution pattern of SG among the plasma lipoproteins is similar to that of cholesterol.

Although the maximal plasma concentration of SG after oral administration was less than 10% of the endogenous concentration, some pharmacological effects were noted. Why are pharmacological effects observed under these conditions? The results presented in Fig. 15 suggest that endogenous SG has no pharmacological activity. Hemostatic test results returned to their original control values 3 h after the administration of SG with chylomicrons and thereafter maintained that level. When SG was administered again with chylomicrons 3 h after the first injection, the hemostatic effect was observed again. When SG was readministered 1 h after the first injection, the hemostatic time did not decrease further, but the effect was prolonged. Thus the maximum effect was obtained by the first injection of SG, and the additional injection had no effect on the hemostatic time but did prolong the effect. These results suggest that the endogenous SG in the plasma before the administration of exogenous SG does not have any pharmacological effect.

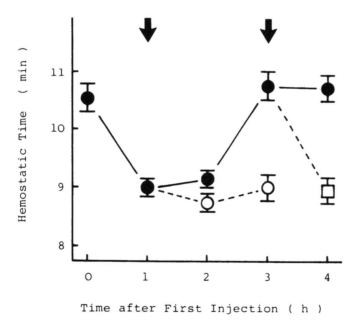

Figure 15 Effect of second injection of SG on hemostatic time. SG complexed with chylomicrons was injected in mice followed by the same dose injected again 1 h (o) or 3 h (□) after the first injection.

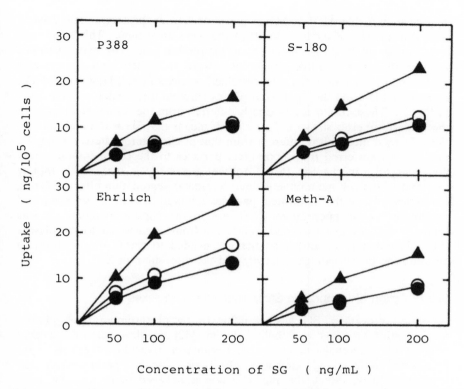

Figure 16 Effect of vehicle on uptake of [³H]SG by cells. The experimental conditions and symbols are the same as for Fig. 1. The lipid nanosphere contains 0.02% SG. Each point shows the mean of duplicate experiments.

However, the endogenous SG isolated from human and mouse plasma had a hemostatic effect when it was injected exogenously into other mice, as seen in Fig. 13. Moreover, the mice fed on an SG-free diet for a week had a longer hemostatic time than normal mice and had negligible amounts of SG in their plasma, as mentioned previously. This result suggests that endogenous SG has some physiological activity.

As stated earlier, the increase in the maximal plasma concentration of SG after oral administration was less than 10% of the plasma concentration of endogenous SG. However, the percentage of increase in the concentration of SG in the active fraction of plasma, in the lower density lipoproteins, was more than that of the total increase in the plasma concentration. Rats and mice have less lower density lipoproteins than HDL in their plasma and their distribution of SG among plasma lipoproteins is similar to that of cholesterol, as shown in Table 11; i.e., SG in lower density lipoproteins might have a sufficient amount of increase to affect the hemostatic time. This hypothesis seems to be the most reasonable one at the present time. However, as shown in Fig. 16, when cultured tumor cells were incubated with lipoproteins containing [³H]SG, all of the cells examined received more [³H]SG from HDL than from LDL and the lipid nanospheres (protein-free analog of LDL seen in Fig. 2). These results are obviously different from the results obtained with the cholesterol ester shown in Fig. 1. This means that SG is internalized into the cells by a mechanism different from that of cholesterol esters. It may be that SG is transferred from lipoprotein particles to the cell membrane by a nonendocytotic mechanism [63]. Another possibility is that SG is rapidly returned to the cell membrane without lysosomal degradation after or during the endocytosis of the lipoprotein particles followed by transfer from the cell membrane to the lipoproteins which are in the culture medium. It was reported that ß-sitosterol is also transferred from erythrocytes to the plasma [51]. In either case, further studies are needed to clarify the relationship between the pharmacological activity and the disposition of SG.

Site of Action of Lipoprotein-Sterylglucoside Complexes

As stated previously, SG has no influence on prothrombin time, plasma recalcification time, and euglobulin lysis time [42] and shows pharmacological effects even in adrenalectomized mice. Because pharmacological effects of SG (hemostatic effect and inhibitory effect on vascular permeability) are considered to be direct, the site of action is believed to be vascular wall, including vascular endothelium and the surrounding tissues. Each class of plasma lipoproteins can be taken up by vascular endothelium via endocytosis and receive lysosomal degradation. To reach various cells, including those of the vessel wall, plasma lipoproteins have to pass through the vascular endothelium. Each class of lipoproteins can also be transported across the vascular endothelium via transcytosis to be delivered to the surrounding tissues. As summarized in Chapter 2, endocytosis is predominantly a receptor-mediated process and transcytosis is mostly receptor-independent. Lower density lipoproteins containing SG can be taken up by endothelial cells and conveyed to the lysosomal compartment (endocytosis), or they can be translocated across the cell, bypassing the lysosome, and discharged into the interstitial fluid (transcytosis). It is known in general that endocytosis of LDL by either microvascular or arterial endothelium is less extensive than transcytosis [64,65]. The transport of macromolecules across the vascular endothelium and the subendothelial wall plays important roles in the regulation of vascular homeostasis. Chien et al. [66] showed that the transendothelial transport of macromolecules depends on the diameter of particles and transmural pressure. In an experiment using lipid nanospheres (protein-free analog of LDL in Fig.

2) containing SG, no significant hemostatic effect was observed 1 h following the intravenous injection (25 μg/kg) to mice (unpublished result). From the result, it is clear that receptor recognition of lipoproteins is an essential factor for the pharmacological effects of SG. Therefore, it is reasonable that the site of action for SG is the endothelial cells of the vascular wall. Lower density lipoproteins that carry SG are taken up the endothelial cells via a receptor-mediated process, and SG subsequently shows the pharmacological effects.

III. CONCLUDING REMARKS

Carrier potential must be demonstrated pharmacologically in the development of a drug delivery system; i.e., pharmacological evaluation is most important for the development of systems. We have shown pharmacologically for the first time the carrier potential of lipoproteins in vivo, using ß-sitosteryl-ß-D-glucoside as a pharmacological agent.

In the case of sterylglucoside, lower density lipoproteins are an excellent carrier to minimize the dose of the sterylglucoside without affecting the maximal pharmacological effect. The hemostatic effect of the sterylglucoside in mice and the inhibitory effect of the sterylglucoside on the vascular permeability in rats were observed only after the intravenous administration of complexes of the sterylglucoside with the lower density lipoproteins. The minimum dose of the sterylglucoside complexed with lower density lipoproteins which effected hemostasis was less than 1% of that when a conventional vehicle was used. The lower density lipoproteins used as drug carriers decreased significantly the dose of the sterylglucoside needed to exhibit pharmacological effects.

In conclusion, sterylglucoside administered orally binds spontaneously to lower density lipoproteins such as chylomicrons, VLDL, or LDL in the absorption process, and the sterylglucoside–lower density lipoprotein complexes are considered to be delivered efficiently from the circulatory system to a site of action by the lipoprotein delivery system. The site is still unclear but some pharmacological studies as well as the results discussed in this chapter suggest that the vascular wall, especially endothelial cells, may be a site of action for SG. In the circulatory system, the sterylglucoside which associates with the lower density lipoproteins is transferred to the high density lipoproteins and becomes inert. Since the bioavailability of the sterylglucoside is extremely low after oral dosing, intravenous administration seems to be more suitable than oral dosing. We are now developing a suitable dosage form for intravenous administration of the sterylglucoside.

The use of plasma lipoproteins as a drug delivery system is very promising, and lipoproteins have some advantages over other microparticles and nanoparticles. For example, plasma lipoproteins are endogenous and stable substances in the body and show less nonspecific uptake by the reticuloendo-thelial system (RES) than other drug carriers such as liposomal carrier systems. It is well known that most other carrier systems show RES-mediated clearance, which is a major factor in determining the biodistribution of colloidal carriers [67]. But LDLs are also trapped by Kupffer cells, which are the mononuclear phagocytic cells responsible for the receptor-dependent catabolism of LDL in the ([68,69]; Chapter 7). When the tissue for targeting does not include the liver or spleen, it is critical for colloidal drug carriers to avoid uptake by the RES [70]. High density lipoprotein may also be a good vehicle for other pharmacologically active agents because it has a specific metabolic fate. Furthermore, progress is being made in the preparation of synthetic HDL particles [71] as well as synthetic LDL particles [72].

There are also problems in the investigation of the carrier potential of lipoproteins, since there are species differences in lipoprotein metabolism. Sometimes we are troubled by this problem when we design experiments. In human LDL is a major carrier of cholesterol, whereas in rats and dogs HDL serves this function. Mouse adrenal glands obtain cholesterol from both of LDL and HDL [73]. However, many papers described in Chapter 5 and our studies illustrated in Fig. 1 show that tumor cells isolated from mice or rats as well as from humans have a greater uptake of LDL than of HDL. There may not be species differences in the leakage of lipoproteins from the capillaries at sites of inflammation and tumor vascularization, because this seems to depend on the size or diameter of lipoprotein particles. Therefore, mice and rats may be a first choice in studies on LDL as a possible vehicle in drug delivery. It is also an advantage that there is less dilution of injected LDL with endogenous LDL in rats and mice. Treatment with hypolipidemic agents such as 4-aminopyrazolopyrimidine also make good experimental models [34]. On the other hand, as human plasma contains much LDL, the dilution of injected LDL may decrease the targeting efficiency. However, there are many patients with hypolipidemia. In any case, further studies on lipoprotein metabolism throughout the body are needed to solve the problems.

It is crucial to incorporate a compound without interfering with the lipoprotein receptor-recognizing properties. Many investigators have developed ways to incorporate exogenous compounds into lipoproteins ([74–76]; Chapters 3 and 4.) In general, drugs that become incorporated into the lipophilic core of lipoproteins appear to participate in receptor-mediated uptake into tissues. On the other hand, compounds that bind to the surface of lipoproteins, especially apoproteins, may interfere with the normal metabolic fate of the plasma lipoproteins. For example, Pohland and Counsell reported drug-induced alterations of the properties of lipoproteins [53]. Mitotane (o-p'-DDD), which associates with lipoproteins, disrupts the normal receptor-mediated uptake process and thereby alters the metabolic fate of lipoproteins. Recently complexes of LDL and a lipophilic derivative of doxorubicin, AD-32, have been produced and their in vivo fate in mice has been determined [77,78], (Chapter 5). It was concluded that it is possible to associate cytotoxic agents with LDL without interfering with its in vivo behavior.

Defective endocytosis of LDL has been observed in various types of malignant cells [79–83], as LDL is considered to be a carrier of drugs to cancer cells. In addition, it is known that some compounds that associate with plasma lipoproteins enter into cells by nonendocytotic mechanisms. For example, benzo[a]pyrene uptake by cells is a simple partitioning phenomenon, controlled by the relative lipid volumes of extracellular donor lipoproteins and of cells, and does not involve lipoprotein endocytosis as an obligatory step [84].

Recently, for activation of antitumor activity of macrophages, a scavenger lipoprotein pathway using a chemically modified lipoprotein (acetyl-LDL) has been used to deliver the lipophilic immunomodulator muramyl tripeptide phosphatidylethanolamine to macrophages [85]. Plasma lipoproteins are considered to be able to carry not only lipophilic compounds but also liposolubilized prodrugs. Plasma lipoproteins also have a stabilizing effect on the degradation of chloroethylnitrosoureas [86] and daunomycin [87].

Studies on the application of plasma lipoproteins in drug delivery and drug disposition are giving important information and strategies not only in cancer treatment but also in the other therapeutic or diagnostic fields. The concept of plasma lipoproteins as physiological carriers of drugs will be applied to clinical therapy and diagnosis in the near future.

REFERENCES

1. Seki J, Okita A, Watanabe M, Nakagawa T, Honda K, Tatewaki N, Sugiyama M. Plasma lipoproteins as drug carriers: pharmacological activity and disposition of the complex of ß-sitosteryl-ß-D-glucopyranoside with plasma lipoproteins. J Pharm Sci 1985; 74:1259–1264.

2. Green PHR, Glickman RM. Intestinal lipoprotein metabolism. J Lipid Res 1981; 22:1153–1173.

3. Brown MS, Kovanen PT, Goldstein JL. Regulation of plasma cholesterol by lipoprotein receptors. Science 1981; 212:628–635.

4. Rudman D, Hollins B, Bixler TJ II, Mosteller RC. Transport of drugs, hormones and fatty acids in lipemic serum. J Pharmacol Exp Ther 1972; 180:797–810.

5. Hobbelen PMJ, Coert A, Geelen JAA, van der Vies J. Interactions of steroids with serum lipoproteins. Biochem Pharmacol 1975; 24:165–172.

6. Keenan RW, Kruczek ME, Fischer JB. The binding of [3H]dolichol by plasma high density lipoproteins. Biochim Biophys Acta 1977; 486:1–9.

7. Vauhkonen M, Kuusi T, Kinnunen PKJ. Serum and tissue distribution of benzo[a]pyrene from intravenously injected chylomicrons in rat in vivo. Cancer Lett 1980; 11:113–119.

8. Chen TC, Bradley WA, Gotto AM Jr, Morrisett JD. Binding of the chemical carcinogen, p-dimethylaminoazobenzene, by human plasma low density lipoproteins. FEBS Lett 1979; 104:236–240.

9. Skalsky HL, Guthrie FE. Binding of insecticides to human serum proteins. Toxicol Appl Pharmacol 1978; 43:229–235.

10. Krinsky NI, Cornwell DG, Oncley JL. The transport of vitamin A and carotenoids in human plasma. Arch Biochem Biophys 1958; 73:233–246.

11. Huang HS, Goodman DS. Vitamin A and carotenoids. I. Intestinal absorption and metabolism of ^{14}C-labeled vitamin A alcohol and ß-carotene in the rat. J Biol Chem 1965; 240:2839–2844.

12. Goodman DS, Blomstrand R, Werner B, Huang HS, Shiratori T. The intestinal absorption and metabolism of vitamin A and ß-carotene in man. J Clin Invest 1966; 45:1615–1623.

13. Yuzuriha T, Takeda M, Katayama K. Transport of [^{14}C]coenzyme Q$_{10}$ from the liver to other tissues after intravenous administration to guinea pigs. Biochim Biophys Acta 1983; 759:286–291.

14. Urien S, Riant P, Albengres E, Brioude R, Tillement JP. In vitro studies on the distribution of probucol among human plasma lipoproteins. Mol Pharmacol 1984; 26:322–327.

15. Naruszewicz M, Carew TE, Pittman RC, Witztum JL, Steinberg D. A novel mechanism by which probucol lowers low density lipoprotein levels demonstrated in the LDL receptor-deficient rabbit. J Lipid Res 1984; 25:1206–1213.

16. McLean LR, Hagaman KA. Probucol reduced the rate of association of apolipoprotein C-III with dimyristoylphosphatidylcholine. Biochim Biophys Acta 1988; 959:201–205.

17. Parthasarathy S, Young SG, Witztum JL, Pittman RC, Steinberg D. Probucol inhibits oxidative modification of low density lipoprotein. J Clin Invest 1986; 77:641–644.

18. Lemaire M, Tillement JP. Role of lipoproteins and erythrocytes in the in vitro binding and distribution of cyclosporin A in the blood. J Pharm Pharmacol 1982; 34:715–718.

19. Sgoutas D, MacMahon W, Love A, Jerkunica I. Interaction of cyclosporin A with human Lipoproteins. J Pharm Pharmacol 1986; 38:583–588.

20. Mraz W, Reble B, Kemkes BM, Knedel M. The role of lipoproteins in exchange and transfer of cyclosporin—results from in vitro investigations. Transplant Proc 1986; 18:1281–1284.

21. Gal D, Ohashi M, MacDonald PC, Buchsbaum HJ, Simpson ER. Low-density lipoprotein as a potential vehicle for chemotherapeutic agents and radionucleotides in the management of gynecologic neoplasms. Am J Obstet Gynecol 1981; 139:877–885.

22. Rudling MJ, Collins VP, Peterson CO. Delivery of aclacinomycin A to human glioma cells in vitro by the low-density lipoprotein pathway. Cancer Res 1983; 43:4600–4605.

23. Iwanik MJ, Shaw KV, Ledwith BJ, Yanovich S, Shaw JM. Preparation and interaction of a low-density lipoprotein:daunomycin complex with P388 leukemic cells. Cancer Res 1984; 44:1206–1215.

24. Yanovich S, Preston L, Shaw JM. Characteristics of uptake and cytotoxicity of a low-density lipoprotein–daunomycin complex in P388 leukemic cells. Cancer Res 1984; 44:3377–3382.

25. Firestone RA, Pisano JM, Falck JR, MacPhaul MM, Krieger M. Selective delivery of cytotoxic compounds to cells by the LDL pathway. J Med Chem 1984; 27:1037–1043.

26. Halbert GW, Stuart JFB, Florence AT. A low density lipoprotein–methotrexate covalent complex and its activity against L1210 cells in vitro. Cancer Chemother Pharmacol 1985; 15:223–227.

27. Vitols SG, Masquelier M, Peterson CO. Selective uptake of a toxic lipophilic anthracycline derivative by the low-density lipoprotein receptor pathway in cultured fibroblasts. J Med Chem 1985; 28:451–454.

28. Ho YK, Smith RG, Brown MS, Goldstein JL. Low-density lipoprotein (LDL) receptor activity in human acute myelogenous leukemia cells. Blood 1978; 52:1099–1114.

29. Vitols S, Gahrton G, Öst Å, Peterson C. Elevated low density lipoprotein receptor activity in leukemic cells with monocytic differentiation. Blood 1984; 63:1186-1193.

30. Vitols S, Gahrton G, Björkhorm M, Peterson C. Hypocholesterolaemia in malignancy due to elevated low-density-lipoprotein-receptor activity in tumor cells: evidence from studies in patients with leukaemia. Lancet 1985; II-8465:1150-1154.

31. Rudling MJ, Ståhle L, Peterson CO, Skoog L. Content of low density lipoprotein receptors in breast cancer tissue related to survival of patients. Br Med J 1986; 292:580–582.

32. Weichert JP, Groziak MP, Longino MA, Schwendner SW, Counsell RE. Potential tumor- or organ-imaging agents. 27. Polyiodinated 1,3-disubstituted and 1,2,3-trisubstituted triacylglycerols. J Med Chem 1986; 29:2457–2465.

33. Damle NS, Seevers RH, Schwendner SW, Counsell RE. Potential tumor- or organ-imaging agents. XXIV. Chylomicron remnants as carriers for hepatographic agents. J Pharma Sci 1983; 72:898–901.

34. Counsell RE, Korn N, Pohland RC, Schwendner SW, Seevers RH. Fate of intravenously administered high-density lipoprotein labeled with radioiodinated cholesteryl oleate in normal and hypolipidemic rats. Biochim Biophys Acta 1983 750:497–503.

35. Isaacsohn JL, Lees AM, Lees RS, Strauss HW, Barlai-Kovach M, Moore TJ. Adrenal imaging with technetium-99m-labelled low density lipoproteins. Metabolism 1986; 35:364–366.

36. Lees RS, Garabedian HD, Lees AM, Schumacher DJ, Miller A, Isaacsohn JL, Derksen A, Strauss HW. Technetium-99m low density lipoproteins: preparation and biodistribution. J Nucl Med 1985; 26:1056–1062.

37. Stemerman MB, Morrel EM, Burke KR, Colton CK, Smith KA, Lees RS. Local variation in arterial wall permeability to low density lipoprotein in normal rabbit aorta. Atherosclerosis 1986; 6:64–69.

38. Seki J, Azukizawa M, Sugiyama M. In vivo evaluation of plasma lipoproteins (LDL) and its protein-free analogue as carriers of pharmacological agents. Abstr

IInd Int ISSX Meeting (ISSX-88), "Xenobiotic Metabolism and Disposition," Kobe, Japan, 1988:53.

39. Roberts DCK, Miller NE, Price SGL, Crook D, Cortese C, La Ville A, Masana L, Lewis B. An alternative procedure for incorporating radiolabelled cholesteryl ester into human plasma lipoproteins in vitro. Biochem J 1985; 226:319–322.

40. Hynds SA, Welsh J, Stewart JM, Jack A, Soukop M, McArdle CS, Calman KC, Packard CJ, Shepherd J. Low-density lipoprotein metabolism in mice with soft tissue tumours. Biochim Biophys Acta 1984; 795:589–595.

41. Norata G, Canti G, Ricci L, Nicolin A, Trezzi E, Catapano AL. In vivo assimilation of low density lipoproteins by a fibrosarcoma tumour line in mice. Cancer Lett 1984; 25:203–208.

42. Nomura T, Watanabe M, Inoue K, Ohata K. Pharmacological studies on steryl-ß-D-glucosides. II. Influence on vascular fragility, permeability and hemorrhage. Jpn J Pharmacol 1978; 28 (suppl):110P.

43. Okuyama E, Yamazaki M. The principles of Tetragonia tetragonoides having an antiulcerogenic activity. I. Isolation and identification of sterylglucoside mixture (Compound A). Yakugaku Zasshi 1983; 103:43–48.

44. Miles DH, Stagg DD, Parish EJ. Investigation of constituents and antitumor activity of Spartina cynosuroides. J Nat Prod 1979; 42:700.

45. King ML, Ling HC, Wang CT, Su MH. Sterols and triterpenoids of Gymnosporia trilocularis Hay. J Nat Prod 1979; 42:701.

46. Hardwick E, Shireman R, Remsen J. Cellular uptake of [^{14}C]sitosterol from low density lipoprotein in vitro. Fed Proc 1982; 41:719.

47. Subbiah MTR, Kuksis A. Differences in metabolism of cholesterol and sitosterol following intravenous injection in rats. Biochim Biophys Acta 1973; 306:95–105.

48. Ikeda I, Sugano M. Comparison of absorption and metabolism of ß-sitosterol and ß-sitostanol in rats. Atherosclerosis 1978; 30:227–237.

49. Nordby G, Norum KR. Substrate specificity of lecithin:cholesterol acyltransferase. Esterification of desmosterol, ß-sitosterol, and cholecalciferol in human plasma. Scand J Clin Lab Invest 1975; 35:677–682.

50. Sugano M, Morioka H, Kida Y, Ikeda I. The distribution of dietary plant sterols in serum lipoproteins and liver subcellular fractions of rats. Lipids 1978; 13:427–432.

51. Sugano M, Kida Y. In vivo exchange of ß-sitosterol between erythrocytes and plasma of rats. Agric Biol Chem 1980; 44:2703–2708.

52. Avigan J. A method for incorporating cholesterol and other lipids into serum lipoproteins in vitro. J Biol Chem 1959; 234:787–790.

53. Pohland RC, Counsell RE. The role of high density lipoproteins in the biodistribution of two radioiodinated probes in the rat. Toxicol Appl Pharmacol 1985; 77:47–57.

54. Massey JB. Kinetics and transfer of α-tocopherol between model and native plasma lipoproteins. Biochim Biophys Acta 1984; 793:387–392.

55. Lund-Katz S, Hammerschlag B, Phillips MC. Kinetics and mechanism of free cholesterol exchange between human serum high- and low-density lipoproteins. Biochemistry 1982; 21:2964–2969.

56. Morton RE. Binding of plasma-derived lipid transfer protein to lipoprotein substrates: the role of binding in the lipid transfer process. J Biol Chem 1985; 260:12593–12599.

57. Zilversmit DB, Hughes LB, Balmer J. Stimulation of cholesterol ester exchange by lipoprotein-free rabbit plasma. Biochim Biophys Acta 1975; 409:393–398.

58. Marcel YL, Vezina C, Teng B, Sniderman A. Transfer of cholesterol esters between human high density lipoproteins and triglyceride-rich lipoproteins controlled by a plasma protein factor. Atherosclerosis 1980; 35:127–133.

59. Quig DW, Zilversmit DB. Disappearance and effects of exogenous lipid transfer activity in rats. Biochim Biophys Acta 1986; 879:171–178.

60. Andersen JM, Dietschy JM. Relative importance of high and low density lipoproteins in the regulation of cholesterol synthesis in the adrenal gland, ovary, and testis of the rats. J Biol Chem 1978; 253:9024–9032.

61. Oliver MF. What do we know and what do we need to know? In:Gotto AM Jr, Miller NE, Oliver MF, eds. High density lipoproteins and atherosclerosis. Amsterdam:Elsevier/North Holland Biomedical Press, 1978:221–225.

62. Eisenberg S, Windmueller HG, Levy RI. Metabolic fate of rat and human lipoprotein apoproteins in the rat. J Lipid Res 1973; 14:446–458.

63. Pittman RC, Knecht TP, Rosenbaum MS, Taylor CA Jr. A nonendocytotic mechanism for the selective uptake of high density lipoprotein-associated cholesterol esters. J Biol Chem 1987; 262:2443–2450.

64. Vasile E, Simionescu M, Simionescu N. Visualization of the binding, endocytosis and transcytosis of low density lipoprotein in the arterial endothelium in situ. J Cell Biol 1983; 96:1677–1689.

65. Vasile E, Popescu G, Simionescu M, Simionescu N. Interaction of low density lipoprotein and β-very low density lipoprotein with the arterial endothelium in normal and hypercholesterolemic animals. Abstr 4th Int Symp Biol Vasc Endoth Cell, Noordwijkerhout, 1986:123.

66. Chien S, Fan F, Lee ML, Handley DA. Effects of arterial pressure on endothelial transport of macromolecules. Biorheology 1984; 21:631–641.

67. Douglas SJ, Davis SS, Illum L. Nanoparticles in drug delivery. Crit Rev Ther Drug Carrier Syst (US) 1987; 3:233–261.

68. Harkes L, van Berkel TJC. Quantitative role of parenchymal and non-parenchymal liver cells in the uptake of [^{14}C]sucrose-labelled low-density lipoprotein in vivo. Biochem J 1984; 224:21–27.

69. Packard CJ, Slater HR, Shepherd J. The reticuloendothelial system and low density lipoprotein metabolism in the rabbit. Biochim Biophys Acta 1982; 712:412–419.

70. Davis SS, Illum L. Colloidal carriers and drug targeting. Acta Pharm Technol 1986; 32:4–9.

71. Pittman RC, Glass CK, Atkinson D, Small DM. Synthetic high density lipoprotein particles. Application to studies of the apoprotein specificity for selective uptake of cholesterol esters. J Biol Chem 1987; 262:2435–2442.

72. Chun PW, Brumbaugh EE, Shiremann RB. Interaction of human low density lipoprotein and apolipoprotein B with ternary lipid microemulsion. Physical and functional properties. Biophys Chem 1986; 25:223–241.

73. Kovanen PT, Schneider WJ, Hillman GM, Goldstein JL, Brown MS. Separate mechanisms for the uptake of high and low density lipoproteins by mouse adrenal gland *in vitro*. J Biol Chem 1979; 254:5498–5505.

74. Krieger M, Smith LC, Andersson RGW, Goldstein JL, Kao YJ, Pownall HJ, Gotto AM Jr, Brown MS. Reconstituted low density lipoprotein: a vehicle for delivery of hydrophobic fluorescent probes to cells. J Supramol Struct 1979; 10:467–478.

75. Craig IF, Via DP, Sherrill BC, Sklar LA, Mantulin WW, Gotto AM Jr, Smith LC. Incorporation of defined cholesteryl esters into lipoproteins using cholesteryl ester-rich microemulsions. J Biol Chem 1982; 257:330–335.

76. Lundberg B, Suominen L. Preparation of biologically active analogs of serum low density lipoprotein. J Lipid Res 1984; 25:550–558.

77. Vitols S, Gahrton G, Peterson C. Significance of the low-density lipoprotein (LDL) receptor pathway for the in vitro accumulation of AD-32 incorporated into LDL in normal and leukemic white blood cells. Cancer Treat Rep 1984; 68:515–520.

78. Masquelier M, Vitols S, Peterson C. Low-density lipoprotein as a carrier of antitumoral drugs: *in vivo* fate of drug–human low-density lipoprotein complexes in mice. Cancer Res 1986; 46:3842–3847.

79. Gal D, Simpson ER, Porter JC, Snyder JM. Defective internalization of low density lipoprotein in epidermoid cervical cancer cells. J Cell Biol 1982; 92:597–603.

80. Clayman RV, Bilhartz LE, Spady DK, Buja LM, Dietschy JM. Low density lipoprotein-receptor activity is lost in vivo in malignantly transformed renal tissue. FEBS Lett 1986; 196:87–90.

81. Tomita K, Ono M, Masuda A, Akiyama S, Kuwano M. Defective endocytosis of low-density lipoprotein in monensin-resistant mutants of the mouse balb/3T3 cell line. J Cell Physiol 1985; 123:369–376.

82. Sainte-Marie J, Vidal M, Philippot JR, Bienvenue A. Internalization of low-density-lipoprotein-specific receptors in leukemic guinea pig lymphocytes. Eur J Biochem 1986; 158:569–574.

83. Cutts JL, Melnykovych G. Defective utilization of cholesterol esters from low-density lipoprotein in a human acute lymphoblastic leukemia T cell line. Biochim Biophys Acta 1988; 691:65–72.

84. Plant AL, Benson DM, Smith LC. Cellular uptake and intracellular localization of benzo[*a*]pyrene by digital fluorescence imaging microscopy. J Cell Biol 1985; 100:1295–1308.

85. Shaw JM, Futch WS Jr, Schook LB. Induction of macrophage antitumor activity by acetylated low density lipoprotein containing lipophic muramyl tripeptide. Proc Natl Acad Sci USA 1988; 85:6112–6116.

86. Weinkam RJ, Finn A, Levin VA, Kane JP. Lipophilic drugs and lipoproteins: partitioning effects on chloroethylnitrosourea reaction rates in serum. J Pharmacol Exp Ther 1980; 214:318–323.

87. Kerr DJ, Hynds S, Wheldon TE, Shepherd J, Kaye SB. Cytotoxic drug targeting to lung cancer cells in vitro with a complex of low density lipoprotein (LDL)–daunomycin. Br J Cancer 1987; 55:335.

10

Key Issues in the Delivery of Pharmacological Agents Using Lipoproteins: Design of a Synthetic Apoprotein-Lipid Carrier

J. Michael Shaw*
Alcon Laboratories, Fort Worth, Texas

Kala V. Shaw
Virginia-Maryland Regional College of Veterinary Medicine, Blacksburg, Virginia

I. INTRODUCTION

Two features common to lipoproteins make them attractive models as carriers of drugs and other therapeutic agents. First, lipoproteins are present as naturally occurring particles in the blood and interstitial fluids and have the capacity for interaction with pharmacological levels of drug molecules. Second, the uptake of lipoproteins by cell types found in numerous tissues and organs occurs by the transport process known as receptor-mediated endocytosis (Chapters 1, 2). Lipoprotein particles such as low density lipoproteins (LDLs) and high density lipoproteins (HDLs) when combined with pharmacological agents represent unique endogenous particles for characterizing cellular uptake processes, extravascular transit, and accompanying biological responses. Consequently, studies concerned with lipoproteins and pharmacological agents are likely to have a direct application on the design and manufacture of pharmaceutically acceptable colloidal–particle systems.

Present Affiliation:
*   Sterling Drug Inc., Malvern, Pennsylvania

This chapter focuses on several key issues or dilemmas which make the study of lipoproteins as drug carriers challenging and directly applicable to other particles and vesicles introduced into biological fluids. In addition, studies are described for the preparation of a synthetic, reconstituted apoprotein peptide–lipid complex and its interaction with fibroblast, endothelial, and epithelial cell types. The complex consists of phospholipid bilayer vesicles which contain anchored amphipathic derivatives of an amino acid peptide derived from the 48,000 dalton apolipoprotein E (apo E). The synthetic peptide fragment is known to be an essential domain for binding of intact apo E to the apolipoprotein B/E receptor and the proteoglycan, heparin. Experimental binding studies with the highly basic apo E peptide–phospholipid vesicle complex and cultured cells show both similarities and differences to intact, native lipoproteins.

II. KEY ISSUES

A. The Complexity of Circulating Levels of Lipoproteins

Lipoprotein Interconversions Within the Blood Pool

Figure 1 is a flow diagram of lipoprotein conversions, many of which occur within the blood pool [1,2]. Although lipoproteins are synthesized and assembled primarily in the liver and intestine, it is well known that processing and exchange occur within the blood compartment through the action of a variety of enzymes, transferases, and interactions at the endothelial wall. Four basic classes of lipoproteins are illustrated in Fig. 1: chylomicrons (CMs), very low density lipoprotein (VLDL), LDL, and HDL. In addition, a number of intermediate species and subclasses include the remnant chylomicron, remnant VLDL, β-VLDL, nascent HDL (HDL$_n$), HDL$_3$, HDL$_2$, apo E–enriched HDL (HDL$_e$), and oxidized LDL (LDL$_o$).

Lipoproteins are in constant flux between the blood pool and tissue compartments like the liver, spleen, adrenal gland, various peripheral tissues, the endothelial wall, and macrophage scavenger cell types (Fig. 1). The presence of a lipoprotein–drug complex in such a dynamic blood pool may lead to rapid metabolism and interconversion prior to its arrival at a biological site of interest. A more detailed discussion of lipoprotein metabolism and interconversions is presented in Chapters 1 and 2.

Alterations in Disease States and During Therapy

Table 1 lists diseases, conditions, and therapeutic treatments that lead to metabolic alterations of specific classes of lipoproteins. The most obvious issue

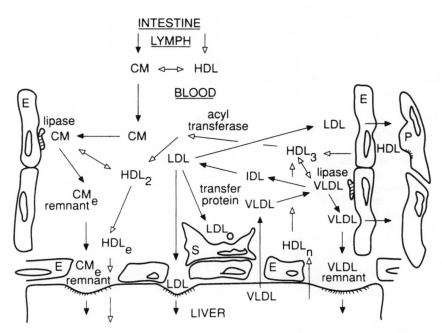

Figure 1 Lipoprotein conversions within the blood pool and transport between tissue compartments. CM, chylomicron; CM$_e$ remnant, chylomicron remnant enriched in apoprotein E; HDL, high density lipoprotein; HDL$_3$, HDL$_2$, subclasses of high density lipoprotein; HDL$_e$, high density lipoprotein enriched in apoprotein E; LDL, low density lipoprotein; LDL$_o$, oxidized low density lipoprotein; IDL, intermediate density lipoprotein; VLDL, very low density lipoprotein; VLDL remnant, very low density lipoprotein remnant; E, endothelial cell; S, scavenger cell of the mononuclear phagocyte system; P, peripheral tissue cell.

or dilemma here is the consequence of fluctuating levels of natural lipoprotein classes and attempting to deliver a drug using an exogenously supplied lipoprotein. Consider the utilization of an LDL–drug complex for therapy in a patient with abnormally high endogenous LDL levels. The LDL–drug complex would compete with the same tissue/cell sites as the endogenous LDL, resulting in a less efficient uptake of drug by the targeted tissues. A second issue of concern is the capacity for interaction of endogenous lipoprotein pools with exogenously supplied free drugs by either parenteral or oral routes (see Chapters 4 and 9). A clinical profile of such a case has been made for cyclosporin A during immunosuppression therapy. Cyclosporin A associates with plasma LDL and exhibits a similar tissue distribution to the LDL receptor

Table 1 Lipoprotein Alteration in Disease and Therapeutic Treatment

Disease or treatment	Ref.
Atherosclerosis	3,4
Diabetes	5,6
Rheumatoid arthritis	7
Systemic lupus erythematosus	8,9
Liver disease	10–12
Obesity	13,14
Diet	15,16
Hypothyroidism	17–19
Certain cancers	20,21
Nephrotic syndrome	22,23
Corticosteroid therapy	24,25
Use of androgenic–anabolic steroids	26,27
Adrenergic beta-blockers	28,29
Cyclosporin	30,31
Oral contraceptives	32,33

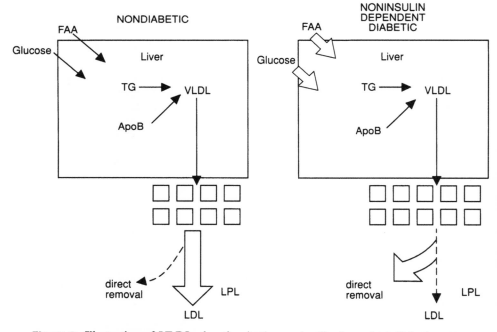

Figure 2 Illustration of VLDL elevation in the noninsulin-dependent diabetic state. FFA, free fatty acid; TG, triglyceride; VLDL, very low density lipoprotein; ApoB, apoprotein B, LDL, low density lipoprotein; LPL, lipoprotein lipase. "Direct removal" indicates direct uptake of VLDL by an apo B receptor–mediated endocytosis pathway. (Adapted from Ref. 5.) □ = VLDL

[34]. An inverse relationship between immunosuppression/toxicity and serum cholesterol/triglyceride values; i.e., high immunosuppression/toxicity occurs in patients with low cholesterol/triglyceride values and vice versa [35]. Consequently, the efficacy of cyclosporin A correlates with levels of its plasma carrier, the circulating lipoprotein, LDL.

Two of the diseases listed in Table 1, diabetes and atherosclerosis, are highlighted here to illustrate key regulatory enzymatic events that affect lipoprotein pools and how lipoprotein–therapeutic agents can be utilized as effectors in the course of disease or as diagnostic tools. Diabetes mellitus is divided into two classes, Type I, which is the less common and involves the autoimmune destruction of pancreatic beta cells, and Type II, known as noninsulin-dependent diabetes, with onset usually at maturity. Both forms of diabetes share hyperglycemia and vascular complications and show elevations in VLDL and changes in the subclasses of HDL [5]. In the nondiabetic, VLDL is processed mainly by the enzyme lipoprotein lipase associated with the endothelial cell wall and further converted to LDL with components of VLDL also contributing to the formation of HDL subclasses (Fig. 2). In the noninsulin-dependent diabetic, the elevated free fatty acid levels lower lipoprotein lipase activity, giving rise to the accumulation of VLDL with decreased processing of VLDL to LDL and other lipoprotein classes. The elevated VLDL in this case is further processed by "direct removal," i.e., uptake via the apo B receptor–mediated endocytosis pathway (Fig. 2). In the noninsulin-dependent diabetic, the delivery of a drug using a lipoprotein carrier such as VLDL would be difficult because of the high levels of native VLDL competing for the same tissue receptor and transport routes.

Alterations in LDL and HDL levels which affect serum cholesterol are a hallmark factor for determining risk in the development of atherosclerosis and eventual coronary heart disease [36]. The atherosclerotic lesions that progress from fatty streaks to intermediate lesions to atheromatous plaques are composed of high percentages of cholesterol, cholesterol ester, and phospholipid. Deposition of mainly LDL in the lesion site is thought to be a major contributing factor in the progression of plaque formation (Fig. 3). Using hot stage polarized microscopy, Small and collaborators [37] examined the thermal properties of arterial wall lesions and were able to accurately predict the composition of the crystalline domains in specimens from cholesterol-fed animals and familial hypercholesterolemic patients. Stratification of lipid deposits is thought to occur during plaque formation (Fig. 3). The early identification and therapy of fatty streak and plaque-forming regions in arterial intima walls is a potential area for utilization of lipoproteinlike carriers. Diagnostic imaging agents have been incorporated into LDL and are being proposed in further studies for distinguishing plaque areas [38,39]. It may eventually be possible to sequester into LDL, agents which stimulate regression of atheromatous plaques or to include therapeutic agents in oral form which associate with LDL and, following long term therapy, reduce or negate the

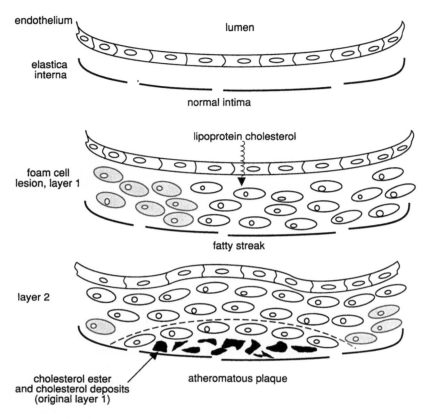

Figure 3 Deposition of lipoprotein cholesterol in the normal arterial intimal wall to give rise to fatty streak and atheromatous plaque. (Adapted from Ref. 37.)

formation of atherosclerotic lesions. The utilization of the antioxidant probucol, which associates with LDL during assembly and in the plasma, is discussed in Chapters 4 and 9. Probucol may reduce the oxidative alteration of LDL and subsequent uptake by macrophage foam cell precursors in the arterial wall [40,41].

B. Lipoprotein Interaction with Hydrophobic and Amphipathic Pharmacological Agents

Examples of Hydrophobic and Amphipathic Molecules

In predicting those drugs, therapeutic agents, and other compounds that interact and partition into lipoproteins, good selections can be made by choosing molecules with high octanol/water partition coefficients or those molecules with distinct lipophilic and hydrophilic regions (i.e., amphipathic). These classes of drugs and agents, and their interaction with lipoproteins, are described in Chapters 3 and 4. In our studies [42–44], we take the characteristic lipophilic derivatization approach used by others in which a water-soluble drug is rendered amphipathic by covalent linkage via acyl ester, amide, or phosphate ester to biologically compatible lipophilic functional groups such as fatty acid, cholesterol, diglyceride, or phospholipid [45–49]. The utilization of muramyl tripeptide phosphatidylethanolamine [50] or hexadecyl methotrexate [51] is a good example of rendering highly water-soluble molecules amphipathic by covalent coupling to a phospholipid or fatty acid (Fig. 4: further discussion in Chapters 3 and 6). Successful formation of the lipoprotein–drug complexes is accomplished by the partition of the amphipathic molecules into intact lipoproteins by the dry-film stir technique [42]. Reconstituted apoprotein–lipid-drug vesicular or micellar systems can be prepared by utilizing high ratios of tertiary butanol/water to solubilize the phospholipids, lipids, and amphipathic molecules followed by lyophilization then rehydration (Section III).

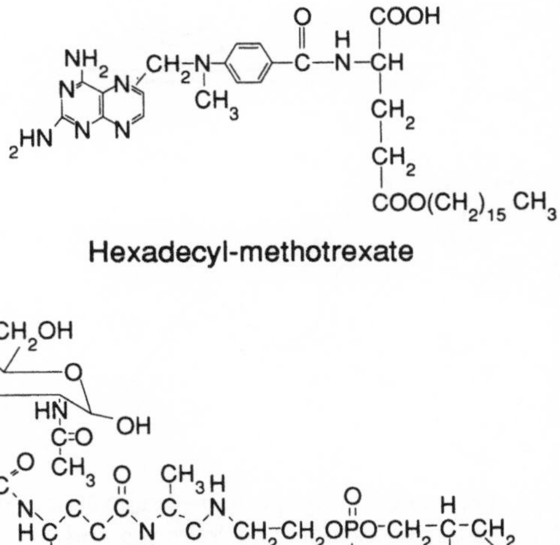

Hexadecyl-methotrexate

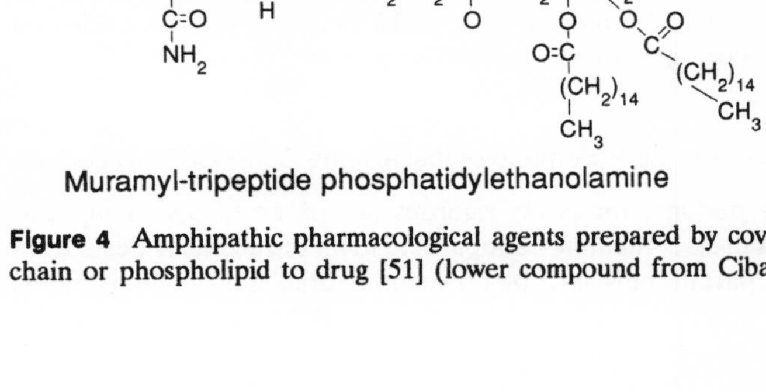

Muramyl-tripeptide phosphatidylethanolamine

Figure 4 Amphipathic pharmacological agents prepared by covalent linkage of alkyl chain or phospholipid to drug [51] (lower compound from Ciba-Geigy Ltd.)

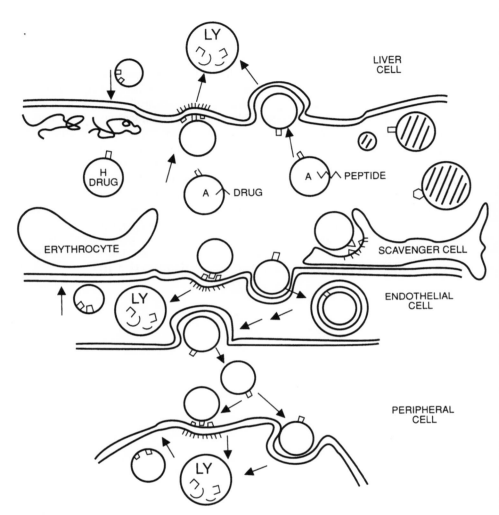

Figure 5 Potential pathways for inactivation of lipoprotein–drug colloidal particles in the blood pool during extravasation across barriers and following endocytotic uptake by peripheral tissues. H-drug, lipoprotein–hydrophobic drug; A-drug, lipoprotein–amphipathic drug; A-peptide, lipoprotein–amphipathic peptide.

Losses of Lipoprotein-Bound Pharmacological Agents During In Vivo Delivery

Figure 5 illustrates perhaps the most rigorous set of problems facing site-specific delivery of a drug-containing nanoparticle to an intracellular destination when administered parenterally into blood or interstitial fluid compartments.

Three different classes of lipoprotein-associated agents are illustrated within the vascular system: (a) hydrophobic drug sequestered in the oily interior domain of the lipoprotein (H-drug); (b) amphipathic drug with the lipophilic region anchored in the surface of the lipoprotein and the hydrophilic region occupying a region exterior to the particle (A-drug); and (c) an amphipathic peptide therapeutic agent aligned much like the amphipathic drug (A-peptide). Within the vascular compartment each of these three classes of lipoprotein–drug complex is subject to inactivation or reduced effectiveness by a number of conceivable events, including (a) transfer of drug between other endogenous lipoprotein particles and cellular components, (b) surface bioadsorption of plasma proteins and glycoproteins to the lipoprotein–drug complex with subsequent loss in site specificity, and (c) competition of lipoprotein–drug complexes with endogenous lipoproteins for receptor sites or pathways of extravasation. In addition, A-drug- and A-peptide–lipoprotein carriers are subject to proteolytic, esterase, or other enzymatic action on those linkages or functional groups external to the particle.

During extravasation one should appreciate the enormous complexity of maintaining drug association with lipoprotein or other potential carriers during movement from blood through sinusoids or fenestrations; transcytosis through the endothelial layer into the subendothelial space; diffusion through extracellular matrix and basement membrane biopolymers such as elastin, collagen, fibronectin, laminin, and proteoglycan networks; and, finally, endocytosis into cells of the peripheral tissues and survival of the therapeutic agent in the low pH environment of endosomes and lysosomes. During this intricate transport process, the lipoprotein–drug complex is in close association with membranes, offering excellent opportunity for amphipathic or lipophilic drug transfer to membrane domains and/or enzymatic modifications.

C. Cellular Uptake of Lipoproteins Occurs at Tissue and Cell Sites by Multiple Pathways

Extravasation and Dominance by the Liver

The last issues we wish to discuss are the occurrence of multiple pathways for uptake or transport of lipoproteins at numerous anatomical sites and the involvement of the liver and small intestine as the major tissues in the synthesis and uptake of lipoproteins. The presence or absence of openings in the form of fenestrated/nonfenestrated capillaries, passage through basement membranes, and the process of transcytosis are the primary means for extravascular transport of lipoproteins across or through endothelial cell barriers. Once extravasation occurs, receptor-mediated and nonreceptor-mediated endocytosis enables entry of lipoproteins into peripheral tissues and cells. An understanding of the regulatory controls over these transport processes is key in understanding both the pharmacokinetics and pharmacodynamics of site delivery for drugs using lipoproteins. The only lipoprotein to be extensively compared for receptor-mediated and nonreceptor-mediated endocytosis in various tissues is LDL [52,53]. Transcytosis of LDL and VLDL has been examined at the arterial endothelial wall in several species of animals ([54,55]; Chapter 2). Very little is known, however, about contrasting patterns of these transport pathways in various disease states or aging with the exception that during physiological conditions in which LDL is elevated, nonreceptor-mediated endocytosis becomes more widespread in tissues.

Of paramount importance is the central role of the liver in the uptake and catabolism of lipoproteins, therefore compromising the utilization of lipoprotein–drug complexes for site delivery to nonhepatic sites. Knowledge is limited in areas concerned with the chemical and biological modifications of lipoproteins and their relative uptake in liver and other tissues. Therapeutic regimens and metabolic patterns that may downregulate the production and expression of lipoprotein receptors in the liver, yet maintain site delivery to other tissues, will make available intriguing avenues for the utilization of lipoproteins as carriers. These topics and issues are covered more extensively in Chapter 1 and 2 with targeting to the liver described in Chapter 7.

Technique for Selective Targeting of Lipoproteins: Monoclonal Antibody-Modified Lipoproteins

We have examined one approach for the chemical modification of lipoproteins which theoretically should enable an almost endless range of cellular specificities to be probed. Monoclonal antibodies (MABs) or their Fab fragments exhibiting specificity toward cell surface components, basement membranes, or extracellular matrix macromolecules were covalently bound to the lipoprotein surface giving rise to MAB-directed targeting. To test the feasibility of this approach, we covalently bound an IgG2a MAB with specificity toward the histocompatibility complex H-2Kk to the oligosaccharide component of apoprotein B-100 of purified LDL [43]. During this covalent modification of apoprotein B-100, all LDL receptor binding activity of the derivatized LDL was lost. The interaction of the LDL [^{125}I]anti-H2Kk was examined in H-2Kk haplotype-positive L-929 cells and haplotype-negative Vero kidney or HeLa cells in vitro (Fig. 6). As illustrated in the figure, LDL[^{125}I]anti-HKk and free H-2Kk bound poorly to antigen-negative cells but strongly to antigen-positive L-929 cells [43]. Clearly, this MAB modification led to a redirecting of the binding specificity of the lipoprotein colloidal particle. Although a targeting advantage was gained, considerable refinement in the technique is required. First, the covalent attachment of the large MAB (150,000 daltons) to the LDL created a particle that tended to aggregate. The use of the smaller Fab fragment as targeting ligand might help in this regard. Second, the chemistry

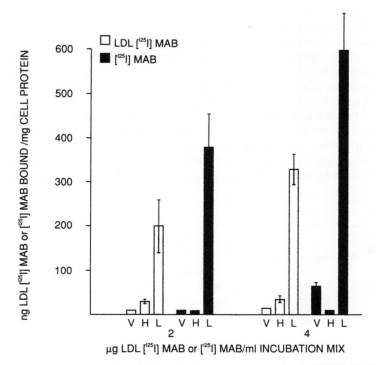

Figure 6 Targeting of LDL–MAB covalent complex to MAB haplotype–positive cultured cells. An [125I]anti-H2K^k showned binding specificity toward H-2K^k positive L-929 cells (L) but not LDL-receptor positive Vero cells (V) or HeLa cells (H). (From Ref. 43.)

utilized for the covalent linking required periodate oxidation/borohydride reduction, which destroyed the natural antioxidants associated with LDL, carotene and vitamin E. The aggregation of the MAB-derivatized LDL nanoparticle is an intolerable property, opposed to receptor-mediated interactions. Furthermore, the loss of LDL-bound antioxidants may render the LDL more subject to oxidative damage, giving rise to a toxic particle [56–59].

Inhibition of Drug-Resistant Cell Lines

It is now well known that cells resistant to the effects of the anthracycline antibiotic daunomycin show cross-resistance to other cytotoxic natural products, and this resistance pattern is described as multidrug resistance. Multidrug-resistant cells consistently overexpress a 170–180 kilodalton membrane glycoprotein, P-glycoprotein, which acts as a transmembrane drug efflux pump with ATPase activity [60–61]. The cellular capacity for active pumping of daunomycin from the cell is known to confer resistance to drug cytotoxicity [62]. In our studies daunomycin-induced transport-resistant P388 leukemia cells were utilized in comparing the growth inhibition properties of LDL-bound daunomycin versus free daunomycin [63]. Resistant P388 leukemia cells were incubated with either LDL–daunomycin complex or high concentrations of free daunomycin for 30 min followed by perfusion and fluorescence video image monitoring of the cell-associated drug as a function of time (Fig. 7). Cells loaded with free daunomycin showed a rapid efflux of drug from the cell, likely via the P-glycoprotein pump. In sharp contrast, LDL-bound daunomycin was retained by the resistant cells for longer times, presumably because of a tight association of the drug with the lipoprotein particle following endocytosis into the cell.

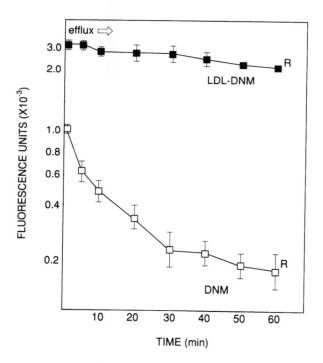

Figure 7 Cellular retention of LDL-bound daunomycin (DNM) in transport-resistant cells. Resistant P388 leukemia cells were loaded with free DNM (□, 1.0 μg DNM/ml) or LDL-DNM (■, 0.1 μg DNM/ml, 21 μg protein/ml). At time zero the drug-loaded cells were perfused with Hanks BSS at 37°C in a flow cell of a fluorescence microscope. Fluorescence units were determined using a digitized video image analysis technique. (From Ref. 63.)

Although the apoprotein B-100 of the LDL particle is rapidly broken down in lysosomes following receptor-mediated endocytosis, the lipid interior domain consisting of the cholesterol ester is reported to hydrolyze slowly, requiring 18–24 h [64]. Only in the case of LDL-bound daunomycin could drug-resistant P388 leukemia cells be growth inhibited (88% in the presence of 0.4 μg/ml LDL–daunomycin after a 2 h incubation). Leukemia colony growth showed no inhibition in the presence of 0.4 μg/ml free daunomycin after a 2 h incubation. To confirm growth inhibition of the resistant cells by LDL-bound daunomycin, a cell cycle analysis profile based on DNA content was performed. Results of the studies illustrate that uptake of LDL–daunomycin by resistant P388 leukemia cells leads to inhibition of cell growth at the G_2-M phase of the cell cycle (Fig. 8). The increase in the G_2-M peak area for LDL-daunomycin–treated cells is indicative of a lack in capacity of the resistant cells to undergo mitosis.

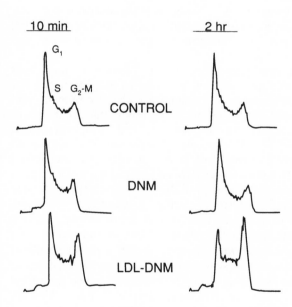

Figure 8 Transport-resistant cells incubated with LDL-DNM are growth inhibited at the G_2-M phase of the cell cycle. Resistant P388 leukemia cells were incubated for 1 min or 2 h with Hanks BSS (control), free DNM (0.4 μg/ml) or LDL-DNM (0.4 μg DNM/ml). Cells were replated for 24 h in drug-free complete media then washed cells stained with propidium iodide. The stained nuclei were analyzed for DNA propidium fluorescence using a cell sorter and DNA distributions analyzed for the proportion of cells in G_1, S, and G_2-M phases of the cell cycle. (From Ref. 63.)

In summary, daunomycin derived from LDL is rapidly endocytosed and remains associated with the slowly hydrolyzable lipid domain of the particle. The daunomycin is maintained at high levels within the intracellular compartments of the transport-resistant leukemia cells, ultimately leading to growth inhibition and cell death.

III. DESIGN OF A BIOADSORPTIVE APOPROTEIN-PEPTIDE-LIPID SYSTEM

A. Rationale

There are several reasons for pursuing research in the development of a lipoproteinlike microemulsion whose components are prepared by chemical synthesis and are reconstituted in a step by step fashion.

First, lipoproteins are complex and pharmaceutically unstable particles. Low density lipoprotein is subject to self-aggregation upon brief vortexing [65] and apo B-100 is a difficult and complex protein to characterize and manipulate [66].

Second, certain lipoproteins have properties in addition to their lipid transport function which make them difficult to utilize as drug carriers. For example, purified LDL is documented to contain associated enzymatic and even hormonal activity [67-69]. More serious is the report that treatment of LDL with the enzyme cholesterol oxidase leads to toxic properties when injected into normal rabbits [70]. We have also observed an LDL-associated esterase activity toward an LDL-bound drug analog, namely, the ester-linked hexadecyl group of hexadecyl methotrexate (HMTX) [44]. Figure 9 illustrates the effect of increasing concentrations of LDL–HMTX complex, freshly prepared (t_0) or aged (t_1), on the activity of purified dihydrofolate reductase. Aging of the complex leads to enhanced inhibition of the reductase, which is caused by hydrolysis of hexadecyl methotrexate to free methotrexate. No methotrexate release was observed with aqueous solutions of phospholipid bilayer–bound hexadecyl methotrexate, suggesting that LDL-associated esterase activity is responsible for the hydrolysis of the gamma ester–linked hexadecyl group (R. Baichwal, J. M. Shaw, A. Rosowsky, and V. Schirch, Virginia Commonwealth University and Dana-Farber Cancer Institute, unpublished results).

The third reason concerns control over drug loading and site-directed delivery. Since apolipoprotein–peptide–lipid complexes are fashioned step by step with a defined composition, the drug load can be varied. Furthermore, control over recognition and bioadsorptive properties can be built into the peptide by variations on the primary sequence, since small apolipoproteinlike

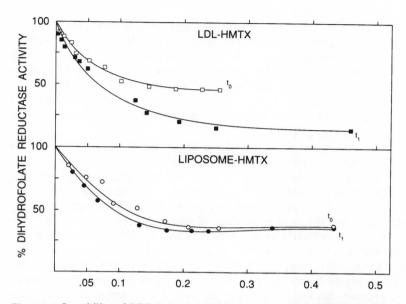

Figure 9 Instability of LDL-bound hexadecyl methotrexate (HMTX); release of free methotrexate and inhibition of dihydrofolate reductase activity. The X axis is represented by 0.05 to 0.5 μM HMTX. LDL–HMTX and liposome–HMTX complexes were freshly prepared using the dry-film stir procedure [44] and immediately incubated with a highly purified dihydrofolate reductase from *Lactobacillus casei* in the presence of dihydrofolate, NADPH, and 50 mM Tris buffer, pH 7.3. Generation of tetrahydrofolate by the enzyme was monitored by following the loss in absorbance of NADPH at 340 nm. Freshly prepared (t_0) LDL–HMTX and liposome–HMTX were both inhibitory toward the enzyme but following aging for 1 day (t_1) at 4°C only LDL–HMTX showed increased inhibition of reductase activity (R. Baichwal, J. M. Shaw, A. Rosowsky, and V. Schirch, Virginia Commonwealth University and DANA-Farber Cancer Institute, unpublished results). In separate HPLC analytical studies, free MTX was identified upon aging of LDL–HMTX in aqueous buffers [44] (L. Matherly and C. Barlow, Virginia Commonwealth University unpublished results).

peptides, i.e., $\leq$ 20 amino acids, are prepared by solid phase synthesis as well as by DNA-engineered cell biology techniques.

B. Receptor- and Heparin-Binding Domain of Apoprotein E

Introduction and the Bioadsorptive Character of Basic Peptides

The gene for apoprotein E is found on chromosome 19 and in addition to a sequence coding for a hydrophobic signal peptide, the gene also contains multiple repeats coding for a 22 amino acid amphipathic helix [71–73]. Apo E is a single chain glycosylated protein consisting of 299 amino acids with no intradisulfide linkages; it occurs naturally as different isoforms [74,75]. From sequence analysis, the secondary structure of apo E consists of 62% alpha helix, 18% random coil, 11% beta turn, and 9% beta sheet [74]. Within the amino acid sequence from residues 134 to 158 a positively charged domain is present, which is critical for binding to the apo B/E receptor (i.e., LDL receptor) and heparin [76,77].

As a first step in designing a synthetic lipoproteinlike microemulsion, we elected to utilize the cationic peptide domain for reconstitution studies. Illustrated in Fig. 10 is the peptide in which we have focused our activities, namely, amino acid residues 140–150. This peptide is enriched in arginine and lysine residues which are accepted to be important in both the apo B binding of intact LDL and the apo E binding of intact apo E-enriched HDL to receptors found on cell surfaces in a variety of tissues [78]. Natural mutations of apo E with amino acid substitutions at residue 145 or 146 lack effective binding to the apo B/E receptor [76,78,79]. In addition, a monoclonal antibody to the 1D7 epitope of apo E, i.e., the immediate vicinity between residues 140–150, prevents binding to the apo B/E receptor on cultured human skin fibroblasts [77]. The rationale for choosing the basic peptide for reconstitution studies relates to its importance for receptor binding and the ability to synthesize the 11 amino acid peptide using solid phase synthesis.

It is evident from the literature that basic peptides, basic sequences, or amphipathic segments in proteins, as well as certain highly basic proteins (pI > 8), play an important role in adsorption to cell and tissue surfaces [80,81]. For example, the membrane pore–forming fragment of colicin A contains a region of eight positively charged side chains which are thought to attract the protein to membrane surfaces and enable the more hydrophobic α-helical regions to spontaneously penetrate the membrane bilayer [82]. Many secretory proteins contain a signal peptide that facilitates interaction of the proteins with

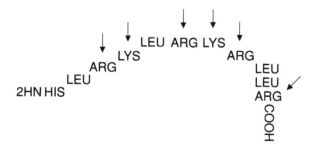

Figure 10 Basic peptide domain from apoprotein E responsible for binding to the apoprotein B/E receptor and heparin. Arrows indicate basic amino acid residues.

membranes. The amino terminus of the signal peptide usually contains several positively charged basic amino acids followed by a hydrophobic stretch which enhances adsorption to the negatively charged membrane surface followed by insertion of the hydrophobic domain [80]. Clustered basic sequence requirements have been identified for a number of proteins that are translocated into the nucleus of cells [83–85]. Conjugation of synthetic peptides, identified to be nuclear targeting regions, with nonnuclear proteins such as albumin and ferritin, leads to transport into the nucleus of cells [86,87]. Certain peptide toxins are composed of basic amino acid residues that are localized in specific domains, as is the case for melittin, or occupy one side of an amphipathic helix, as is found in bombolitin I and mastoparan [80,81]. Finally, cationized albumin and ferritin (pI > 8) are reported to undergo adsorptive-mediated endocytosis across capillary endothelial barriers, being initiated by the electrostatic interaction of the positively charged proteins with anionic sites on the endothelial surface [88,89]. The common action of these different types of basic domains and the basic apo E peptide is to facilitate interaction with cell membranes.

Apoprotein E Peptide and Acetylated Apoprotein E Peptide Anchored in Lipid Domains

Hydrophobic interactions play an important role in the interaction of apoproteins with the lipid domain of the lipoprotein [90,91]. The highly basic apo E peptide is very water soluble and must first be modified on the N- or C-terminal end with a lipophilic functional group suitable for anchoring the basic peptide in a lipid domain. The lipid system chosen for our first series of studies consisted of phospholipid liposomes.

Two forms of the apo E peptide were prepared: (a) the 11 amino acid peptide with a dipeptide spacer, phenylalanine-alanine, at the N-terminus and (b) the 11 amino acid peptide acetylated at both lysines with the phenylalanine-alanine spacer as in (a). To prepare amphipathic phenylalanine-alanine-apo E or phenylalanine-alanine-acetylated apo E, activated stearoyl groups were coupled in amide linkage to the N-terminal phenylalanine (A. Day, R. Freer, and J. M. Shaw, Virginia Commonwealth University and Alcon Laboratories, unpublished results). The acetylation was performed on the free blocked lysines and prior to assembly of the peptide from appropriately "blocked" amino acids during solid phase synthesis. The amphipathic peptide and amphipathic acetylated peptide are illustrated in Fig. 11. Isoelectric points were estimated at 12.96 and 19.95, respectively (H. T. Wright, Virginia Commonwealth University, unpublished results).

The amphipathic peptides were solubilized in tertiary butanol and mixed with 1-palmitoyl-2-oleoylphosphorylcholine and the lipophilic fluorescent dye dioctadecylindocarbocyanine at a mol ratio of 99.34% phospholipid/0.41% apo

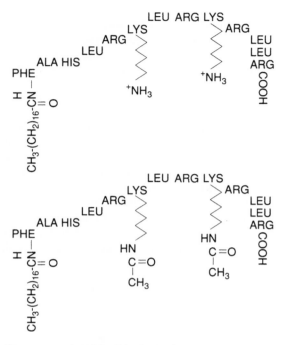

Figure 11 Amphipathic derivatives of apoprotein E basic peptide and acetylated apoprotein E basic peptide. The lipophilic anchor is a stearoyl group in amide linkage through a Phe-Ala spacer to the *N*-terminal end of the peptide (A. Day, R. Freer, and J. M. Shaw, Virginia Commonwealth University and Alcon Laboratories, unpublished results).

E fragment/0.25% dye. The solution in tertiary butanol was freeze-dried to remove solvent followed by the addition of phosphate buffered saline and pH adjustment to ~7.2. Peptide–liposome complexes were formed by brief bath sonication and incubation at 37°.

C. Interactions of Apoprotein E-Peptide-Lipid Systems with Cultured Cells

Flow Cell Fluorescence Microscopy

To examine the interaction of the fluorescent-labeled amphipathic apo E–liposome complexes with cultured cells, we utilized fluorescence microscopy equipped with a flow cell for maintaining cells viable in the presence of defined media and a video image analysis system (J. M. Shaw and S. Browder, Alcon Laboratories). The basic procedure for performing experiments is outlined here. Fibroblast, endothelial, or epithelial cells are first cultured on coverslips then exposed to apo E–liposome complexes at different concentrations or time periods. The coverslips with attached cells are inverted and placed in a specially designed flow cell in which four different coverslips are individually examined during perfusion with defined media. The fluorescent image from the microscope is projected onto a silicon intensifier target camera and then displayed on a video monitor. A software driven IBM-AT computer is used to collect the image during a specified time frame with the aid of an electronic shutter, and the image is rapidly digitized with the use of image processing computer boards. During an experiment, digitized images are transferred to the computer hard disc and ultimately stored on magnetic tape at ~200 video images per tape. A typical digitized image of fibroblasts magnified ~800 times following incubation with stearoyl-Phe-Ala-apo E–peptide in fluorescent-labeled liposomes is illustrated in Fig. 12. The intensity of the pixels (single picture elements) shown in the image vary in gray level on a scale of 0 (black, minimal fluorescence) to 256 (white, maximal fluo-

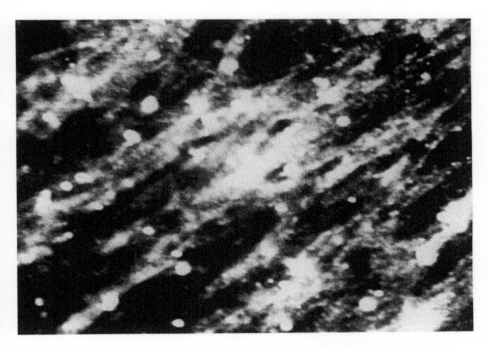

Figure 12 Digitized fluorescent image of a washed fibroblast monolayer after incubation with stearoyl-Phe-Ala-apo E peptide in fluorescent-labeled liposomes for 30 min at 37°C. Magnification 800x.

rescence).   The highlighted pixels defined by a "user set" threshold are determined for each full field and expressed as "video area of bioadsorption" (μm2).   These procedures and analyses represent a semiquantitative approach to monitoring the interactions of fluorescent-labeled particles with different cell types.

Bioadsorption to Fibroblast, Endothelial, and Epithelial Cell Types

We set out to investigate the kinetic and quantitative nature of the interaction of stearoyl-Phe-Ala-apo E–liposome complex with different cell types (J. M. Shaw, S. Browder, and W. Howe, Alcon Laboratories).   The following cell classes isolated from one tissue source, the human eye, were utilized in these studies: fibroblasts from the stromal region of the cornea;   trabecular meshwork endothelial cells known to be involved in the filtration of intraocular fluid in the eye; and epithelial cells from the corneal surface.   Figures 12–14 show digitized fluorescent images of each cell type cultured to near confluency

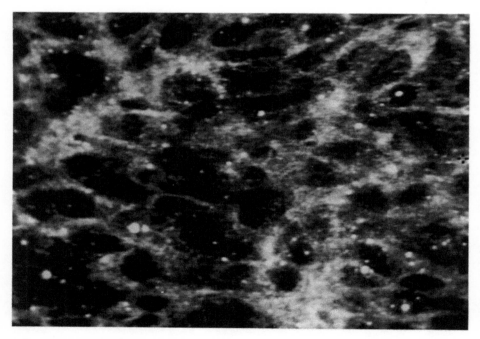

Figure 13 Digitized fluorescent image of a washed endothelial cell monolayer after incubation with stearoyl-Phe-Ala-apo E peptide in fluorescent-labeled liposomes for 30 min at 37°C. Magnification 800x.

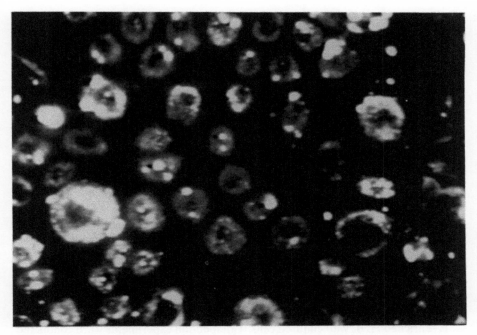

Figure 14 Digitized fluorescent image of a washed epithelial cell monolayer after incubation with stearoyl-Phe-Ala-apo E peptide in fluorescent-labeled liposomes for 30 min at 37°C. Magnification 400 x.

on coverslips following incubation with the stearoyl-Phe-Ala-apo E–liposome complex. A striking fluorescent pattern was shown for each cell type upon incubation with the formulation. The interaction was especially rapid for the fibroblast (Fig. 12) and endothelial cell types (Fig. 13), occurring at rates of 15–30 video area units per minute, as compared to the epithelial cells (Fig. 14), 1.5–2.5 video area units per minute. The interaction with all cell types was largely extracellular in nature, this being obvious with the epithelial cells, which were "ringed" with binding at the cell surface and showed little intra-cellular uptake (Fig. 14). The "dark holed" areas observed in the endothelial cells represent the cell body and intracellular compartment (Fig. 13).

We can only speculate as to the nature of the extracellular and cell surface components involved in the interaction. For example, it is reported that he apo E peptide (residues 141–150) and similar basic peptides from apo B bind to heparin and to proteoglycans enriched in chondroitin-6-sulfate with intact LDL acting as a competitive inhibitor [92–94]. Proteoglycans are highly anionic polymeric species containing uronic acid and sulfated groups. These glycosaminoglycans occur on the extracellular surface and matrix regions and are subdivided into four families, chondroitin sulfate proteoglycan, heparin sulfate proteoglycan, dermatin sulfate proteoglycan, and keratin sulfate proteoglycan [95]. There is considerable interest in observations that LDL interacts with arterial proteoglycans and may play an important role in the development of atherosclerosis [96–98]. Specific proteoglycans are secreted by a variety of cell types, including arterial endothelial cells, fibroblasts from a variety of tissues, and the trabecular meshwork endothelial cells utilized in our studies. Charged regions of other extracellular matrix proteins such as collagen or fibronectin secreted by the fibroblast or endothelial cells could also be involved in the binding to the apo E peptide–lipid complex. For example, similarities of fluorescence labeling patterns were apparent between the stearoyl-Phe-Ala-Apo E–liposome complex (Figs. 12, 13) and the pattern of fluorescence from fibroblasts or endothelial cells following interaction with MAB to collagen IV or fibronectin (data not shown). Finally, the apo B/E receptor present on fibroblast and endothelial cells may be involved in the binding; however, evidence is lacking for the presence of this receptor on the apical surfaces of epithelial cells.

To learn more about the interactions of the stearoyl-Phe-Ala-apo E–liposome complex with cell types, we carried out two standard assays that have been utilized for examining specificity of interaction of LDL or apo E–enriched HDL with cultured cells. Competition experiments were conducted using an excess of purified human LDL. Also used in the studies was apo E peptide acetytated at the epsilon amino groups on the lysine residues resulting in a neutralization of the normal positive charge of the basic lysine residues at physiological pH. In fibroblasts, both competition of the stearoyl-Phe-Ala-apo E–liposome complex with excess native LDL and acetylation of lysine residues of the stearoyl-Phe-Ala-apo E resulted in almost total loss of interaction with cells (Fig. 15). Both competition with LDL and acetylation of key lysine residues on apoprotein B of LDL or apoprotein E of HDL were shown to inhibit interactions with the apo B/E receptor and heparin [4,94]. Our competition results can be explained as either (a) genuine competition between the LDL and stearoyl-Phe-Ala-apo E–liposome complex for the cell binding site or (b) fusion of the stearoyl-Phe-Ala-apo E–liposome complex with the excess LDL in the incubation medium, resulting in losses in cell binding. In addition to fibroblasts, endothelial cells also show strong interactions with the stearoyl-Phe-Ala-Apo E–liposome complex and negligible interactions with the stearoyl-Phe-Ala-acetylated ApoE–liposome complex (Fig. 16).

The isoelectric points for the apo E peptide and the acetylated apo E peptide show little difference, suggesting that overall charge between the two peptides plays little role in the cell binding. The large abundance of arginines in the apo E peptide serve to maintain a high isoelectric point regardless of the presence of the lysine residues. Of more interest, however, is the finding

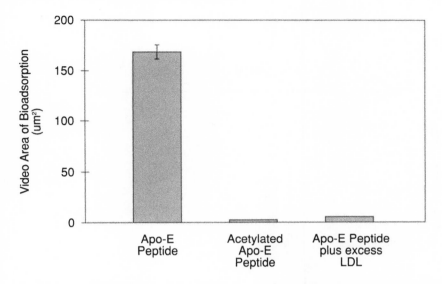

Figure 15 Video area of bioadsorption to fibroblasts of stearoyl-Phe-Ala-apo E peptide liposomes (Apo-E peptide), stearoyl-Phe-Ala-acetylated apo E peptide lipsomes (acetylated apo-E peptide) and stearoyl-Phe-Ala-apo E peptide liposomes plus excess LDL (apo-E plus excess LDL) following a 30 min incubation at 37 °C. Video area of bioadsorption (μm^2) to the cell monolayer was determined for each formulation by image analysis of the highlighted pixels. Cell monolayers cultured on coverslips were incubated in a total volume of 5 ml consisting of 80% Ham's F-10 medium and 20% Dulbecco's phosphate-buffered saline with calcium and magnesium. The formulation consisted of 12.5 nmol peptide, 3000 nmol phosphatidylcholine, and 7.5 nmol dye in a total volume of 5 ml. Prior to image analysis, cells were washed a total of four times, once with 5 ml Ham's F-10, once with 5 ml Ham's F-10 containing 1% albumin, then twice with 5 ml Ham's F-10.

that secondary structure modeling of the N-terminal–blocked apo E peptide (Phe-Ala plus residues 140 through 146) supports the presence of an α-helix in which lysines 6 and 9 play an important role in creating a positively charged amphipathic face on the peptide (Fig. 17; H. T. Wright, Virginia Commonwealth University, unpublished results). The same peptide with acetylated lysine residues loses the positively charged amphipathic face (Fig. 17). The importance of this positively charged α-helical amphipathic face in the apo E peptide and similar peptides from apoprotein B have been noted by others and were thought to be important in the affinity of lipoproteins for chondroitin-6-sulfate–rich aortic proteoglycans and the apo B/E receptor [77,92]. An alternative explanation for the role of lysine acetylation is that these modified amino acid residues may lead to an altered anchoring, asymmetry, or aggregation of

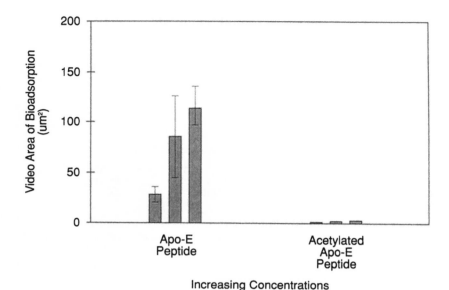

Figure 16 Video area of bioadsorpotion to endothelial cells of increasing concentrations of stearoyl-Phe-Ala-apo E peptide liposomes (apo-E peptide) and stearoyl-Phe-Ala-acetylated apo E peptide liposomes (acetylated apo-E peptide) following a 30 min incubation at 37°C. Video area of bioadsorption (μm^2) to the cell monolayer was determined for each formulation by image analysis of the highlighted pixels. Cell monolayers cultured on coverslips were incubated in a total volume of 5 ml consisting of 80% Ham's F-10 medium and 20% Dulbecco's phosphate-buffered saline with calcium and magnesium. Each concentration indicated in the bar graph represented 6.25 nmol peptide, 1500 nmol phosphatidylcholine, and 3.75 nmol dye; 12.5 nmol peptide, 3000 nmol phosphatidylcholine, and 7.5 nmol dye; or 25 nmol peptide, 6000 nmol phosphatidylcholine, and 15 nmol dye. Prior to image analysis, cells were washed a total of four times, once with 5 ml Ham's F-10, once with 5 ml Ham's F-10 containing 1% albumin, then twice with 5 ml Ham's F-10.

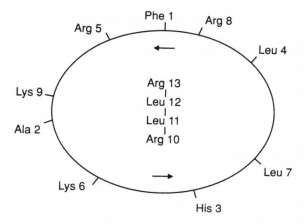

Figure 17 Helical wheel representation of the Phe-Ala-apo E peptide in which residues 1–9 are predicted to exhibit an α-helix conformation (H. T. Wright, Virginia Commonwealth University unpublished results).

the peptide in the phospholipid bilayer, making the charged arginine regions of the peptide unavailable for interaction with the fibroblast and endothelial cell monolayers. We should be able to distinguish between the two hypotheses by examining other peptide analogs as well as surface charge of the anchored stearoyl-Phe-Ala-apo E– or stearoyl-Phe-Ala-acetylated-apo E–liposome complexes.

D. Summary: Apoprotein E Peptide–Lipid Complex and Bioadsorption

The apo E peptide–liposome complex in many respects behaves as a highly bioadhesive particle to cell surfaces and extracellular matrix. The fact that the cellular binding of the vesicular particle can be competitively inhibited with excesses of LDL and by acetylation of the lysine residues suggests that the complex behaves like native LDL, which is well known to interact with such cellular macromolecules as the apo B/E receptor, heparin, and chondroitin-6-sulfate–rich proteoglycans. From a chemical-physical perspective, however, the present apo E peptide–lipid complex is not a nanoemulsion like native LDL but a heterogeneous, multibilayered liposome. The large complex does not appear to undergo endocytosis and, if studied in vivo, could be subject to adverse interactions and fusions with tissues and blood plasma–containing components. Furthermore, the 11 amino acid apo E peptide most certainly lacks many necessary structural and functional properties of the ~48,000 dalton apoprotein E or the ~500,000 dalton apoprotein B found in native lipoproteins. The large concentration of basic amino acids in the small peptide may render the peptide–lipid complex highly susceptible to trypsinlike protease activity. Nevertheless, the apo E peptide–liposome complex would appear to satisfy the following criteria set out in the "Rationale" section: (a) complete preparation by chemical synthesis and reconstitution in a step by step fashion; (b) no aberrant enzymatic or hormonal activities; and (c) drug loading in a controlled fashion during reconstitution.

IV. CONCLUDING REMARKS

Lipoproteins normally are envisioned as carriers for cholesterol and triglyceride, with aberrant behavior in metabolism leading to the development of atherosclerosis and heart disease. Pharmacological and dietary intervention in the metabolism of lipoproteins is primarily aimed at the modulation of different lipoprotein classes, and the composition and physicochemical properties of the lipids typically found in the particles, i.e., cholesterol, cholesterol ester, triglyceride, phospholipid, and the degree of saturation/-unsaturation. This chapter and earlier chapters entertain a different perspective with respect to lipoproteins, namely, (a) certain drugs and other agents can associate endogenously with lipoproteins, and (b) pharmacological or diagnostic agents can be complexed with purified lipoproteins and delivered to various tissue compartments within the body.

From a pharmaceutical standpoint, we feel that two approaches are especially attractive for the delivery of drugs and other therapeutic agents via lipoproteins. First, certain pharmacological agents taken by the oral route would by design naturally transport and incorporate into the lipoprotein assembly process much like cholesterol, triglycerides, vitamin E, and carotene [1,99,100]. These pharmacological agents subsequently would be transported to tissue compartments by lipoprotein-mediated pathways. Chapter 9 describes such a process for the sterylglycosides. Chylomicrons would be the major carrier for drugs from intestine via the lymph to the liver with drug secretion from the liver "in complex" with VLDL followed by drug–LDL associations occurring within the blood pool. High density lipoprotein would play a role as a carrier in the blood and interstitial fluid pools.

Second, synthetic drug–containing microemulsions with anchored apoproteinlike peptides acting as bioadhesive, targeting, and/or transport signaling agents could be administered by parenteral routes into appropriate tissue compartments. Distinct advantages of the peptide microemulsion–drug complex over native lipoproteins for drug delivery include control over the preparation, properties, and drug loading of the complex. Even though these two approaches are radically different, each is subject to the constraints of a number of "key issues" which attest to (a) the complex metabolism of lipoproteins and the altered levels of lipoproteins in disease states, (b) the necessity of hydrophobic and amphipathic pharmacological agents for interaction with lipoproteins, (c) the transport and uptake of lipoproteins by at least four different pathways, and (d) the dominance of lipoprotein uptake by certain organs such as the liver.

REFERENCES

1. Gotto AM, Pownall HJ, Havel RJ. Introduction to the plasma lipoproteins. Methods Enzymol 1986; 128:3–41.
2. Shepherd J, Packard CJ. Pharmacological approaches to the modulation of plasma cholesterol. Trends Pharmacol Sci 1988; 9:324–329.
3. Lipid Research Clinics Program. The lipid research clinics coronary primary prevention trial. I. Reduction in incidence of coronary heart disease. JAMA 1984; 251:31–364.
4. Goldstein JL, Brown MS. The low-density lipoprotein pathway and its relation to atherosclerosis. Annu Rev Biochem 1977; 46:897–930.
5. Howard BV. Lipoprotein metabolism in diabetes mellitus. J Lipid Res 1987; 28:613–628.
6. Reaven GM, Chen YD. Role of insulin in regulation of lipoprotein metabolism in diabetes. Diabetes Metab Rev 1988; 4:639–652.
7. Lakatos J, Harsagyi A. Serum total, HDL, LDL cholesterol and triglyceride levels in patients with rheumatoid arthritis. Clin Biochem 1988; 21:93–96.
8. Ilowite NT, Samuel P, Ginzler E, Jacobson MS. Dyslipoproteinemia in pediatric systemic lupus erythematosus. Arthritis Rheum 1988; 31:859–863.
9. Ettinger WH, Hazzard WR. Elevated apolipoprotein-B levels in corticosteroid-treated patients with systemic lupus erythematosus. J Clin Endocrinol Metab 1988; 67:425–428.
10. Seidel D. Lipoproteins in liver disease. J Clin Chem Clin Biochem 1988; 25:541–561.
11. Avgerinos A, Kourti A, Chu P, Harry DS, Raptis S, McIntire N. Plasma lipid and lipoprotein response to carbohydrate feeding in cirrhotic patients. J Hepatol 1988; 6:315–324.
12. Sabesim SM. Lipid and lipoprotein abnormalities in alcoholic hepatitis. Gastroenterology 1978; 74:283–285.
13. McNamara PJ. Effects of fat-modified diets on cholesterol and lipoprotein metabolism – human obesity. Annu Rev Nutr 7:273–290.
14. Cominacini L, Zocca I, Garbin U, Davoli A, Micciolo R, DeBastiani P, Bosello O. High density lipoprotein composition in obesity: interrelationships with plasma insulin levels and body weight. Int J Obesity 1988; 12:343–352.
15. Havel RJ. Dietary regulation of plasma lipoprotein metabolism in humans. Prog Biochem Pharmacol 1983; 19:111–122.

16. Knuiman JT, West CE, Katan MB, Hautvast JG. Total cholesterol and high density lipoprotein cholesterol levels in populations differing in fat and carbohydrate intake. Arteriosclerosis 1987; 7:612–619.

17. Althaus BO, Staub JJ, Ryff-De Leche A, Oberhansi A, Stahelin HB. LDL/HDL – changes in subclinical hypothyroidism: possible risk factors for coronary heart disease. Clin Endocrinol 1988; 28:157–163.

18. Tenenbaum D, Gambert P, Fischbach M, d'Athis P, Nivelon JL, Lallemant C. Alterations of serum high-density lipoproteins and hepatic lipase activity in congenital hypothyroidism. Biol Neonate 1988; 54:241–254.

19. Davidson NO, Powell LM, Wallis SG, Scott J. Thyroid hormone modulates the introduction of a stop condon in rat liver apolipoprotein B messenger RNA. J Biol Chem 1988; 263:13482–13485.

20. Vitols S, Gahrton G, Björkholm M, Peterson C. Hypocholesterolemia in malignancy due to elevated LDL receptor activity in tumor cells: evidence from studies in leukemic patients. Lancet 1985; 23:1150–1154.

21. Budd D, Ginsberg H. Hypocholesterolemia and acute myelogenous leukemia. Association between disease activity and plasma low density lipoprotein cholesterol concentrations. Cancer 1986; 58:1361–1365.

22. Querfield V, Gnasso A, Aaberbosch W, Augustin J, Scharer K. Lipoprotein profiles at different stages of the nephrotic syndrome. Eur J Pediatr 1988; 147:233–238.

23. Kohtoku N, Okuda F, Shimizu K, Takagi H, Fujii Z, Fujii H, Kusukawa R. Serum lipoprotein abnormalities in patients with nephrotic syndrome and effects of steroid therapy. Nippon Jinzo Gakkai Shi 1989; 31:335–344.

24. Ettinger WH, Hazzard WR. Prednisone increases very low density lipoprotein and high density lipoprotein in healthy men. Metabolism 1988; 37:1055–1058.

25. Becker, DM, Chamberlain B, Swank R, Hedgwald MG, Girardet R, Baughman KL, Kwiterovich PO, Pearson TA, Ettinger WH, Renlund D. Relationship between corticosteroid exposure and plasma lipid levels in heart transplant recipients. Am J Med 1988; 85:632–638.

26. Kleiner SM, Calabrese LH, Fielder KM, Naito HK, Skibinski CI. Dietary influences on cardiovascular disease risk in anabolic steroid-using and non-using body builders. J Am Coll Nutr 1989; 8:109–119.

27. Cohen JC, Noakes TD, Benade AJ. Hypercholesterolemia in male power lifters using anabolic-androgenic steroids. Physician Sportsmed 1988; 16:49–56.

28. Brun D, Moorjani S. The effect of beta-blockers and diuretics on lipoprotein metabolism. Union Med Can 1980; 109:1763–1764.

29. Pollare T, Lithell H, Morlin C, Prantare H, Hvarfner A, Ljunghall S. Metabolic effects of diltiazem and atenolol: results from a randomized, double-blind study with parallel groups. J Hypertens 1989; 7:551–559.

30. Ballantyne CM, Podet EJ, Patsch WP, Harati Y, Appel V, Gotto AM, Young JB. Effects of cyclosporin therapy on plasma lipoprotein levels. JAMA 1989, 262:53–56.

31. Vathsala A, Weinberg RB, Schoenberg L, Grevel J, Goldstein RA, Van Buren CT, Lewis RM, Kahan BD. Lipid abnormalities in cyclosporine-prednisone-treated renal transplant recipients. Transplantation 1989; 48:37–43.

32. Boonsiri B, Kolkijkovinda S, Chutivongse S, Crona N, Medberg M, Gundersen G, Samsioe G, Garza-Flores J, Valles de Bourges V, Juarez-Ayala J. A multicentre comparative study of serum lipids and lipoproteins in four groups of oral combined contraceptive users and a control group of IUD users. World Health Organization. Task Force on Oral Contraceptives. Contraception 1988; 38:605–629.

33. Lindberg UB, Enk L, Crona N, Silfverstolpe G. A comparison of the effects of ethinyl estradiol and estradiol valerate on serum and lipoprotein lipids. Maturitas 1988; 10:343–352.

34. Mraz W, Zink RA, Graf A, Preis D, Illner WD, Land W, Siebert W, Zottlein H. Distribution and transfer of cyclosporin among the various human lipoprotein classes. Transplant Proc 1983; 15:2426–2429.

35. DeGroen PC. Cyclosporin low-density lipoprotein and cholesterol. Mayo Clin Proc 1988; 63:1012–1021.

36. Havel RJ. Lowering cholesterol: rationale, mechanisms and means. J Clin Invest 1988; 81:1653–1660.

37. Small DM. Progression and regression of atherosclerotic lesions: insights from lipid physical biochemistry. Arteriosclerosis 1988; 8:103–129.

38. Lees AM, Lees RS, Schoen FJ, Isaacsohn JL, Fischman AJ, McKusick KA, Strauss HW. Imaging human atherosclerosis with technetium-99M labeled low density lipoproteins-endarterectomy scintigraphy. Arteriosclerosis 1988; 8:461–470.

39. Sinzinger H, Angelberger P. Imaging and kinetic studies with radiolabelled autologous low-density lipoproteins (LDL) in human atherosclerosis. Nucl Med Commun 9:859–866.

40. Baker SG, Joffe BI, Mendelsohn D, Seftel HS. Treatment of homozygous familial hypercholesterolaemia with probucol. S Afr Med 1982; 62:7–11

41. Parthasarathy S, Young SG, Witztum JL, Pittman RC, Steinberg D. Probucol inhibits oxidative modification of low density lipoprotein. J Clin Invest 1986; 77:641–644.

42. Iwanik MJ, Shaw KV, Ledwith BJ, Yanovich S, Shaw JM. Preparation and interaction of a low density lipoprotein–daunomycin complex with P388 leukemia cells. Cancer Res 1984; 44:1206–1215.

43. Shaw JM, Shaw KV, Schook LB. Drug delivery particles and monoclonal antibodies. In:Schook LB, ed. Monoclonal antibody production techniques and applications. New York:Marcel Dekker, 1987:285–310.

44. Shaw JM, Shaw KV, Yanovich S, Iwanik M, Futch WS, Rosowsky A, Schook LB. Delivery of lipophilic drugs using lipoproteins. Ann NY Acad Sci 1987; 507:252–271.

45. Lipidot Y, Rappoport S, Wolman Y. The use of fatty acid anhydrides to acylate. J Lipid Res 1967; 8:142–145.

46. Bermel MS, Turcotte JG, Steim JM, Notter RH. Interfacial properties of arabinofuranosyl diphosphate diacylglycerol, a semi-synthetic anticancer liponucleotide. J Colloid Interfac Sci 1987; 115:167–175.

47. Rosowsky A, Kim SH, Ross J, Wick MM. Lipophilic 5'-(alkyl phosphate) esters of 1-ß-D-arabinofuranosylcytosine and its N^4-acyl and 2,2'-anhydro-3'-0' acyl derivatives as potential prodrugs. J Med Chem 1982; 25:171–178.

48.   Shashoua VE, Jacob JN, Ridge R, Campbell A, Balessarini RJ. Circumventing
       the blood brain barrier using drug covalently bound to cholesterol or glyceryl
       di-fatty acid ester. J Med Chem 1984; 27:659–664.
49.   Firestone RA, Pisano LM, Falck JR, McPhaul MM, Krieger M. Selective
       delivery of cytotoxic compounds to cells by the LDL pathway. J Med Chem
       1984; 27:1037–1043.
50.   Koff WC, Showalter SD, Hamper B, Fidler IJ. Protection of mice against fatal
       herpes simplex 2 infection by liposomes containing muramyl tripeptide. Science
       1985; 228:495–496.
51.   Rosowsky A, Forsch RA, Yu CS, Lazarus H, Beardsley GP. Methotrexate
       analogues. 21. Divergent influence of alkyl chain length on the dihydrofolate
       reductase affinity and cytotoxicity of methotrexate monoesters. J Med Chem
       1984; 27:605–609.
52.   Spady DK, Bilheimer DW, Dietschy JM. Rates of receptor-dependent and
       independent low density lipoprotein uptake in the hamster. Proc Natl Acad
       Sci USA 1983; 80:3499–3503.
53.   Steinbrecher VP, Witztum JL, Kesaniemi YA, Elam RL. Comparison of
       glucosylated low density lipoprotein with methylated or cyclohexanedione-
       treated low density lipoprotein in the measurement of receptor-independent low
       density lipoprotein catabolism. J Clin Invest 1983; 71:960–964.
54.   Vasile E, Simionescu N. Transcytosis of low density lipoprotein through
       vascular endothelium. In:Copley AL, Hamashima Y, Seno S, Venkatachalam
       MA, eds. Glomerular dysfunction and biopathology of vascular wall. New
       York:Academic Press, 1985:87–101.
55.   Nistor A, Simionescu M. Uptake of low density lipoproteins by the hamster
       lung. Am Rev Respir Dis 1986; 134:1266–1272.
56.   Henriksen T, Evensen SA, Carlander B. Injury to human endothelial cells in
       culture induced by low density lipoprotein. Scand J Clin Invest 1979;
       39:361–368.
57.   Hessler JR, Morel DW, Lewis JL, Chisolm GM. Lipoprotein oxidation and
       lipoprotein-induced cytotoxicity. Arteriosclerosis 1983; 3:215–222.
58.   Cathcart MK, McNally AK, Morel DW, Chisolm GM. Superoxide anion
       participation in human monocyte-mediated oxidation of low density lipoprotein
       and conversion of low density lipoprotein to a cytotoxin. J Immunol 1989;
       142:1963–1969.
59.   Steinbrecher VP, Parthasarathy DS, Leake DS, Witztum JL, Steinberg D.
       Modification of low density lipoprotein by endothelial cells involves lipid
       peroxidation and degradation of low density lipoprotein phospholipids. Proc
       Natl Acad Sci USA 1984; 81:3883–3887.
60.   Riordan JR, Deuchars K, Kartner N, Alon N, Trent J, Ling V. Amplification
       of P-glycoprotein genes in multidrug resistant mammalian cell lines. Nature
       1985; 316:817–819.
61.   Hamada H, Tsuruo T. Purification of the 170- to 180- kilodalton membrane
       glycoprotein associated with multidrug resistance. J Biol Chem 1988;
       263:1454–1458.
62.   Imba M, Johnson RK. Active efflux of daunorubicin and adriamycin in

sensitive and resistant sublines of P388 leukemia. Cancer Res 1979; 39:2200–2203.

63. Yanovich S, Preston L, Shaw JM. Characteristics of uptake and cytotoxicity of a low density lipoprotein–daunomycin complex in P388 leukemic cells. Cancer Res 1984; 44:3377–3382.

64. Kovanen PT, Faust JR, Brown MS, Goldstein JL. Low density lipoprotein receptors in bovine adrenal cortex. I. Receptor mediated uptake of low density lipoprotein and utilization of its cholesterol for steroid synthesis in cultured adrenocortical cells. Endocrinology 1979; 104:599–609.

65. Khoo JC, Miller E, McLaughlin P, Steinberg D. Enhanced macrophage uptake of low density lipoprotein after self-aggregation. Arteriosclerosis 1988; 8:348–358.

66. Kane JP. Apolipoprotein B: structural and metabolic heterogeneity. Annu Rev Physiol 1983; 45:637–650.

67. Steinbrecher VP, Pritchard PH. Hydrolysis of phosphatidylcholine during LDL oxidation is mediated by platelet activating factor acetyl-hydrolase. J Lipid Res 1989; 30:305–315.

68. Block LH, Pletscher A. Low density lipoprotein: an old substance with a new function. Trends Pharmacol 1988; 9:214–216.

69. Scott-Burden T, Resink TJ, Hahn AW, Baur U, Box RJ, Buhler FR. Induction of growth-related metabolism in human vascular smooth muscle cells by low density lipoprotein. J Biol Chem 1989; 264:12582–12589.

70. Bernheimer AW, Robinson WG, Linder R, Mullins D, Yip YK, Cooper NS, Seidman I, Uwajima T. Toxicity of enzymatically-oxidized low-density lipoprotein. Biochem Biophys Res Commun 1987; 148:260–266.

71. Olaisen B, Teisberg P, Gedde-Dahl T. The locus for apolipoprotein E is linked to the complement component C3 locus on chromosome 19 in man. Hum Genet 1982; 62:233–236.

72. McLean JW, Eishourbagy NA, Chang DJ, Mahley RW, Taylor JM. Human apolipoprotein E mRNA cDNA cloning and nucleotide sequencing of a new variant. J Biol Chem 1984; 259:6498–6504.

73. Luo CC, Li WH, Moore MN, Chan L. Structure and evolution of the apolipoprotein multigene family. J Mol Biol 1985; 187:325–340.

74. Rall SC, Weisgraber KH, Mahley RW. Human apolipoprotein E; the complete amino acid sequence. J Biol Chem 1982; 257:4171–4178.

75. Davignon J, Gregg RE, Sing CF. Apolipoprotein E polymorphism and atherosclerosis. Arteriosclerosis 1988; 8:1–21.

76. Innerarity TL, Friedlander EJ, Rall SC, Weisgraber KH, Mahley RW. The receptor-binding domain of human apolipoprotein E: binding of apolipoprotein E fragments. J Biol Chem 1983; 258:12341–12346.

77. Weisgraber KH, Innerarity TL, Harder KJ, Mahley RW, Milne RW, Marcel YL, Sparrow JT. The receptor-binding domain of human apolipoprotein E: monoclonal antibody inhibition of binding. J Biol Chem 1983; 258:12348–12354.

78. Mahley RW, Innerarity TL. Lipoprotein receptors and cholesterol homeostasis. Biochim Biophys Acta 1983; 737:197–222.

79. Mahley RW, Innerarity TL, Rall SC, Weisgraber KH. Plasma lipoproteins: apolipoprotein structure and function. J Lipid Res 1984; 25:1277–1294.

80. Kaiser ET. Peptides with affinity for membranes. Ann Rev Biophys Chem 1987; 16:561–581.

81. Degrado WF. Design of peptides and proteins. Adv Protein Chem 1988; 39:51-115.

82. Parker MW, Pattus F, Tucker AP, Tsernoglou D. Structure of the membrane-pore-forming fragment of colicin A. Nature 1989; 337:93–96.

83. Chelsky D, Ralph R, Jonak G. Sequence requirements for synthetic peptide-mediated translocation to the nucleus. Mol Cell Biol 1989; 9:2487–2492.

84. Adam SA, Loll TJ, Mitchell MA, Gerace L. Identification of specific binding proteins for a nuclear location sequence. Nature 1989; 337:276–279.

85. Dang CV, Lee WM. Nuclear and nucleolar targeting sequences of c-erb-A, C-myb, N-myc, p53, HSP70 and HIV tat proteins. J Biol Chem 1989; 264:18019–18023.

86. Goldfarb DS, Gariepy J, Schoolnik G, Kornberg RD. Synthetic peptides as nuclear localization signals. Nature 1986; 322:641–644.

87. Landford RE, Kanda P, Kennedy RC. Induction of nuclear transport with a synthetic peptide homologous to the SV40 antigen transport signal. Cell 1986; 46:575–582.

88. Kumagai AK, Eisenberg JB, Pardridge WM. Absorptive-mediated endocytosis of cationized albumin and a 3-endorphin-cationized albumin chimeric peptide by isolated brain capillaries. J Biol Chem 1987; 262:15214–15219.

89. Simionescu M, Simionescu N, Silbert JE, Palade GE. Differentiated microdomains on the luminal surface of the capillary endothelium. II. Partial characterization of their anionic sites. J Cell Biol 1981; 90:614–621.

90. Yang CY, Gu ZW, Weng S, Kim TW, Chen SH, Pownall HJ, Sharp PM, Liu SW, Li WH, Gotto AM, Chan L. Structure of apolipoprotein B-100 of human low density lipoproteins. Arteriosclerosis 1989; 9:96–108.

91. Knott TJ, Rall SC, Innerarity TL, Jacobson SF, Urdea MS, Wilson BL, Powell LM, Pease RJ, Eddy R, Nakai H, Byers M, Priestley LM, Robertson E, Rall LB, Betsholtz C, Shows TB, Mahley RW, Scott J. Human apolipoprotein B: structure of carboxy-terminal domains, sites of gene expression, and chromosomal localization. Science 1985; 230:37–43.

92. Camjeo G, Olofsson SO, Lopez F, Carlsson P, Bondjers G. Identification of apo B-100 segments mediating the interaction of low density lipoproteins with arterial proteoglycans. Arteriosclerosis 1988; 8:368–377.

93. Weisgraber KH, Rall SC, Mahley RW, Milne RW, Marcel YL, Sparrow JT. Human apolipoprotein E. Determination of the heparin binding sites of apolipoprotein E3. J Biol Chem 1986; 261:2068–2076.

94. Mahley RW, Weisgraber KH, Innerarity T. Interactions of plasma lipoproteins containing apolipoproteins B and E with heparin and cell surface receptors. Biochim Biophys Acta 1979; 575:81–91.

95. Martin GR, Timpl R, Kuhn K. Basement membrane proteins: molecular structure and function. Adv Prot Chem 1988; 39:1–50.

96. Skeele RH, Wagner WD, Rowe HA, Edwards IJ. Artery wall derived proteoglycan plasma lipoprotein interaction: lipoprotein binding properties of extracted proteoglycans. Atherosclerosis 1987; 65:51–62.

97.   Hoff HF, Wagner WD. Plasma low density lipoprotein accumulation in aortas of hyper-cholesterolemic swine correlates with modifications in aortic glycosaminoglycan composition. Atherosclerosis 1986; 61:231–236.

98.   Wight TN.   Cell biology of arterial proteoglycans.   Arteriosclerosis 1989; 9:1–20.

99.   Traber MG, Ingold KU, Burton GW, Kayden HJ. Absorption and transport of deuterium-substituted 2R 4'R 8'R-alpha tocopherol in human lipoproteins. Lipids 1988; 23:791–797.

100. Ashes JR, Burley RW, Sidhu GS, Sleigh RW. Effect of particle size and lipid composition of bovine blood high density lipoprotein on its function as a carrier of ß-carotene. Biochim Biophys Acta 1984; 797:171–177.

Index

A

For Product Safety Concerns and Information please contact our EU representative GPSR@taylorandfrancis.com
Taylor & Francis Verlag GmbH, Kaufingerstraße 24, 80331 München, Germany

www.ingramcontent.com/pod-product-compliance
Ingram Content Group UK Ltd.
Pitfield, Milton Keynes, MK11 3LW, UK
UKHW040927180425
457613UK00004B/46